Band Structure Engineering in Semiconductor Microstructures

NATO ASI Series

Advanced Science Institutes Series

A series presenting the results of activities sponsored by the NATO Science Committee, which aims at the dissemination of advanced scientific and technological knowledge, with a view to strengthening links between scientific communities.

The series is published by an international board of publishers in conjunction with the NATO Scientific Affairs Division

A	**Life Sciences**	Plenum Publishing Corporation
B	**Physics**	New York and London
C	**Mathematical**	Kluwer Academic Publishers
	and Physical Sciences	Dordrecht, Boston, and London
D	**Behavioral and Social Sciences**	
E	**Applied Sciences**	
F	**Computer and Systems Sciences**	Springer-Verlag
G	**Ecological Sciences**	Berlin, Heidelberg, New York, London,
H	**Cell Biology**	Paris, and Tokyo

Recent Volumes in this Series

Volume 184—Narrow-Band Phenomena—Influence of Electrons with Both
Band and Localized Character
edited by J. C. Fuggle, G. A. Sawatzky, and J. W. Allen

Volume 185—Nonperturbative Quantum Field Theory
edited by G. 't Hooft, A. Jaffe, G. Mack,
P. K. Mitter, and R. Stora

Volume 186—Simple Molecular Systems at Very High Density
edited by A. Polian, P. Loubeyre, and N. Boccara

Volume 187—X-Ray Spectroscopy in Atomic and Solid State Physics
edited by J. Gomes Ferreira and M. Teresa Ramos

Volume 188—Reflection High-Energy Electron Diffraction and
Reflection Electron Imaging of Surfaces
edited by P. K. Larsen and P. J. Dobson

Volume 189—Band Structure Engineering in Semiconductor Microstructures
edited by R. A. Abram and M. Jaros

Volume 190—Squeezed and Nonclassical Light
edited by P. Tombesi and E. R. Pike

Volume 191—Surface and Interface Characterization by
Electron Optical Methods
edited by A. Howie and U. Valdrè

Series B: Physics

Band Structure Engineering in Semiconductor Microstructures

Edited by

R. A. Abram

School of Engineering and Applied Science
University of Durham
Durham, United Kingdom

and

M. Jaros

Department of Theoretical Physics
University of Newcastle upon Tyne
Newcastle upon Tyne, United Kingdom

Plenum Press
New York and London
Published in cooperation with NATO Scientific Affairs Division

Proceedings of a NATO Advanced Research Workshop
on Band Structure Engineering in Semiconductor Microstructures,
held April 10–15, 1988,
in Il Ciocco, Italy

Library of Congress Cataloging in Publication Data

NATO Advanced Research Workshop on Band Structure Engineering in
Semiconductor Microstructures (1988: Il Ciocco, Italy)
    Band structure engineering in semiconductor microstructures / edited by R.
A. Abram and M. Jaros.
        p.   cm.—(NATO ASI series. Series B, Physics; v. 189)
    "Proceedings of a NATO Advanced Research Workshop on Band Structure
Engineering in Semiconductor Microstructures, held April 10–15, 1988, in Il
Ciocco, Italy"—T.p. verso.
    "Published in cooperation with NATO Scientific Affairs Division."
    Includes bibliographies and indexes.
    ISBN 978-1-4757-0772-4   ISBN 978-1-4757-0770-0 (eBook)
    DOI 10.1007/978-1-4757-0770-0
    1. Semiconductors—Congresses. 2. Microstructure—Congresses. 3. Energy-
band theory of solids—Congresses. I. Abram, R. A. II. Jaros M. III. North Atlantic
Treaty Organization. Scientific Affairs Division. IV. Title. V. Series.
QC610.9.N36   1988   88-29430
530.4'1—dc19   CIP

© 1989 Plenum Press, New York
Softcover reprint of the hardcover 1st edition 1989
A Division of Plenum Publishing Corporation
233 Spring Street, New York, N.Y. 10013

SPECIAL PROGRAM ON CONDENSED SYSTEMS OF LOW DIMENSIONALITY

This book contains the proceedings of a NATO Advanced Research Workshop held within the program of activities of the NATO Special Program on Condensed Systems of Low Dimensionality, running from 1983 to 1988 as part of the activities of the NATO Science Committee.

Other books previously published as a result of the activities of the Special Program are:

Volume 148   INTERCALATION IN LAYERED MATERIALS
              edited by M. S. Dresselhaus

Volume 152   OPTICAL PROPERTIES OF NARROW-GAP LOW-DIMENSIONAL
              STRUCTURES
              edited by C. M. Sotomayor Torres, J. C. Portal, J. C. Maan, and
              R. A. Stradling

Volume 163   THIN FILM GROWTH TECHNIQUES FOR LOW-DIMENSIONAL
              STRUCTURES
              edited by R. F. C. Farrow, S. S. P. Parkin, P. J. Dobson,
              J. H. Neave, and A. S. Arrott

Volume 168   ORGANIC AND INORGANIC LOW-DIMENSIONAL CRYSTALLINE
              MATERIALS
              edited by Pierre Delhaes and Marc Drillon

Volume 172   CHEMICAL PHYSICS OF INTERCALATION
              edited by A. P. Legrand and S. Flandrois

Volume 182   PHYSICS, FABRICATION, AND APPLICATIONS OF
              MULTILAYERED STRUCTURES
              edited by P. Dhez and C. Weisbuch

Volume 183   PROPERTIES OF IMPURITY STATES IN SUPERLATTICE
              SEMICONDUCTORS
              edited by C. Y. Fong, Inder P. Batra, and S. Ciraci

Volume 188   REFLECTION HIGH-ENERGY ELECTRON DIFFRACTION AND
              REFLECTION ELECTRON IMAGING OF SURFACES
              edited by P. K. Larsen and P. J. Dobson

PREFACE

This volume contains the proceedings of the NATO Advanced Research Workshop on Band Structure Engineering in Semiconductor Microstructures held at Il Ciocco, Castelvecchio Pascoli in Tuscany between 10th and 15th April 1988.

Research on semiconductor microstructures has expanded rapidly in recent years as a result of developments in the semiconductor growth and device fabrication technologies. The emergence of new semiconductor structures has facilitated a number of approaches to producing systems with certain features in their electronic structure which can lead to useful or interesting properties. The interest in band structure engineering has stimulated a variety of physical investigations and novel device concepts and the field now exhibits a fascinating interplay between pure physics and device technology. Devices based on microstructures are useful vehicles for fundamental studies but also new device ideas require a thorough understanding of the basic physics.

Around forty researchers gathered at Il Ciocco in the Spring of 1988 to discuss band structure engineering in semiconductor microstructures. A broad view was taken of the field, but the emphasis was put on physical understanding with much consideration given to methods of calculation and experimental investigation of band structure as well as to new ideas and practical achievements in engineering bandstructure. The collection of papers in this volume is a record of most of the talks given at the Workshop. The meeting programme allowed substantial time for discussion but unfortunately it was not possible to provide a written record of the contributions from a lively and informed audience, that interspersed the talks and became an integral part of the Workshop. Because of their considerable overlap of subject matter, the papers have been arranged under three broad headings (Electronic Structure, Band Offsets and Stability; Transport Properties; and Optical Properties) rather than attempting a division into a number of smaller groupings.

The editors are grateful to all those concerned with the organisation of the Workshop, and to the speakers who have met a tight schedule in providing manuscripts for the proceedings. Thanks are also due to Miss Jane Wilson and Mrs. Pauline Morrell for their assistance in the preparation of material for the meeting and the proceedings.

CONTENTS

ELECTRONIC STRUCTURE, BAND OFFSETS AND STABILITY

Comments on "Can band offsets be modified controllably?"   1
    R. M. Martin

The pressure dependent band offset in a type II superlattice, a
    test for band line-up theories.   7
    L. M. Claessen

Electronic properties of semiconductor interfaces : the control of
    interface barriers   21
    F. Flores

Polar/polar, covalent/covalent and covalent/polar semiconductor
    superlattices   33
    S. Ciraci

Band offsets at semiconductor heterojunctions: bulk or interface
    properties?   51
    S. Baroni, R. Resta and A. Baldereschi

The physics of Hg-based heterostructures   61
    M. Voos

Valence band discontinuities in HgTe-CdTe-ZnTe heterojunction
    systems   81
    J. P. Faurie

Exact envelope function equations for microstructures and the
    particle in a box model   99
    M. G. Burt

A method for calculating electronic structure of semiconductor
    superlattices: perturbation   111
    H. M. Polatoglou, G. Kanellis and G. Theodorou

The effects of ordering in ternary semiconductor alloys:
    electronic and structural properties   119
    K. E. Newman, D. Teng, J. Shen and B. L. Gu

Ab-initio molecular dynamics studies of microclusters   129
    W. Andreoni, G. Pastore, R. Car, M. Parrinello and P. Giannozzi

TRANSPORT PROPERTIES

Quantum interference in semiconductor devices   137
    M. Pepper

A review of developments in resonant tunnelling   149
    L. Eaves, F. W. Sheard and G. A. Toombs

Observation of ballistic holes   167
    M. Heiblum, K. Seo, H. P. Meier and T. W. Hickmott

Quantum transport theory of resonant tunnelling devices   177
    W. R. Frensley

Hot electron effects in microstructures   187
    P. Lugli

Models for scattering and vertical transport in microstructures
    and superlattices   201
    D. C. Herbert

Electron beam source molecular beam epitaxy of $Al_xGa_{1-x}As$ graded
    band gap device structures   217
    R. J. Malik, A. F. J. Levi, B. F. Levine, R. C. Miller, D. V.
    Lang, L. C. Hopkins and R. W. Ryan

Future trends in quantum semiconductor devices   225
    M. J. Kelly

OPTICAL PROPERTIES

Novel optical properties of InGaAs-InP quantum wells   233
    M. S. Skolnick, K. J. Nash, S. J. Bass, L. L. Taylor, A. D.
    Pitt, L. J. Reed and M. K. Saker

Time resolved spectroscopy of GaAs/AlGaAs quantum well structures 247
    E. O. Göbel

Recombination mechanisms in a type II GaAs/AlGaAs superlattice   253
    T. W. Steiner, D. J. Wolford, S. W. Tozer, T. F. Kuech
    and M. Jaros

Interface recombination in GaAs-GaAlAs quantum wells   259
    B. Sermage

The interface as a design tool for modelling of optical and
    electronic properties of quantum well devices   269
    J. Christen and D. Bimberg

Characterization and design of semiconductor lasers using strain 279
    A. R. Adams, K. C. Heasman and E. P. O'Reilly

Photoreflectance and photoluminescence of strained $In_xGa_{1-x}As/GaAs$
    single quantum wells   303
    D. J. Arent, K. Deneffe, C. Van Hoof, J. De Boeck, R. Mertens
    and G. Borghs

Excitons in quantum well structures   311
    K. K. Bajaj, G. D. Sanders and R. L. Greene

Fourier determination of the hole wavefunctions in p-type
    modulation doped quantum wells by resonant Raman scattering   325
    G. Fasol, T. Suemoto, U. Ekenberg and K. Ploog

Optical properties of superlattices   341
    Y. C. Chang, H. Chu and G. D. Sanders

Ab-initio calculated optical properties of [001] $(GaAs)_n - (AlAs)_n$
    superlattices   359
    R. Eppenga and M. F. H. Schuurmans

Effect of a parallel magnetic field on the hole levels in
    semiconductor superlattices   367
    A. Fasolino and M. Altarelli

Profit from heterostructure engineering   375
    G. J.Rees

Participants   383

Index   385

COMMENTS ON "CAN BAND OFFSETS BE CHANGED CONTROLLABLY?"

Richard M. Martin

Department of Physics
University of Illinois
Urbana, IL 61801

Abstract

Offsets of valence and conduction bands at interfaces are among the key design parameters which allow engineering of electronic properties of semiconductor heterostructures. In this paper are discussed some current theoretical ideas on the causes of the offsets and the extent to which the interface dipole can be changed by atomic scale control of the chemical composition at the interface. The primary conclusion is that significant variations appear possible by the dipoles due to oriented pairs of polar atoms at the interface. Conditions where this can occur are discussed.

Introduction

"Band Gap Engineering" of semiconductor materials can be accomplished if high quality materials can be produced with controlled variations in the active electronic states -- the highest states at the top of the valence bands and the lowest states at the bottom of the conduction bands. In ordinary bulk semiconductors the variables which have been used to engineer the desired properties are the composition of alloys and doping. Modern methods of preparation of materials, such as molecular beam epitaxy and chemical vapor deposition, make it possible to vary the composition of the materials on atomic length scales to create heterostructures. As is well known, these artificial structures exhibit a rich variety of electronic states different from any that exist in ordinary crystals. Therefore, the engineering which we address here is the atomic scale control of the composition of heterostructures and the resulting electronic properties.

The key aspect of heterostructures that lead to novel electronic properties is the spatial variation of characteristic electronic energies.[1] This is commonly described in terms of variations in the band edges with position. As illustrated in Fig. 1, this often can be divided into slowly varying band bending plus rapid variations at abrupt interfaces. The band bending results from charged donor or acceptors which create dipoles which shift the bands, as in ordinary p-n junctions and in n-i-p-i structures. The abrupt jump is due to chemical composition variation immediately at the interface. The net result is that Fermi energies are properly equalized at large distances by the combined effects.

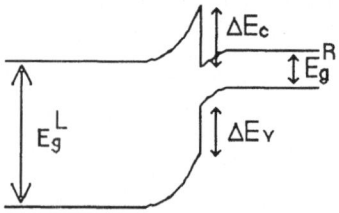

Interface Offset & Band Bending

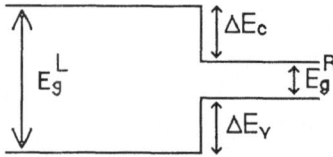

Band Offset Very Near Interface

Figure 1.

The focus of the present talk is theoretical ideas and results on the band offsets in semiconductor heterostructures. We will first review briefly the extensive work[2,3] that has been done on the magnitudes of the abrupt offsets at interfaces. We consider in particular the much-discussed issue of whether or not these offsets are intrinsic, i.e. whether or not they are properties of the bulk materials. If they are intrinsic, then they cannot be changed by variations at the interface. If they are not intrinsic, then control of the interface could provide ways to "engineer" the magnitudes of the offsets.

The crucial point is whether or not changes in the interface can shift the electronic states far from the interface. Two effects of the interface extend to the regions far from the interface: 1) strain induced by matching (or partially matching) the lattice constants at the interface and 2) electrostatic potentials resulting from dipoles at the interface. The effects of strain have been shown to change the offsets between the lowest conduction bands and highest valence bands significantly[4-6]. However, work to date has found effects that could be understood in terms of bulk band deformation potentials. Thus in the present paper we will focus upon the question of whether or not the electrostatic dipole can be varied. In one sense the answer is trivially "yes". An example is the dipole introduced by the charged donors and acceptors in the band bending in Fig. 1. Thus we must rephrase the question: Are there cases where the dipole can be changed over a small distance range ("abruptly") so that its effects are indistinguishable from other contributions to the abrupt offset?

Recent Theoretical Work on Band Offsets

There has been extensive work recently on offsets of bands at semiconductor interfaces. Review of this work was given recently by Tersoff[2] and by the present author;[3] a summary given here. There are three categories into which recent work can be grouped: 1) full self-consistent *ab initio* calculations for particular interface structures[4-11]; 2) simple models for the offsets[4,6,7,12-19]; and 3) calculations on simplified electronic hamiltonians.[20,21] We will not consider the last category here. Even though the simplified hamiltonians are very important and useful, they introduce approximations which make it difficult to compare in detail with the other categories.

Among the full calculations, two groups have carried out extensive work[4-8] on many different interfaces and have examined carefully the question of whether the band offsets change with the interface orientation and structure. The result of careful comparisons is that for a large number of interfaces, the offsets depend only upon the two materials forming the interface and are essentially independent of the interface itself. This conclusion is based upon comparing (100), (110), and (111) interfaces, variations with volume, and transitivity relations.[4,6,8] Thus, we note that for many cases the full *ab initio* calculations support the use of simple models which give the offsets as differences between bulk quantities for the two materials.[2-4,6,7,12-19]

However, if we consider the models of electronic screening at the interface,[14-16] we find that they do <u>not</u> predict the offset to be completely independent of the interface. These models suggest that the offset will tend toward a universal value, but will differ from that value by amounts of order some interface dipole divided by the dielectric constant ε.[15] Since ε is large in semiconductors the change in offsets is reduced, but it sould be non-zero.

The above analysis suggests that the most likely place to search for changes in the offsets are <u>polar</u> interfaces, where changing polarity can change the dipole at the interface. Almost all *ab initio* calculations have been limited to <u>non-polar</u> interfaces, i.e. (110) and other non-polar cases. In the remainder of the paper we discuss recent work on prototype polar interfaces.

Variations of the Band Offsets at Polar Interfaces

In order to test whether or not the band offset can be changed, we now turn to cases in which there are significant changes in polarity of the interface. As is suggested by the name, these are the cases where one would most likely expect changes in the interface dipole. There has been only a small amount of work on such cases; we review points made before and present new quantitative results from very recent work.

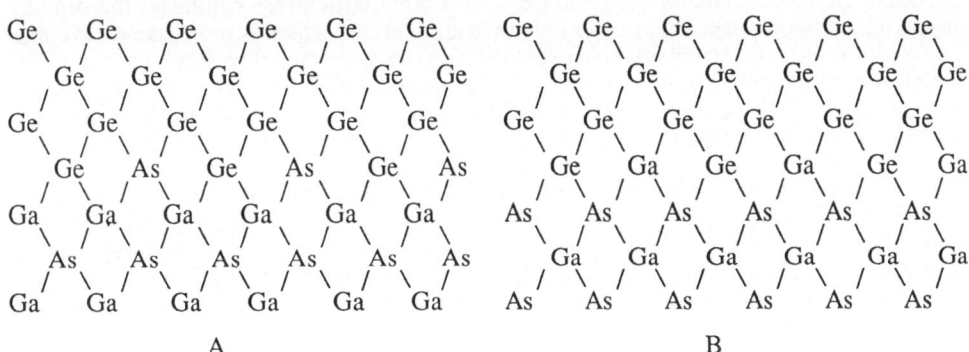

Figure 2.
GaAs/Ge (100) interfaces with mixed Ge/Ga layer (A) and Ge/As layer (B).

It was pointed out by Harrison[22] and by the author[23] that different structures should produce variations in the dipole at polar interfaces such as GaAs/Ge (100). It was shown that interfaces with mixed stoichiometry should be most stable, and conditions were given for cases in which there should be filled bands with no deep interface states.[23] The extreme cases with largest difference in polarity are shown in Fig. 2 A and B. Simple arguments suggest changes in dipole between these two cases to be given by assigning charges $\pm |e|$ to Ga and As screened by the dielectric constant $\varepsilon \sim 10$, giving approximately 1 eV shift.

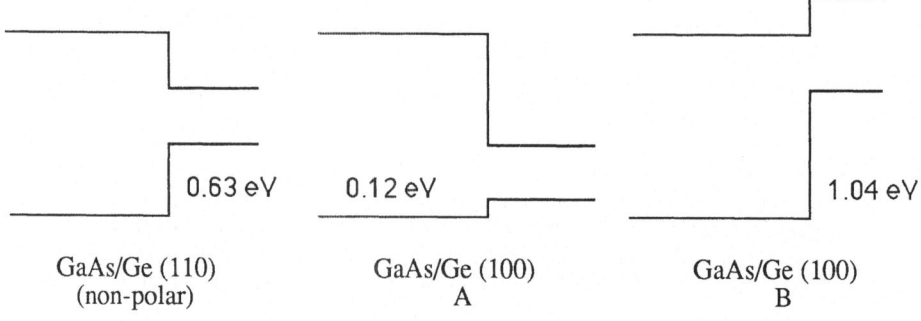

Figure 3.
Cases A and B for the different interfaces in Figure 2.

Subsequently, Kunc and Martin[10] carried out self-consistent calculations on two such interfaces with the finding that the dipole changed by ~0.6 eV. This shows clearly the possibility of large offsets, but the calculations at that time used empirical potentials and made an uncontrolled "alloy" approximation for the interface. Very similar results have been found in self-consistent tight binding calculations.[24] Thus, the quantitative changes have never been calculated with *ab initio* self-consistent methods.

In unpublished work,[25] Satpathy and the author have carried out calculations for exactly the interfaces described in Fig. 2. The result is the large change in offsets shown in Fig. 3, compared to the offset for the non-polar (110) interface.[4,6] By this variation, the offset can be changed from near zero in the valence band to negative values in the conducting band! Thus, if these interfaces can be made, the full calculations predict large changes. The next question is whether or not these polar interfaces are stable under growth conditions. Entropy argues against their stability because entropy will increase if the atoms are disordered forming mixtures of atomic structures; similarly, the dipolar energy would likely make them less stable than an interface which is more like an average (i.e. with an average band offset close to the (110) value!). Nevertheless, even if a partially ordered interface could be made, the dipole could shift by a fraction of the amount shown, which could be significant. Further investigation of energies and entropies are needed to better understand these interfaces.

It is interesting to note that interface A can be transformed into B merely by removing pairs of non-polar atoms (Ge-Ge) and replacing them with oriented pairs of polar atoms (Ga-As). The sign of the change in the dipole corresponds to As positive and Ga negative. This is qualitatively the same effect as produced by charged donor-acceptor pairs, showing the relation to the usual p-n junction. Thus the interface maybe considered as thin sheets of donors and acceptors.

Another polar case that has been investigated recently is Si/CaF_2 where self-consistent calculations have been carried out using the LMTO method.[26] For (111) interfaces, it was shown that interfaces with different polarity (i.e. different ratios at Si-Ca to Si-F bonds) have band offsets differing by ~3 eV.[26] In this case, experiments[27] indicate one interface is found predominantly, and it would be interesting to find ways to control the interface and change the offset experimentally.

In conclusion, theory shows that large changes in interface dipoles - i.e. band offsets - are possible. The requirement is to create interfaces of different polarity. Further theoretical and experimental work is needed to determine the extent to which these can actually be created under realistic growth conditions.

Acknowledgements

The author is indebted to many people for collaborations and discussions - especially C. G. Van de Walle, C. Herring, K. Kunc, S. Satpathy, D. J. Chadi, W. A.

Harrison, J. Tersoff, A. Muñoz, and N. Chetty. This work was done in part at Xerox Palo Alto Research Center supported by ONR Contract N00014-82-C0244. Work at the University of Illinois Materials Research Laboratory was supported by Department of Energy contract DE-AC02-76ER01198.

References

[1] Recent reviews of experimental work and interpretations have been given by G. Margaritondo, Surf. Sci. 168, 439 (1986); R. S. Bauer and G. Margaritondo, Physics Today, 40, 26 (1987).

[2] For a recent review of theory, see J. Tersoff, in *Heterojunctions: A Modern View of Band Discontinuities and Applications*, G. Margaritondo and F. Capasso, eds. (North Holland, Amsterdam, in press).

[3] A recent review comparing various models and full calculations is given in R. M. Martin, "Asian Pacific Symposium on Surface Physics", ed. by Xie Xide (World Scientific, Singapore, 1987), p. 14.

[4] C. G. Van de Walle, Ph.D. thesis, Stanford University, Palo Alto, California, 1986 (unpublished); most results are published in Refs. 5-6.

[5] C. G. Van de Walle and R. M. Martin, J. Vac. Sci. Technol. B 3, 1256 (1985), and Phys. Rev. B 34, 5621 (1986).

[6] C. G. Van de Walle and R. M. Martin, J. Vac. Sci. Technol. B 4, 1055 (1986), and Phys. Rev. B 35, 8524 (1987); Phys. Rev. B 37, 4108 (1987).

[7] M. Cardona and N. E. Christensen, Phys. Rev. B 35, 6182 (1987).

[8] N. E. Christensen, Phys. Rev. B 37, 4528 (1988).

[9] D. M. Bylander and L. Kleinman, Phys. Rev. B 36, 3229 (1987).

[10] S. B. Massidda, B. I. Min, and A. J. Freeman, Phys. Rev. B 35, 987 (1987).

[11] K. Kunc and R. M. Martin, Phys. Rev. B 24, 3445 (1981).

[12] R. L. Anderson, Solid-State Electron. 5, 341 (1962).

[13] W. A. Harrison, J. Vac. Sci. Technol. 14, 1016 (1977).

[14] W. R. Frensley and H. Kroemer, Phys. Rev. B 16, 2642 (1977).

[15] C. Tejedor, F. Flores, and E. Louis, J. Phys. C 10, 2163 (1977); F. Flores and C. Tejedor, J. Phys. C 12, 731 (1979).

[16] J. Tersoff, Phys. Rev. Lett. 52, 465 (1984); J. Tersoff, Phys. Rev. B 30, 4874 (1984); see also W. A. Harrison and J. Tersoff, J. Vac. Sci. Technol. B 4, 1068 (1986).

[17] C. Mailhiot and C. B. Duke, Phys. Rev. B 33, 1118 (1986).

[18] J. A. Van Vechten, J. Vac. Sci. Technol. B 3, 1240 (1985).

[19] M. Jaros, Phys. Rev. B, to be published.

[20]F. Guinea, J. Sanchez-Dehesa, and F. Flores, J. Phys. C 16, 6499 (1983); G. Platero, J. Sanchez-Dehesa, C. Tejedor, and F. Flores, Surf. Sci. 168, 553 (1986); A. Muñoz, J. C. Duran, and F. Flores, Surf. Sci. 181, L200 (1988).

[21]C. Priester, G. Allan, and M. Lanoo, Phys. Rev. 33, 7386 (1986).

[22]W. A. Harrison, J. Vac. Sci. Technol. 16, 1492 (1979).

[23]R. M. Martin, J. Vac. Sci. Technol. 17, 978 (1980).

[24]A. Muñoz, private communication.

[25]S. Satpathy and R. M. Martin, unpublished.

[26]S. Satpathy and R. M. Martin, Bull. Am. Phys. Soc. 33, 374 (1988), and to be published in Phys. Rev.

[27]M. A. Olmstead, R.I.G. Uhrbers, R. D. Bringans, and R. Z. Bachrach, Phys. Rev. B 35, 7526 (1987).

THE PRESSURE DEPENDENT BAND OFFSET IN A TYPE II SUPERLATTICE,

A TEST FOR BAND LINE-UP THEORIES

L.M. Claessen

Max-Planck-Institut für Festkörperforschung

Hochfeld Magnetlabor

25, av. des Martyrs

F-38042 Grenoble

I. INTRODUCTION

The heterojunction band offset, i.e. the position of the bandedges in one semiconductor relative to those in another one in close contact, presents a problem in solid state physics which is neither experimentally nor theoretically well understood. Yet this quantity is of growing and crucial interest for the characterisation and design of novel heterostructure devices, which can now be grown with near perfection by modern growth techniques such as MBE) and MOCVD)[1-3].

Materials between which at the interface a staggered or broken-gap[2,3] lineup is present are particularly suited for accurate optical determination of band offsets. In this context hydrostatic pressure is an interesting external parameter, because it changes the band offsets almost solely by changing Γ-Γ energy gaps of the involved bulk materials. The dependence of the resulting bandoffset on pressure yields information on the way band gaps influence band offsets, and comparisons with theory can be easily made. This will be illustrated in an experiment on an InAs-GaSb type II superlattice under hydrostatic pressure.

II. THE BAND OFFSET

Many attempts have been made to determine the band offsets in a vast variety of heterostructures. Basically a division can be made in optical[4], and electrical[5] (transport) methods. The experimental problem is that the band offsets at the interface are often determined in an indirect way. The experimentally observed quantities are fitted to a model in which one of the parameters is the desired band offset. If the sensitivity to the exact value of the band offset of this model is limited, the results can be unreliable. This makes it often difficult to obtain experimental values on band offsets within a hundred meV. A very notorious example of this problem is the widely studied interface between GaAs and

$Al_xGa_{1-x}As$, at which the GaAs energy gap falls completely in the much wider $Al_xGa_{1-x}As$ gap. The ratio in which the difference in bandgap energy is divided at the interface between the conduction bands and the valence bands has varied over the years from 85:15 to[6-8] 70:30, which spans a difference in the lineup of some hundreds of meV.

From experimental point of view the most suited systems to measure band offsets are those in which there exists a pair of energy levels of which one level is strongly correlated to the bandedges in one material, and the other level to the bandedges in the adjacent material. Measuring the energy separation between such a level pair reveals in a very direct way the band offset. A recently developed method based on this principle is x-ray or UV-ray photoemission spectroscopy[9] (respectively XPS or UPS) in which the core levels of the individual atoms are being probed at the interface. For optical band offset determinations, staggered or broken-gap line-ups[2,3] can be more suited than the usual straddling line-up. These three types of line-up are schematically depicted in figure 1. In a multi-layered structure, such as a multiquantum well (MQW) or a superlattice (SL), con-

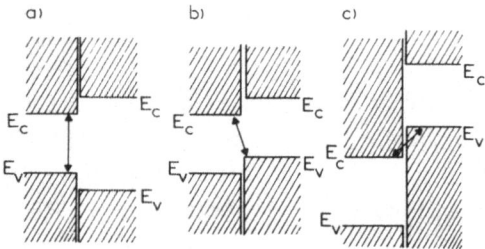

Fig.1. Three possible band line-up configurations: a) straddling, b) staggered, and c) broken-gap. The arrows suggest possible transitions sensitive to band offsets.

finement of electrons and holes will in the case of a staggered or a broken-gap line-up take place in different adjacent layers (because the lowest conduction bandedge and the highest valence bandedge are in different materials). In this situation a level-pair as discussed above can be present in the form of an electron-like, and a hole-like level. In figure 1 possible transitions between such levels sensitive to the band offsets are suggested by arrows.

SL or MQW structures having staggered or broken-gap line-ups are often referred to as type II structures. Interfaces that can lead to type II structures are for instance GaAs or AlAs against $Al_xGa_{1-x}As$ for sufficiently high Al concentrations, and InAs against GaSb. In the first case, the $Al_xGa_{1-x}As$ conduction band (CB) has an indirect (Γ-X) energy gap for x> 0.43. At the GaAs-$Al_xGa_{1-x}As$ interface the $Al_xGa_{1-x}As$ CB X-point can become, under the influence of pressure, the lowest CB-bandedge, leading to a staggered lineup between the GaAs VB Γ-point and the $Al_xGa_{1-x}As$ CB X-point. The staggered lineup at this interface has been investigated by Wolford and coworkers[7,8] in order to measure the band offset very accurately using luminescence on GaAs-Al_xGa_{1-x}As SL and MQW structures.

The second interface, which will be the subject of the rest of this paper, is that of InAs against GaSb, two narrow-gap III-V semiconductors. This widely studied[12-18] interface exhibits a broken-gap line-up (see figure 1c) in which the (Γ) InAs CB bandedge is lower in energy than the (Γ) GaSb VB bandedge. Several optical measurements can be explained with a value of 150±50 meV for Δ. Applying hydrostatic pressure to an InAs-GaSb heterostructure increases the energy gaps of the two materials (10.2 meV/kBar, and 14.5 meV/kBar for InAs and GaSb respectively[10,11]), and therefore immediately affects the overlap Δ, although not in a trivially predictable way (see figure 3). The pressure dependence of Δ therefore can yield information on how a band lineup depends on the relative magnitudes of the involved energy gaps. Moreover, if a theory on band line-ups is able to predict correctly the zero pressure band offset for InAs against GaSb, it should also be able to predict the effects of pressure on this offset if the pressure induced changes in the bulk materials are correctly accounted for. The merit of this method is that the parameters of the bulk materials are changed in a known and controlled 'neat' way, which means another interface is created in the same sample with in principle the same materials. Therefore the InAs-GaSb heterostructure under pressure provides a very interesting system to test band line-up theories.

III. EXPERIMENTAL

We have determined the pressure dependence of the overlap Δ between the InAs CB and the GaSb VB in a superlattice (SL) made of alternating layers of 12 nm InAs and 8

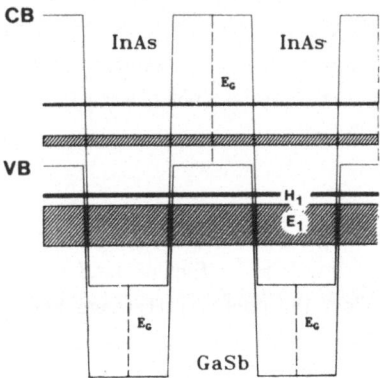

Fig.2. Spatial bandedge variation and subbands in a type II superlattice at the Γ-point. The two most important subbands for the experiments are the electron-like band E_1 and the hole-like band H_1.

nm GaSb grown on a $\langle 100 \rangle$ GaSb substrate. Extensive reviews of the electronic properties of these kind of superlattices can be found in references 15 and 17. The spatial bandedge variation and the main features of the electronic band structure of this SL are illustrated in figure 2. By the additional superlattice periodicity subbands are formed, which due to confinement have different energies than the bulk band edges, hence creating a new effective SL-energy gap defined by the lowest electron-like band (E_1) and the highest hole-like band (H_1). Due to a reduction of the confinement with increasing layer

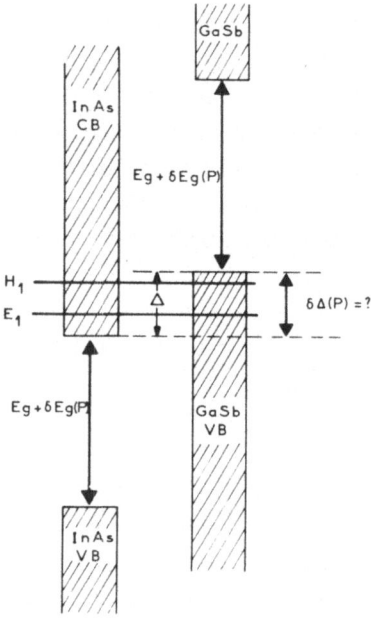

Fig.3. One period of a semimetallic-like InAs-GaSb superlattice showing the relative positions of the conduction bands (CB) and valence bands (VB) as well as the band offset Δ. The increase of the energy gaps with pressure is indicated by $\delta E_g(P)$, resulting in a unknown change in Δ. The involved subbands E_1 and H_1 are also drawn (not to scale).

thicknesses, E_1 sinks below H_1 for SL periodicities exceeding 18 nm. Maan[15] has shown that charge transfer across the interface leading to band bending can be neglected in these SL for thin enough layers. Then the energy difference between the E_1 and the H_1 subband edge at zero wave vector (i.e. neglecting any subband dispersion in the growth direction) is given by the InAs-CB GaSb-VB discontinuity Δ minus the confinement energy for the electrons minus the hole confinement energy. Therefore a measurement of E_1-H_1 as a function of pressure provides direct information about the pressure dependence of the band line-up Δ. Previous measurements[14,15] on the same sample by use of far-infrared magnetotransmission at zero pressure have determined E_1 to be 40meV lower than H_1, and can be explained using a value of 150 meV for Δ. Here we report on results of the same[19-21] experiment done as a function of hydrostatic pressures up to 12 kBar. The idea of this experiment is presented in figure 3 at the hand of one period of the SL. This figure immediately shows that although the pressure dependences of the bulk materials are well known, no unique solution exists for the pressure dependence of Δ.

The experimental principle followed to obtain the subband separation E_1-H_1 is shown in figure 4a, in which the hole-like and the electron-like Landau levels are plotted as a function of a magnetic field perpendicular to the layers. As usual, the continuum of states for motion in the plane of the layers is split into a set of equidistant linearly field dependent levels, with hole levels moving downward and electron levels moving upward in energy. Transitions which can be observed in the present experiment are suggested by the arrows. In this picture, coupling between the different subbands due to non-parabolicity leading to anticrossing[18,19] behaviour as sketched in the figure can be neglected. Then in a magnetic field perpendicular to the layers the subbands E_1 and H_1 evolve in a set of

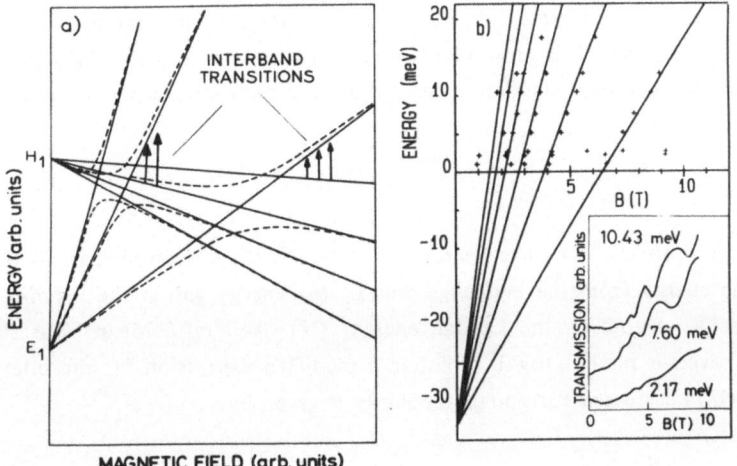

Fig.4. a) Landau level structure of the electron (E_1), and the hole (H_1) sub-bands. The dashed lines suggest a different level structure in the presence of intersubband coupling. The arrows indicate observable transitions. b) Measured transition energies between Landau levels (interband) at p=1.8 kBar. The drawn lines are a fit according to eq.3 and show the idea of linear extrapolation to zero magnetic field in order to obtain the subband separation. The inset shows some experimental spectra for different wave-lengths.

equidistantly spaced Landau levels. Applying the selection rule $\Delta N=0$ for interband transitions from the hole to the electron subband involving Landau levels with index N, the absorption of a photon with energy $\hbar\omega$ is given by[20]

$$\hbar\omega = E_1 - H_1 + (N + \tfrac{1}{2})\, \hbar eB[m_e^{-1} - m_h^{-1}] \qquad (1)$$

in which m_e and m_h are respectively the effective electron and hole mass. The transition energy given by eq. 1 increases linear with field and extrapolates at zero field to the negative energy gap E_1-H_1. In figure 4a this extrapolation is demonstrated by the drawn lines which however have been calculated using a slightly different expression then presented in eq.1 (see eqs.3 and 4).

The measurements are performed by observing the sample transmission of fixed frequency far-infrared radiation coming from an optically pumped molecular gas laser as a function of a swept magnetic field in the Faraday configuration at T=4.2 K. Laser frequencies were chosen from λ=57 μm to 1223 μm. Pressures up to 11.2 kBar have been generated using a CuBe liquid cell[10] equipped with a sapphire window in order to transmit the radiation onto the sample. The transmission was detected with a bolometer made from an Allen Bradley carbon resistor placed in the pressure cell. Some typical transmission curves are presented in the inset of figure 4b for a pressure of 1.8 kBar.

IV. ANALYSIS

Ignoring non-parabolicity in the InAs-CB as done in eq.1. leads to an energy independent effective electron mass, and hence affects strongly the extrapolation to zero magnetic field. Therefore eq.1 is only qualitatively correct. For InAs the CB non-parabolicity can be accounted for in a standard way given by a two-band k.p model, leading to the dispersion relation

$$\frac{\hbar^2 k^2}{2m_o^*} = E(1 + \frac{E}{E_g})$$ (2)

in which k presents the 3D wave vector, m_o^* the effective electron mass at the CB bandedge, E the energy from the bandedge and E_g the energy gap at the Γ-point. Replacing the lefthandside of eq.2 by the Landau energy $(N+\frac{1}{2})\hbar eB/m_o^*$, Maan[15] has shown that for zero wavevector in the growth direction a modified expression for the interband transitions (see eq.1) corrected for non-parabolicity is given by

$$\hbar\omega = -\frac{E_g}{2} + \frac{E_g}{2}\left[1 + \frac{4\hbar eB}{m_o^* E_g}(N+\frac{1}{2}) + \frac{4E_1}{E_g}(1 + \frac{E_1}{E_g})\right]^{1/2} - H_1 - (N+\frac{1}{2})\frac{\hbar eB}{m_h}$$ (3)

The main difference between eqs.1 and 3 is that in the latter the Landau levels do not form straight lines anymore. Moreover, due to an increasing band-gap, pressure also affects the effective electron mass according to[21,22]

$$\frac{m}{m_e^*} = 1 + \frac{2P^2}{3}\left[\frac{2}{E_g} + \frac{1}{E_g + \Delta_{so}}\right]$$ (4)

which relates the effective electron mass at the bandedge to the bandgap, m presents the free electron mass, E_g the bandgap, and Δ_{so} the spin-orbit splitting. P is a matrix element determining the CB-VB interaction[23]. P and Δ_{so} can be assumed to have a very small pressure dependence compared to E_g. According to eq.4 the bandedge mass increases over 20% in InAs going from 0 to 10 kBar. Therefore, E_g and m_o^* were for every experimental pressure adjusted according to eq.4 and substituted in eq.3 with the subband energy separation E_1-H_1 as the fitting parameter. The experimentally observed transition energies are plotted as a function of the magnetic field in fig.4b together with a theoretical fit conform eqs.3 and 4 for N=1 to N=6 at p=1.8 kBar. For the highest field interband transitions (N=1) the experimental points and the best fit to the experiment are plotted in a single picture in fig.5 for several pressures. The dotted lines present a theoretical fit for the two highest pressures in which Landau level broadening due to collision damping[24] has qualitatively been taken into account. The influence of this kind of damping can be

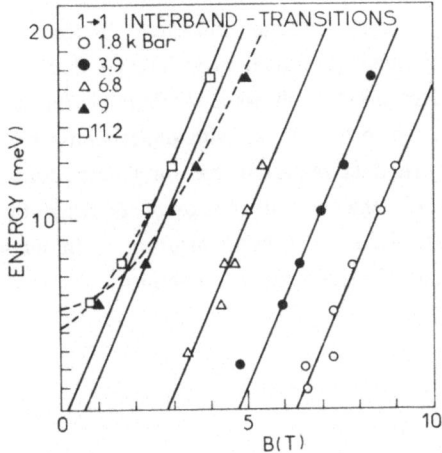

Fig.5. The experimental N=1 interband transitions for various pressures. the drawn lines are calculated as explained in the text. The dotted lines present a calculation including line-broadening due to collision damping.

expected to be more pronounced at high pressures due to possible distortions in the lattice. The obtained dependence of $E_1 - H_1$ on pressure is depicted in fig.6 for all pressures by the open circles. A decrease of $E_1 - H_1$ at a rate of 3.4 meV/kBar is found. According to this picture the previously suggested[19-21] crossover of E_1 and H_1, i.e. the semimetal-semiconductor transition appears not to occur at experimentally attainable pressures.

In zeroth order the rate of change of Δ with pressure should equal the rate of change of $E_1 - H_1$. However confinement is affected by pressure through the increase of E_g which increases the bandedge mass (eq.4) and which decreases the non-parabolicity (eq.2). These two effects work differently on the confinement, resulting in a net decrease with pressure. Therefore the overlap Δ is expected to decrease more with pressure than $E_1 - H_1$. J.C. Maan[21] has shown the confinement to be basically independent on Δ, and has calculated its dependence on pressure in the envelope wave function approximation at zero wave vector. These confinement values have been used to derive Δ from the measured $E_1 - H_1$. The results are plotted in figure 6 as the squares. A rate of change for Δ equal to 4.2 meV/kBar is found, which is indeed higher than the rate of change of $E_1 - H_1$. A more fundamental analysis using the envelope wave function approximation in a magnetic field to calculate directly the transition energies in order to fit Δ has been performed on a limited amount of experimental data and is presented elsewhere[19-21].

V. RESULTS AND VALIDITY

The essential results of the experiment presented in figure 6 have been obtained using very straightforward methods which are from theoretical point[17,18] of view strongly oversimplified. Effects of intersubband coupling have been totally neglected. It has been shown however that introducing any intersubband coupling by a two-band k.p model yield results growing increasingly worse with increasing coupling strength. Such calculations have been used to explain suggested changes[21] in transition slopes which however were not seen in later obtained additional experimental data. An analysis as correctly as possible using a six-band model[19-21] shows a result deviating little from figures 5 and 6, however having overall a worse agreement with experiment. Therefore in view of attainable accuracies in this work data reduction has only been done according to equations 3 and 4, with an outcome for the rate of change of Δ equal to 4.5±1.0 meV/kBar. This value and error have been chosen such as to incorporate upper and lower possible values as can be determined using different analysis methods.

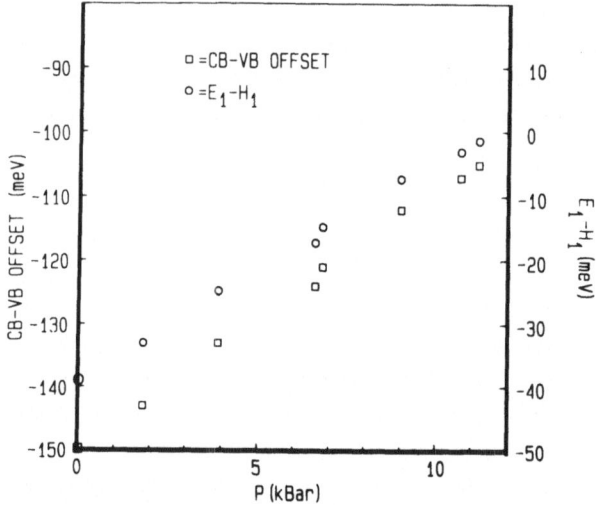

Fig.6. Experimentally obtained results for the subband separation E_1-H_1, and the from this value deduced band overlap Δ.

Moreover, due to a slight lattice mismatch[15] between InAs and GaSb (0.6%), the bandoffset will be strain affected[19]. Because of the sample geometry this strain is in the InAs layer and affects mostly the degenerated valence band[25], which is not crucial for this experiment. The main effect is a very small increase of Δ due to a hydrostatic dilatation of the InAs[25]. However, as InAs and GaSb have nearly the same compressibility[11] no additional strain will be induced by pressure and hence the dependence of Δ on pressure is expected to be not strain affected.

VI. THEORETICAL PREDICTIONS

In the following we like to discuss some band line-up theories applicable to the InAs-GaSb system under pressure. In general[2] band lineup theories can be called "absolute", or "relative". The absolute theories, often called "linear" theories, assume that there is a specific absolute energy associated with the bandedges of a bulk semiconductor. Therefore line-ups should simply be formed by the differences in these absolute bulk energies, discarding interface and surface effects. The second group (relative theories) takes into account the interface effects which means that the line-up is not solely determined by intrinsic bulk parameters.

Absolute theories

Most absolute theories yield socalled natural line-ups, i.e. line-ups that directly correlate the band offset to the valence[26] bands or the conduction bands. The most widely acknowledged and relatively reliable method to yield natural line-ups is the Harrison atomic orbital[2,8,26,27] (HAO) method, historically preceded by the electron affinity rule[2,28,29] (EAR).

The electron affinity rule. The EAR asserts that the conduction band offsets are equal to the difference in electron affinities (EA) of the two semiconductors involved. This means that the vacuum energy levels of the two materials is put at an equal level. Semiconductors can exhibit strong band bending which makes the experimental EA[29] strongly surface and interface dependent, and in such a way obscuring an objective measurement. However, in spite of various arguments against the EAR it works surprisingly well yielding for InAs-GaSb a value of 140 meV for Δ, which is correct within experimental error, and ranks among the best theoretical predictions. The EAR is however not suited to predict pressure dependent band offsets because all EAR values stem from experimental data which are on the average already suspect at ambient pressures, and to our knowledge not even available at high pressures.

The Harrisons atomic orbital theory. The Harrison atomic orbital (HAO) method calculates absolute energies for the valence bandedges in bulk materials and takes the differences in the thus obtained E_v as the valence band offsets. The idea behind the HAO is that linear combinations of s- and p-orbitals (in energy represented by the free-atom energies of the s- and p-states[30], respectively $-\epsilon_s$ and $-\epsilon_p$) form respectively conduction and valence bands in the solid. The absolute VB bandedge energy of a tetrahedral binary semiconductor is in this way given by a suitable combination of the p-state free-atom wave-function energy values ϵ_p^c and ϵ_p^a for respectively the cation and the anion according to[26,27,30]

$$E_v = (\epsilon_p^c + \epsilon_p^a)/2 - [((\epsilon_p^c - \epsilon_p^a)/2)^2 + V^2]^{1/2} \qquad (5)$$

The term V^2 presents the overlap between the p-orbitals of the different atom species, and is given by $V = \eta_{ll'm}\hbar^2/md^2$, with l, l,' and m the usual quantum numbers of angular momentum of the two involved orbitals, η a matrix element, and d the internuclear distance. η can be exactly calculated for atomic wave functions, but to obtain reasonable band parameters from the HAO slightly different values for η seem to be better. The power and validity of the HAO method stands with finding accurate values for the p-state energies to substitute in eq.2 and with finding a suitable value for η. As can be seen in the cited references, there is a rather large spread in available data for the energies ϵ_p and for η. At present it seems that the Herman-Skillman values for the neutral atom energies ϵ_p as given by Kraut[27], in combination with $\eta = 2.16$ yield the best results as far as HAO band line-ups are concerned. For the InAs-GaSb heterointerface the bandoverlap Δ is in such a way determined as $\Delta = 92$ meV, which is not far of the experimentally accepted value of 150 ± 50 meV. Although taking ϵ_p values from other sources will shift E_v considerably, the effect on the band line-up stays within several tens of meV the same.

Under the assumption that the atomic-like states have no pressure dependence, pressure dependences in band offsets can be predicted using the HAO in a simple way. According to eq.5, the only pressure dependence in the absolute energy position of the VB bandedge E_v comes from the compressibility of the lattice and can be easily substituted in this equation. Although the results of the HAO strongly depend on the data source which is used, it can be shown that the dependence of ΔE_v on pressure is rather independent on this choice. Depending on the value of the nearest neighbours coupling matrix element η (see eq.5), the HAO predicts a lowering in energy of the VB bandedges ranging from 1.5 to 3 meV/kBar for both InAs and GaSb. Using the values for η yielding the best zero pressure band offset ΔE_v gives a decrease with pressure of this offset of 0.2 meV/kBar. This means that virtually the VB offsets should be pressure independent in the InAs-GaSb system. This is however in strong disagreement with the experimental results.

The pseudopotential theory. The basic idea of the Frensley-Kroemer[2,31] pseudopotential (FKP) theory is to calculate the bandstructure of the bulk materials relative to some effective interstitial potential V_i which resembles an effective zero potential at infinity. Therefore lining up V_i in order to obtain the band offset is a method closely related to the EAR. Just as the HAO, the FKP predicts several heterojunction band offsets in good agreement with experimental values. An extension of the FKP is the inclusion of a dipole potential V_d in order to allow for interface dipoles caused by charge transfer along the bonds across the interface. This effect causes a discontinuity V_d in the interstitial potential V_i at the interface. For InAs-GaSb a band overlap $\Delta=130$ meV is found by lining-up the interstitial potentials as calculated by the FKP directly, whereas including the dipole potential V_d the InAs CB is found 10 meV above the GaSb VB which means a staggered lineup instead of a broken-gap.

From the FKP it should principally be feasible to extract predictions on pressure dependence of band offsets, if one assumes the atomic-like states to have a pressure independent energy. The calculations using the FKP however lie outside the scope of this work, and have not yet been performed.

Relative theories

  The charge neutrality level concept. An approach conceptually different from the EAR, HAO, and the FKP methods was proposed by Flores and Tejedor[32,33], and elaborated by Tersoff[34]. Here, the influence of interface dipoles, caused by gap-states is supposed to have the strongest influence on the exact value of the band offsets. Gap-states close to the VB are more valence-like, and gap-states close to the CB are more conduction-like. A filled valence-like state will correspond to a charge excess at the interface, whereas an empty conduction-like state will cause a charge deficit. Filling or emptying these states introduces interface dipoles leading to a potential which tends to rearrange the band line-up in such a way as to minimise the interface dipole. The problem therefore resides in finding a line-up suffering very little or not at all from interface dipoles. This can be translated in finding a specific energy level serving as a sort of Fermi-level for the gap-states, called the charge neutrality level, at which gap-states are equally conduction- and valence-like. This level therefore presents an effective midgap (E_b) energy[34]. Ideally, neutrality at the interface is achieved by filling up all gap-states up to E_b and leaving them empty above. It then seems reasonable to put the charge neutrality levels at the heterointerface at the same position in order to establish a band line-up suffering least of interface dipoles. In this way[34] the value of the band overlap Δ at the InAs-GaSb interface is found to be 140 meV. In this calculation E_b is calculated relative to the effective Γ-valence bandedge given by $\frac{1}{3}(2\Gamma_8+\Gamma_7)$, corrected for spin-orbit coupling.

  The pressure dependence of the charge neutrality level (E_b) has not yet been investigated in detail. However, E_b matches the middle of the effective dielectric energy-gap (Penn-gap, E_p) averaged over the first brillouin zone. Therefore an estimate of the pressure dependence of E_p yields a first approximation for the pressure dependence of the charge neutrality level. Instead of calculating E_p by really averaging over the whole zone it can be calculated by evaluating the energies of conduction and valence bands at the Baldereschi points[35] as done in reference 36. For an fcc lattice the first Baldereschi point lies close to the X-point, and therefore the major contribution to the Penn-gap comes from the $X_{6,7}$ gap ($\propto 4$ eV for III-V materials). Although the actual values of E_b and the Penn-gap are not given by the X-gap alone, it is a reasonable to assume that their pressure dependences do follow the X-gap in first order approximation. Within the experimental and theoretical accuracy[11,34] (1-2meV/kBar) the middle of the X-gap moves down with respect to the Γ_8 valence bandedge at a rate of 4 meV/kBar for both InAs and GaSb, hence predicting a valence band offset which is virtually pressure independent. In a different way, the gap density of states can, instead of summing over the Baldereschi points, also be approximated[33] by coming for 25% from the Γ-gap and for 75% from the X-gap. If according to this rule the position of E_b and its pressure dependence are following the average of the middle points of both Γ- and X-gaps with a weight facor of 1:3 respectively, the valence band offset is also found to be pressure independent to within 1 meV/kbar.

The dielectric midgap energy. A method strongly related to the charge neutrality concept is presented by Cardona and Christensen[36]. The idea is that potentials acting on energy bands have in general to be screened by the dielectric constant. If lattice perturbations are caused not by an external potential, but by deformations due to e.g. strain, the deformation potentials which act on the bands have to be screened in a different way in each band, and it is not anymore correct to use the same dielectric constant for this screening. The DME is calculated as the (virtual) energy level at which the screening is given by division through the dielectric constant $\epsilon(k)$, just as would be the case for an external potential acting on all the bands in the same way. The DME is in the middle of the Penn-gap just as the charge neutrality level. In this method bandoffsets are firstly calculated as in the HAO, but the thus resulting differences in DME at the interface are screened by an effective "interface" dielectric constant. The results using the DME concept on the InAs-GaSb interface yield a band overlap $\Delta \simeq 130$ meV.

Using this method, the pressure dependence of the Γ_8-VB offset at the InAs-GaSb interface has been calculated by evaluating the pressure dependence of the Penn-gap at the first Baldereschi pont, as discussed above. An increase of 1.5 meV/kBar is found, which in view of the accuracies[36] can be translated in a nearly pressure independent valence band offset.

The matching of hybrids. If interface dipoles are formed, the potential steps at the interface between the energybands will be screened[37], thus changing the canonical line-up as predicted by the HAO. Using the atomic energy values of the s- and p-states an average hybrid energy $\epsilon_h = \frac{1}{4}\langle\langle\epsilon_s + 3\epsilon_p\rangle\rangle$ can be defined, where the averaging is over the two atom species of the compound. It can be shown on relatively simple grounds[26,37] that screening the discontinuity in the different ϵ_h (in the canonical HAO line-up) at the interface by the optical dielectric constant ϵ_∞ yields a line-up which induces least possible interface dipoles. The hybrid energy is therefore related to the DME and to Tejedors charge neutrality level. For InAs-GaSb this method leads however to a staggered line-up in which the GaSb VB is 160 meV lower than the InAs CB. Because the positions of the bandedges relative to the hybrid energy are calculated in the same way as in the original HAO, the pressure dependence also equals that predicted by the HAO, and is subsequently not compatible with experiments.

VII. RELATED WORK

Beerens[38] et al. have reported magnetotransport measurements on an InAs quantum well sandwiched between GaSb layers under pressure. From an observed reduction of hole-states at the InAs-GaSb interfaces with pressure, the band overlap Δ has been found to decrease indeed with pressure, at a rate of 6,7 meV/kBar. However, their results might be influenced by a strong band bending in the outer GaSb layer, leading to an increase in the number of electrons due to an effective rise of the Fermi level in the well. This situation could lead to an erroneous interpretation of the experimental results as a function of pressure. The presence of such a band bending has recently been calculated and experimentally verified by Altarelli and Maan[39].

In the experiments on GaAs-Ga$_{1-x}$Al$_x$As MQW and SL (see sec.II) Wolford and coworkers[7,8] found the ratio between the CB-offset and the VB offset to be pressure independent. If ad hoc the assumption is made that this ratio is also pressure independent at the InAs-GaSb interface, a decrease of the overlap Δ at a rate of 4 meV/kBar is calculated, which is surprisingly close to the experimental value. Recalling that in the GaAs-Ga$_{1-x}$Al$_x$As system this ratio is almost independent on the composition parameter x, it can be suggested that if at a heterointerface the energy gaps are changed by either small changes in composition or by pressure, the lineup changes in such a way that the ratio between CB and VB discontinuities stays constant.

VIII. CONCLUSION

Hydrostatic pressure has be used to change the band offset of a type II interface in a controlled way by changing the band gaps of the bulk materials. At the InAs-GaSb interface it is observed that the overlap between the InAs CB and the GaSb VB Δ reduces with a rate of 4.5 meV/kBar, which indicates that both CB and VB offsets change with pressure. This result therefore constitutes an interesting test for different theories on band offsets. Many theories predict qualitatively the trend as observed in experiment. However, the valence band discontinuity is predicted to vary only very little with pressure which is not compatible with the experiment. Finally it is found that the ratio between CB and VB offsets is nearly pressure independent, as is the case for the GaAs-Al$_x$Ga$_{1-x}$As interface. To my knowledge there is however no theoretical background for this phenomenon.

Acknowledgements

In this place I would like to acknowledge some people whose efforts has lead to this work. First of all L.L. Chang and L. Esaki who initiated the investigation of the InAs-GaSb system. I like to thank J.C. Maan for solving all kinds of experimental and theoretical problems, A. Fasolino and M. Altarelli for many discussions on the bandstructure problems, and F.Flores for clarifying the charge neutrality level concept. Finally I am indebted to G. Martinez and J.C. Portal for helping with pressure related problems, and to prof. P. Wyder for his interest in this work

References

1. H.Kroemer, Surf.Sci. **132**, 543 (1983)

2. H.Kroemer, J.Vac.Sci.Technol. **B2**, 433 (1984)

3. L.Esaki. IEEE J.Quantum Electron. **QE-22**, 1611 (1986)

4. G.Duggan, J.Vac.Sci. Technol. **B3**, 1224 (1985)

5. T.W.Hickmott in Two-Dimensional Systems: Physics and New Devices, proc. of the Winterschool Mautendorf, February 24-28, 1986, ed. by G.Bauer, F.Kucher, and H. Heinrich, p.72

6. B.A.Wilson,P.Dawson, C.W.Tu, and R.C.Miller, J.Vac.Sci. Technol. **B4**, 1037 (1986)

7. D.J.Wolford,T.F.Kuech,J.A.Bradley,M.A.Gell,D.Ninno, and M.Jaros, J.Vac.Sci. Technol. **B4** , 1043 (1986)

8. M.A.Gell,D.Ninno,M.Jaros,D.J.Wolford,T.F.Kuech, and J.A.Bradley, Phys.Rev.B **35**, 1196 (1987)

9. S.P.Kowalczyk,J.T.Cheung,E.A.Kraut, and W.Grant, Phys.Rev.Lett. **56**, 1605 (1986)

10. G.Martinez, in Optical Properties of Solids, ed. M.Balkanski, Handbook on Semiconductors vol.2, (North Holland, Amsterdam 1980), p.181

11. Landolt-Bornstein: numerical Data and Functional relationships in Science and TEchnology, ed. by D.Bimberg (Springer Verlag New York 1982) vol.17.

12. G.A.Sai-Halasz, L.L.Chang,J.M.Welter,C.A.Chang, and L.Esaki, Solid State Commun. **27**, 935 (1980)

13. Y.Guldner,J.P.Vieren,P.Voisin,M.Voos,L.L.Chang, and L.Esaki, Phys.Rev.Lett. **45**, 1719 (1980)

14. J.C.Maan,Y.Guldner,J.P.Vieren,P.Voisin,M.Voos,L.L.Chang, and L.Esaki, Solid State Commun. **39**, 683 (1981)

15. J.C.Maan, "Infrared and Millimeter Waves", ed. by K.J.Button, Ac.Press, New York **8**, 387 (1982)

16. L.L.Chang, J.Phys.Soc.Jpn, **49**, Suppl.A, 997 (1980)

17. M.Altarelli, Phys.Rev.B **28**, 842 (1983)

18. A.Fasolino, and M.Altarelli, Surface Sci. **142** 322 (1984)

19. L.M. Claessen,J.C.Maan,M.Altarelli,P.Wyder,L.L.Chang, and L.Esaki, Phys.Rev.Lett. **57** 2556 (1986)

20. L.M. Claessen,J.C.Maan,M.Altarelli,P.Wyder,L.L.Chang, and L.Esaki, Superlattices and Microstructures **2**, 551 (1986)

21. J.C.Maan, in Optical Properties of Narrow-gap low-dimensional Structures Ed. by C.M. Sotomayor Torres, J.C.Portal,J.C.Maan, and R.Stradling, (Plenum Press New York 1986)

22. L.M.Roth, ref.10, vol.1, p.474

23. C.R.Pidgeon, ref.10, vol.2, p.223

24. K.H.Seeger: "Semiconductor Physics", ed. by M.Cardona,p.Fulde, and H,-J.Queisser, (Spriner Verlag New York 1982), p.358

25. G.Platero, and M.Altarelli, Phys.Rev.B **36**, 6591 (1987)

26. W.A.Harrison, ref.5, p.62

27. E.A.Kraut, J.Vac.Sci.Technol. **B2** , 486 (1984)

28. J.L.Shay,S.Wagner, and J.C.Phillips, Apll.Phys.Lett. **28**, 31 (1976)

29. G.W.Gobeli, and F.G. Allen, in Semiconductors and Semimetals, ed. by R.K. Willardson, and A.C.Beer, vol.2, (Ac. Press New York 1960), p.263

30. W.A.Harrison: "Electronic Structure and the Properties of Solids. (W.H.Freeman and Co. san Francisco, 1980)

31. W.R.Frensley, and H.Kroemer, Phys.Rev.B **16**, 2642 (1977)

32. F.Flores, and C.Tejedor, J.Phys.C. **12**, 731 (1979)

33. C.Tejedor, F.Flores, and E.Louis, .Phys.C. **10**, 2163 (1977)

34. J.Tersoff, Phys.Rev.B **30**, 4874 (1984)

35. A.Baldereschi, Phys.Rev.B 7, 5212 (1973)

36. M.Cardona, and N.E.Christensen, Phys.Rev.B **35**, 6182 (1987)

37. W.A.Harrison, and J.Tersoff, J.Vac.Sci.Technol. **B4**, 1068 (1986)

38. J.Beerens,G.Gregoris,J.C.Portal,E.E.Mendez,L.L.Chang, and L.Esaki, Phys.Rev.B **35**, 3039 (1987), and Phys.Rev.B **36**, 4742 (1987)

39. M.Altarelli, J.C.Maan,L.L.Chang, and L.Esaki, Phys.Rev.B **35**, 9867 (1987)

ELECTRONIC PROPERTIES OF SEMICONDUCTOR INTERFACES:

THE CONTROL OF INTERFACE BARRIERS

Fernando Flores

Departamento de Física de la Materia Condensada
Facultad de Ciencias. Universidad Autónoma de
Madrid, 28049 Madrid, Spain

ABSTRACT

The electronic properties of metal-semiconductor interfaces and heterojunctions are discussed in relation to
the formation of interface barriers. The theoretical results
presented in this communication suggest that the semiconductor heights are basically determined by the intrinsic properties of the two crystals forming the interface. Small changes, up to 0.3eV, can be introduced, however, in the barriers
by means of adequate intralayers deposited at the interface.
This opens interesting applications from the point of view
of the band structure engineering in semiconductor
microstructures.

1. INTRODUCTION

Semiconductor device characteristics are crutially
dependent on their barrier heights [1] : the purpose of basic
physics is to explain how these barriers are formed in
order to control their values for technological necessities.

The first semiconductor interface was prepared by
F.Braun [2] in 1874. Since then, a lot of experimental and
theoretical work has been done in this field. Let us mention,
from the theoretical point of view, that the first models
about the formation of semiconductor interfaces were proposed
in the late thirties [3,4] , and that the pioneering work of
Bardeen [5] and Heine [6] prepared the field for our current
understanding of these interfaces.

The perspective on semiconductor interfaces changed
in the late seventies with the new experimental tools available to analyse interfaces [7,8]. Due to the great amount of
new information accumulated in the last ten years, a new
and deeper understanding of the semiconductor interfaces
is emerging. For the first time, we start having a real

microscopic image of what is happening at a semiconductor interface, how these microscopic characteristics are related to the interface barriers and, finnally, how we can use all this knowledge to control and modify the different semiconductor interfaces and their barriers.

The purpose of this contribution is to present the actual status of the theory about semiconductor interfaces, considering simultaneously metal-semiconductor and semiconductor-semiconductor junctions. In this discussion, things shall be presented emphasizing some details that can be of importance for controlling semiconductor barriers: in particular, we shall also analyse how the barriers may depend on the detailed geometry of the interface and on the different intralayers that can be deposited at the interface. The aim of this discussion is to show how the interface semiconductor barriers can be controlled, within given limits, by an appropiate microscopic preparation of the interface.

No attempt is made in this contribution to present the experimental evidence on the semiconductors interfaces and their barriers. The interested reader is referred to other papers of this School or to some review papers[7-11].

The structure of this communication is the following: in sect. 2 we discuss some general models for the metal-semiconductor and semiconductor-semiconductor interfaces. The emphasis is put on the Induced Density of Interface States (IDIS) model and the related charge neutrality level of each semiconductor[12-13]. A more detailed analysis of the semiconductor interface barriers is presented in section 3. Most of the results discussed in this section have been obtained using a selfconsistent tight-binding approach that is briefly introduced in section 3.1 In section 3.2 main results for metal-semiconductor interfaces are presented,while in section 3.3 heterojunctions are discussed, and in section 3.4 we analyse the effect of intralayers on metal-semiconductor and semiconductor-semiconductor junctions. Finnally, in section 4 we present our conclusions from the point of view of the engineering of the semiconductor interfaces.

2. GENERAL MODELS FOR SEMICONDUCTOR INTERFACES

In this section we present some general models for different semiconductor interfaces. In section 2.1 we consider the metal-semiconductor junctions, while in section 2.2 heterojunctions are discussed. The aim of this section is to present the IDIS model, which seems to offer the simplest and most adequate approach to the semiconductor interfaces formation. In order to put this model in perspective, other related models are also presented.

2.1. Metal-semiconductor interfaces

The Schottky model[3-4] appears if it is assumed that

no electron can be transfered from the metal to the semiconductor. The Bardeen model [5] appears when there is a high density of surface states, and it is assumed that these states control the Fermi energy level.

In these two models it is assumed that the density of states at the interface is given by the superposition of the density of surface states for each independent crystal. The IDIS model[12,13] tries to take into account the effects of coupling both crystals. Fig. 1 illustrates how the interface density of states changes in the 1-dimensional model of a metal covalent semiconductor junction[14] for different degrees of coupling between the metal and the semiconductor. The important point about the results[11] shown in fig.1 is that the interface density of states in symmetric with respect to the semiconductor mid-gap, and that a charge neutrality level (the mid-gap in a 1-dimensional model) can be defined for the semiconductor, in such a way that this level plays the role of an effective Fermi level for the crystal at the interface. This means that, when forming the junction, the metal Fermi level tends to equalize the charge neutrality level of the semiconductor.

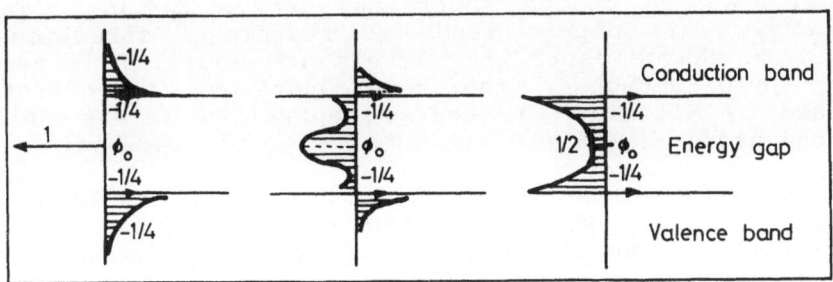

Fig. 1.Induced density of states around the main gap of a one-dimensional metal-semiconductor model (a). For a decoupled interface a surface state appears at the middle of the gap. (b) As the metal approaches the semiconductor the surface state broadens. (c) For an intimate contact the resonance extends to the whole gap[14].

Similar arguments can be given for ionic semiconductors. In general, one can introduce a charge neutrality level for each semiconductor independent of the face orientation. In table I we show the results calculated by Tersoff for different semiconductors.

In a zeroth-order approximation, the IDIS model predicts a Schottky barrier that is metal independent and defined by the difference between the energy gap and the charge neutrality level of the semiconductor as given by table I. Small corrections to this value can be calculated taking into account [14] the transfer of charge between the metal and the semiconductor; in the zeroth-order approximation discussed above it is assumed that there is no transfer of charge between the two crystals: when the Fermi level and the charge neutrality level are aligned the semiconductor remains neutral.

TABLE I

Si	Ge	GaP	InP	AlAs	GaAs	InAs	GaSb
0.36	0.18	0.81	0.76	1.05	0.50	0.50	0.07

Charge neutrality levels for different semiconductors measured from the top of the valence band (in eV).

2.2. Heterojunctions

Anderson model[15] is the equivalent of the Schottky model for metal-semiconductor interfaces. In the IDIS model for heterojunctions[16], it is assumed that a strong interface dipole is induced between the two semiconductors when their respective charge neutrality levels do not coincide in energy. This dipole tends to restore a situation for which both charge neutrality levels are equal. In a zeroth-order approximation [14], the heterojunction band-offset is obtained by aligning the charge neutrality levels of both semiconductors.

The best evidence supporting this model has been given by Margaritondo[17] who has shown that the predictions of this model for the heterojunctions of Si and Ge with other semiconductors are in good agreement with the experimental evidence. Katnani and Margaritondo[18] also found that for many semiconductors the transitivity rule holds[11]. This can be easily explained by means of fig. 2, where all the charge neutrality levels of different semiconductors have been aligned. Notice that a similar argument can be applied to Schottky-barriers by aligning the metal Fermi energies and the semiconductors charge neutrality level (fig.2).

Summarizing the comments of this section, in fig.2 we find the main implications of the IDIS model. According with this figure, we can define an absolute value for the position of the electronic energy levels of a crystal: this is defined by either the Fermi energy of metals or the charge neutrality level of semiconductors. This level defines all the possible barrier heights between one crystal and other metal or semiconductor.

Taken at its face value, the IDIS model imposes strong restrictions to the possibility of changing or controlling the barrier heights of different interfaces. As fig.2 shows, if a metal (say, a monolayer) is deposited between two semiconductors, the band offset between the last two crystals cannot be changed. We should comment that this result can

only be taken as a zeroth order approximation: more accurate
calculations show that things are a little more complex
at the different interfaces. Indeed, changes of a few tenths
of eVs can be induced with respect to the results of the
IDIS model, depending on the different interface interactions. In this sense, notice that a few tenths of eV is
a small quantity when compared with most of the semiconductors optical gaps (the charge neutrality level is an average
of the different mid-gaps defined over the 2-dimensional
Brillonin Zone[14]); thus, we can say that although the

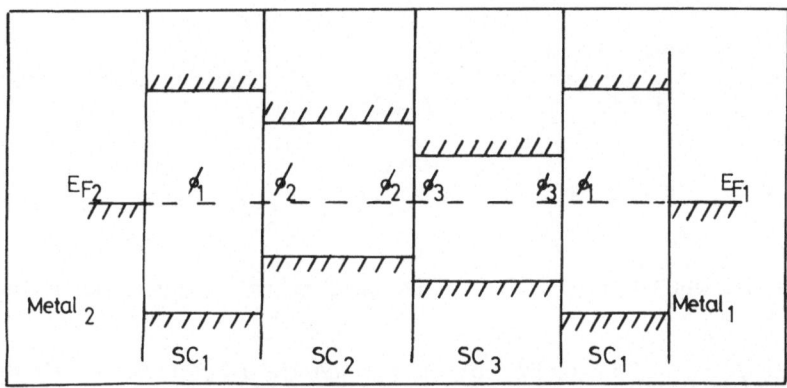

Fig. 2 Shows the alignment of the charge neutrality levels
and Fermi levels of different metals and semiconductors.

IDIS model explains in a simple way the physics operating
at different interfaces, accurate results can only be obtained by means of more complex models. This observation brings
us to the next section, where a first-principle self-consistent tight-binding model is used to analyse Schottky barriers
and heterojunctions.

3. BEYOND THE IDIS MODEL

The IDIS model applies strictly only to the case of
an interface between two thick crystals. What happens in
the case of thin films, say a monolayer of a metal on a
thick semiconductor, has to be analysed using a different
approach. This case, as well as the case of intralayers
between two crystals, and the effect of the specific geometry
of the interface, i.e, where the atoms sit on and how this
position affects the interface barrier, will be considered
in this section. The general approach we have used to discuss
these effects will be presented in section 3.1: this is
a first-principle selfconsistent approach with no adjustable
parameter [19]. The main results obtained using this approach
will be presented in the rest of the section.

3.1. A Self-consistent tight-binding approach for interfaces

In this approach[19] we describe the electronic properties of any junction by means of a linear combination of

the atomic orbitals, ψ_{in}, of each atom. We assume ψ_{in}, and its eigenvalue in the atom, E_{in}, to be known. Starting with this basis, one can use a Linear Combination of Atomic Orbitals and show[19] that the effective Tight-Binding hamiltonian associated with the crystal built up by the single atoms is defined by the following parameters:

$$(Heff)_{\alpha\alpha} = E_\alpha + \frac{1}{4} \sum_\beta (S\alpha\beta)^2 (E_\alpha - E_\beta) - S\alpha\beta \, T\alpha\beta \qquad (1a)$$

$$(Heff)_{\alpha\beta} = T\alpha\beta \qquad (1b)$$

where $\alpha = (i,n)$, $\alpha\beta = \langle \psi_\alpha | \psi_\beta \rangle$ and:

$$T_{\alpha\beta} = -\frac{1}{2} \int_{\sigma_{\alpha\beta}} (\psi_\alpha \nabla \psi_\beta - \psi_\beta \nabla \psi_\alpha) \cdot d\underset{\sim}{s} \qquad (2)$$

here $\sigma_{\alpha\beta}$ is a particular surface chosen such that the following condition is satisfied:

$$\int_{\Omega_i} \psi_{in} \psi_{i'n'} \, d\underset{\sim}{r} = \int_{\Omega_{i'}} \psi_{in} \psi_{i'n'} \, d\underset{\sim}{r} = \frac{1}{2} \int_\Omega \psi_{in} \psi_{i'n'} \, d\underset{\sim}{r} = \frac{1}{2} S_{in,i'n'} \qquad (3)$$

where the whole space, Ω, is split into the two subspaces Ω_i and $\Omega_{i'}$.

Equ.(2) holds only for short-range potentials. Wavefunctions ψ_α and ψ_β are, however, the solution of long range atomic potentials. In that case, it can be proved that equ.(2) has to be modified to:

$$T\alpha\beta = -\frac{\gamma}{2} \int_{\sigma_{\alpha\beta}} (\psi_\alpha \nabla \psi_\beta - \psi_\beta \nabla \psi_\alpha) \cdot d\underset{\sim}{s} \qquad (4)$$

where γ is typically about 1.4 for most semiconductors and normal metals.

Equ (4) yields the off-diagonal elements of the effective hamiltonian, Heff, and equ.(1a) can then be used to obtain how the overlap between orbitals, $S\alpha\beta$, and the hopping interaction, $T\alpha\beta$, change the diagonal component.

In our approach, instead of calculating all the different elements of the tight-binding hamiltonian from equs(1a) and (4), we only use these equs to obtain the interface interactions between the atoms of the crystals forming the interface. As regards the bulk semiconductor parameters of each crystals, we have chosen to use Vogl et al's parameters[20] wich are known to give rather accurate electronic band structures.

Once we have defined the different tight-binding elements of the total hamiltonian, we proceed to calculate the electronic charge at the interface. From this charge, we obtain the induced electrostatic potential on each atom and correct for its energy levels with the induced potentials (see ref.21). Then, a selfconsistent calculation is performed with the energy levels of each atom being related to the transfer of charge between the two crystals.

Notice that the Hartree-selfconsistency introduced in the calculation takes into account in an appropriate way the charge neutrality conditions related to the charge neutrality levels of the IDIS model: this explains why this tight-binding calculation goes beyond the discussion of section 2.

3.2. Metal-semiconductor junctions

The self-consistent tight-binding (SCTB) method has been applied to analyse different metal-semiconductor inter-faces. Here, we mention three speciific cases; (a)Ag on GaAs (110): (b) K on GaAs (110) and (c) Al on Si(111).

First of all, we have considered the early stages in the formation of Schottky barriers. The crutial point is to find out how the barrier height depends on the number of metal layers deposited on the semiconductors. Ortega et al [22] have analysed the effect of depositing 1,2...and 5 monolayers of Ag on GaAs (110): their results show that the Fermi level is practically pinned by only a monolayer although small changes (± 0.05eV) appear with the deposition of more metal layers.This result seems to be in good agreement with the experimental evidence [7] .

The case of alkali metals on semiconductors is of great interest: in this case, the metal atoms have a large size and there is no interdiffusion between the semiconductor and the deposited metal. On the other hand, recent experimental data [23] show no interface reactivity and a Schottky barrier formation that is completed with the deposition of an alkali metal monolayer.

We have considered the case of K on GaAs (110), and have analysed the Schottky barrier as a function of the adsorbed site for the deposited alkali metal[24] . The chemisorption energy has been also calculated using the SCTB approach. Our results show that the Ga-site is the most favourable for K, and that the Fermi energy, ~ 0.57eV, has a slight dependence on the site position (the experimental value[23] for Ef is 0.55eV). We should comment that the slight dependence shown by Ef on the adsorption site is mainly due to the large atomic size of K.

Things appear slightly different for smaller metal atoms; this is the case of Al deposited on Si (111). It has been found [25] that a monolayer of Al on Si (111) pins the Fermi level at the same position as a thick Al-layer, in agreement with previous results for Ag. More interestingly, in refs.25 and 26 it has been found that the barrier height presents slight changes as a function of the sites of the chemisorbed metal-atom: thus, the barrier height for an Al-Si (111) interface is shifted by ± 0.2eV with respect to the jellium model of Al when the atoms are located on a top position or on a three-fold coordinated site of the semiconductor surface. This result shows that the semiconductor charge neutrality level can be slightly changed with the chemisorbed metal site, thus, each specific metal-semiconductor interface has to be considered by itself if one is interested in calculating the barrier height with high accuracy (say, with an error less than ±0.2eV).

3.3. Heterojunctions

Heterojunctions offer a great advantage over metal-semiconductor junctions from the point of view of a theoretical model, since the geometry of the interface is rather well known, at least for those semiconductors having a good lattice matching. Accordingly, they present a good case to which apply the SCTB model discussed above. Table II shows our results for the band-offsets of different heterojunctions [11]. In the same table we include the results of the IDIS model and of the selfconsistent local density (SCLD) approach [27].

TABLE II

	SCTBM	IDISM	SCLDM	EXPERIMENTAL (22)
GaP-Si	0.64	0.58	0.61	0.80
GaAs-Ge	0.61	0.37	0.63	0.3-0.55
AlAs-Ge	0.94	0.87	1.05	0.95
AlAs-GaAs	0.32	0.50	0.37	0.38-0.58
ZnSe-Ge	1.70	1.61	2.17	1.40-1.52
ZnSe-GaAs	1.01	1.21	1.59	1.10

Band offsets (in eV) for different heterojunctions as calculated with different models.

It is worth mentioning the striking agreement between the results of the SCTB model and the SCLD approach. Notice that even the IDIS model yields reasonable results, although in some cases some differences of 0.2eV appear when comparing with the other models for heterojunctions of low ionic semiconductors.

As regards the early stages in the formation of hetero-junctions, we have also found [28], for the deposition of Ge on GaAs (110) or (100), that a monolayer of Ge is enough to form completely the barrier. This is in agreement with the results for metal-semiconductor interfaces, suggesting that in all the cases the semiconductor barriers are mainly controlled by the few layers deposited around the interface. This result and the posibility of modifying slightly the barriers with the geometry of the interface, suggest that it could be worth looking at the effect of intralayers deposited at the interface of a junction. This brings us to the next paragraph.

3.4. Intralayers

The SCTB method has been applied to the analysis of two different cases: (i) the first one corresponds to a Ge-ZnSe interface [29] ; Niles et al [30] have found experimentally that the deposition of Al at this interface can introduce important changes in the band-offset. (ii) The second case we have analysed is the Si-GaP heterojunction [31] , with different intralayers [32] of Cs, Al and H. This case allows us to see how the electronegativity of the deposited species can change the heterojunction band-offset.

As regards the Ge-Al-ZnSe (110) interace, let us comment that the theoretical results [29] for the deposition of only a monolayer of Al at the interface can introduce changes in the barrier offset of $\sim$ 0.2eV, in good agreement with the experimental evidence [30] . This result is mainly due to the slight changes introduced in the charge neutrality level of each semiconductor, as induced by the metal monolayer, and shows how metal intralayers can induce shifts in the heterojunction band-offsets.

More interestingly, we have also analysed [32] the effect of different intralayers in the band-offset of the Si-GaP heterojunction. Table III shows our results for a monolayer of Cs, Al or H. We find changes that depend clearly on the electronegativity of the chemisorbed species.

Table III

Cs	Al	H
0.4 ±0.15eV	0.15 ± 0.06eV	−1.15 ± 0.05eV

Change in the band-offset of the GaP-Si (110) heterojunction as a function of the deposited intralayer.

We should comment that the error bars introduced in table III are related to the different geometries we have assumed for the chemisorbed monolayer.

Summarizing this section, we conclude that intralayers of different ionicity produce changes in the bandoffset of a heterojunction that are related to the electronegativity of the adsorbed species.

4. CONCLUSIONS

We have presented a discussion of the main physical mechanisms controlling the barrier formation of semiconductor interfaces. Our analysis shows that the barrier heights between a metal and a semiconductor or between two semiconductors are basically determined by the intrinsic properties of the two crystals. This result, however, is only valid with an accuracy which is around a small fraction of the semiconductors optical gap. This implies that small changes in the barrier heights of semiconductor junctions, 0.3eV, can be introduced by means of adequate intralayers deposited at the interface. These changes are important enough, however, to have interesting implications as regards the engineering of semiconductor interfaces.

We conclude that an appropriate use of an adequate intralayer opens the possibility of controlling the barrier heights of different semiconductor interfaces.

Acknowledgment

This work has been partially supported by the Scientific Cooperation Contract N5.ST2J-0254-7-E (EDB) between the European Communities and the Universidad Autónoma de Madrid (Spain).

REFERENCES

1.- E.H. Rhoderick, Metal-Semiconductor Contact (Oxford: Oxford University Press 1987).
2.- F.Braun, Papp.Ann. **153**,556 (1874)
3.- N.F. Mott, Proc.Cam.Phyl.Soc. **34**,568 (1938).
4.- W. Schottky, Z.Phys. **113**,367 (1939).
5.- J. Bardeen, Phys.Rev. **71**, 717 (1947).
6.- V. Heine, Phys.Rev. **138**, 1689 (1965).
7.- T. Kendelewicz and I. Lindau, Crit. Rev, **13**, 27 (1986)
8.- L. Brillson, Handbook of Synchorotron Radiation, Vol II, ed. G.V. Marr (Amsterdam: North-Holland, 1985)
9.- G. Le Lay, J.Vac.Sci.Technol. **2**, 354 (1983)
10.-C.Calandra, O.Bisi and G.Ottaviani, Surf.Sci.Rep, **4**, 271 (1984).
11.-F.Flores and C.Tejedor, J.Phys.C **20**, 145 (1987).
12.-C.Tejedor, F.Flores and E.Louis, J.Phys.C **11**, L19 (1978)
13.-J. Tersoff, Phys.Rev.Lett, **32**, 465 (1984)
14.-F.Flores, Semiconductor Interfaces: Formation and Properties (ed. G.Le Lay, J.Derrien and N.Baccara) Springer-Verlag (1987).
15.-R.L.Anderson, Ph.D.Thesis Syracuse University, New York (1960).
16.-C.Tejedor and F.Flores, J.Phys C **11**,L19 (1978).
17.-G.Margaritondo, Phys.Rev.B **31**, 2526 (1985).
18.-A.D. Katnani and G.Margaritondo, Phys.Rev.B **28**,1944 (1983).
19.-F.Flores, A.Martín-Rodero, E.C.Goldberg and J.C.Durán Il Nuovo Cimento D (1988) in press.
20.-P.Vogl, H.P.Hjalmaron and J.D.Dow, J.Phys.Chem.Solids **44**, 365 (1983).

21.-F.Guinea, J.Sánchez-Dehesa and F.Flores, J.Phys.C **17**, 2039 (1983).

22.-J.Ortega, J.Sánchez-Dehesa and F.Flores Phys.Rev.B (1988) in press.

23.-M.Prietsh et al (to be published)

24.-J.Ortega and F.Flores (to be published).

25.-G.Platero, J.A.Vergés and F.Flores, Surface Sci.**168**, 100 (1986).

26.-H.I.Zhang and M. Schlüter, Phys.Rev.B **18**, 1928 (1978)

27.-C.G.Van del Walle and R.M.Martín, J.Vac.Sci.Techonol.B **4**,1055 (1986).

28.-A.Muñoz, J.Sánchez-Dehesa and F.Flores, Europhysics Lett **2**, 335 (1986)

29.-J.C. Durán, A. Muñoz and F.Flores, Phys.Rev.B **35**,772 (1987)

30.-D.W.Niles, E.Colavita, D.Perfetti, C.Quaresima and M.Capozi. J.Vac.Sci.Technol. A**4**, 962 (1986)

31.- P.Perfetti, Surface Sci. **189/190**, 362 (1987)

32.-J.C.Durán, A.Muñoz, R.Pérez and F.Flores (to be published).

31

POLAR/POLAR, COVALENT/COVALENT and COVALENT/POLAR SEMICONDUCTOR SUPERLATTICES

S.Ciraci
Department of Physics,
Bilkent University,
Bilkent 06533 Ankara, Turkey

A comparative study of polar/polar, covalent/covalent, and covalent/polar semiconductor superlattices is presented. Based on the calculated formation energy, charge density and electronic structure, the effects of the superlattice parameters on the overall properties (stability, quantum well structure, indirectness of the band gap etc.) are discussed. Geometry optimized total energy calculations indicate that the formation energy of all superlattices studied is positive implying that the separation (or segregation) into constituent crystals are favored. The contribution of the superlattice periodicity, lattice mismatch, and polarity of sublattices in the formation energy vary depending upon the type of the superlattice. The electronic structure is found to depend strongly on the sublattice periodicity when it is small.

1. INTRODUCTION

Advances in epitaxial growth tehcniques have led to the production of highly perfect synthetic semiconductor heterostructures[1,2]. Alternatingly and epitaxially grown semiconductor crystals, whose electronic properties in particular fundamental band gaps are different, form the semiconductor superlattices. The semiconductor superlattice can be viewed as a multiple quantum well separated by the potential barriers of the band discontinuities. Carriers confined in these quantum wells display a quasi 2-D character leading to a new regime of quantum effects[3]. They have a quantization different from that of the 3-D systems. Their transport, screening and scattering properties are also significantly different. High carrier mobilities, and novel electronic properties observed in the modulatingly doped[4] $(GaAs)_n/(Ga_{1-x}Al_xAs)_m$ superlattices have initiated many fundamental and applied research on the new devices in microelectronics and photonics. Progress made in the heterostructures of compound semiconductors has stimulated the idea of increasing carrier mobility in the Si/Ge quantum well structure. In an effort to compensate for the deficiencies of silicon, and to further upgrade this well established technology the epitaxial growth of the pseudomorphic $Si/Si_{1-x}Ge_x$ superlattice has been achieved[5,6]. Furthermore, the growth of pure Ge pseudomorphically resricted to the Si(001) substrate has been realized[7] in spite of the large lattice mismatch of 4%. More importantly, direct optical transitions observed[7] in such a system (Si_4/Ge_4) are found neither in constituent crystals, nor in the $Si_{0.5} Ge_{0.5}$ alloy.

Clearly, semiconductor superlattices present new conceptual ideas about synthetic semiconductors and new device applications. Apart from the unusual electronic properties (high mobility etc.) stemming from the 1-D quantum wells a semiconductor superlattice $(S_1)_m/(S_2)_n$ provides many degrees of freedom for controlling the electronic properties of this system. These are basically the repeat periods (m,n and m+n), and the sublattices (S_1 and S_2) themselves. The repeat period n+m (and also the sublattice thickness n and m) determine the subband strucuture and the confinement of states for a given depth of the quantum well. Normally, S_1 and S_2 have similar lattice parameters, but different electronic structures. Sometimes, they can be replaced by the alloy $(S_1)_{1-x} (S_2)_x$, which yields continuously varying properties. This way the band gap and the band-offset of the superlattice can be monitored to engineer the quantum well structure. In the growth and overall character of a heterostructure, the polarity of the sublattices, S_1 and S_2, play a crucial role, which leads to a more fundamental classification. These are polar/polar, covalent/covalent and covalent/polar heterostructures. Consequently, major theoretical and experimental effort has gone into understanding the effect of the polarity differences, such as the stability, interface defects, and charge transfer.

This paper present a comparative study of the polar/polar, covalent/covalent and covalent/polar semiconductor superlattices. The issues addressed are the stability and electronic properties. In addition, the δ-doping in Si_n/Ge_n, which gives rise to marked changes in the character of the superlattice, is discussed. It is shown that as far as the formation energy and the stability is conserned the difference of polarity in Si/GaAs is much more important than the large lattice mismatch. The interface charging is the major cause of the large atomic rearrangement at the interface. While the formation energy of $(GaAs)_n/(AlAs)_n$ decreases with increasing n, the formation energy of a strained, covalent/covalent Si_n/Ge_n superlattice increases with increasing n. Calculated formation energies of all superlattices are positive, favoring the separation (or segregation) into constituent crystals, or defect formation in the strained sublattices. The electronic structure, especially the location of the confined carriers are strongly dependent upon the size of the sublattices. In contrast to $(GaAs)_n/(AlAs)_n$ $(n>\sim 10)$, Si_n/Ge_n is an indirect band gap semiconductor, but the energy difference between the direct and indirect band gap recedes with increasing n.

2. METHOD

This review is based on results of ab-initio pseudopotential calculations [8-12]. For large repeat periods, $n + m$ the empirical tight binding (ETB) method[13] is used to explore the trends. Self-consistent pseudopotential calculations have been performed within the framework of the local density-functional theory[14] applied in momentum space[15] Scalar relativistic effects were included by the use of nonlocal,norm-concerving ionic pseudopotentials[16], but the spin-orbit interaction has not been considered. Ceperly- Alder exchange and correlation potential in a parametrized form has been used[17]. Bloch states are expanded with the kinetic energy cut-off correponding to $|\vec{k}+\vec{G}|^2 = 12$ Ry, which leads to a basis set consisting of ≈ 1200 plane waves in large unit cells. The energy cut-off is raised up to 18 Ry for calculating electronic properties. The equilibrium structures are determined by the minimization of the total energy with respect to the structural degrees of freedom. Then these equilibrium configurations are tested by calculating the interatomic forces[18].

3. $(GaAs)_n/(AlAs)_n$ SUPERLATTICE

3.1 Energetics

High-quality $GaAs/Ga_{1-x}Al_xAs$ superlattices with layer thickness of the order of atomic dimensions have already been achieved. Despite the excellent growth properties, the interest has focused recently on the energetics and stability of these superlattices. Kuan et al.[19] made an important observation indicating a long-range order in $Ga_{1-x}Al_xAs$ alloys grown on the GaAs(110) and (100) substrates. The separation (or segregation) into constituent crystals with a thin layer alternation, $(GaAs)_n/(AlAs)_n$ $n < 3$, was the primary cause of the observed long-range order. This implies that $Ga_{1-x}Al_x As$ is unstable at high temperature, inspite of the almost perfect lattice match of GaAs and AlAs. Concomitant with this observation the local-density, total- energy minimization calculations[20] have also favored the stability of the $GaInP_2$ ordered structure not only relative to the alloy, but also relative to disproportionation into constituent crystals, GaP and InP, which have 7% lattice mismatch.

The stability analysis of the ordered $GaAlAs_2$ phase is carried out by calculating the formation energy (or enthalpy of formation at $T = 0^\circ K$)

$$\Delta E^f = E_T^o(GaAlAs_2)\{-E_T^o[(GaAs)_2] - E_T^o[(AlAs)_2]\}/2$$

in terms of the total energies, E_T^o, of each crystal in their equilibrium structure. To reduce computational errors, all calculations were performed using the tetragonal supercell containing four atoms. The ordered $GaAlAs_2$ structure in this supercell corresponds to $(GaAs)_1/(AlAs)_1$ superstructure in the [001] orientation. E_T^o for GaAs and AlAs is found by minimizing the total energy with respect to the cubic lattice constant. For $(GaAs)_1/(AlAs)_1$ an extensive geometry optimization of the structure was carried out by allowing the bands to relax, and the lattice constant to vary. The formation energy ΔE^f is calculated to be 2.6 mRy/cell or (35 meV/cell), implying the instability of the superstructure relative to the disproportionation into constituent compounds. Bylander and Kleinman[21] ($\Delta E^f = 9.2$ meV) and Wood et al.[22] ($\Delta E^f = 21$ meV) reported also positive formation energies. Furthermore, calculations by Kunc and Batra[23] showed that the order $(GaAs)_1/(AlAs)_1$ phase is favored with respect to the alloy, but becomes less stable with respect to the $(GaAs)_n/(AlAs)_n$ for n=2,3. Hence, the conclusion drawn from all these results is that $\Delta E^f[(GaAs)_n/(AlAs)_n]$ is positive, but decreases as n increases, favoring the segregation. That the segregation is favored not only relative to the alloy phase, but also relative to the ordered ternay phase has to be valid for other III-V compounds. An assertion[24] made is that small ΔE^f gives rise to the excellent epitaxial growth properties of $(GaAs)_n/(AlAs)_n$ for n ranging from one to large values.

3.2 Electronic Structure

Earlier empirical pseudopotential calculations[25] for $(GaAs)_n/(AlAs)_n$ have revealed that a Kronig-Penney-type analysis within the effective-mass approximation is appropriate for heterostructures with large superlattice perodicities. Then the experimental and theoretical consensus is that for large n the band alignment is the type-I leading to the electron and hole states both are confined in GaAs. Whereas, the appropriateness of the simple effective mass approach becomes questionable when n is small. As a matter of fact, the observation of Finkman et al.[26]. points out a different band-lineup for very small n, *i.e.* a staggered type-II alignment. This observation has been confirmed by the empirical calculations[27], and later by the first principle calculation of Ciraci and Batra[8,9,24].

The band edge discontinuity is a macroscopic concept and is meaningful only when well-defined bulk regions is identified in the adjacent sublattices. For small n, for example $(GaAs)_1/(AlAs)_1$, the band-edge discontinuity, can not be a matter of subject, since this structure is considered as a new crystal, rather than a heterostructure. In fact the self-consistent field calculations reveals neither interface nor quantum well states[9]. In contrast to the superlattice with large period of alternation it has an indirect band gap with the lowest conduction band occuring at the R-point. The band structure of $(GaAs)_1/(AlAs)_1$ oriented in the [001] direction and the tetragonal Brillouin zone (SBZ) are shown in Fig.1.

The band offset and the electronic structure of $(GaAs)_n/(AlAs)_n$ superlattices with large periodicity have been treated in a number of earlier studies[28]. The SCF- pseudopotential calculations by Batra, Ciraci and Nelson[24] yield the valence band discontinuity for $(GaAs)_4/(AlAs)_4$ to be ≈ 0.3 eV indicating the hole and electron quantum wells located in GaAs-sublattice. In accordance with this finding the charge distribution of the highest valence band state has larger weight in the GaAs-region. Whereas, the lowest three conduction band are localized in the AlAs sublattice. The minimum of the conduction band in AlAs occurs at the six X-points of the fcc BZ. In the superlattice structure the four X-points in the (001) plane coincide with the M-points of SBZ. The remaining two X-points (or X_z) along the superlattice direction are folded to the Γ-point. Consequently, the lowest conduction band state of $(GaAs)_4/(AlAs)_4$ which occurs at the Γ-point originates from the X-state of AlAs. The second and third lowest conduction band states, which are only 70 meV above the first one occur at the M-point of SBZ. The localization of these states indicates that electrons and holes are separated both in direct and momentum space. Only the fourth lowest conduction band state (or second lowest state at Γ) originates from the lowest conduction band state of the bulk GaAs. Clearly, the localization of the conduction band states are in contrast with the calculated band discontinuity. This paradoxical situation emerges from the definition of the band-alignment. Even though the calculated band-offset of $(GaAs)_4/(AlAs)_4$ is the type-I, the relative energy position of the sublattice states determines the confinment. A state in one sublattice $\psi(E, k_{//})$, can match to the state in the other sublattice $\Phi(E', k'_{//})$, as long as their momentum and energies are conserved. If $V_o = E - E'$ differs from zero near a band minimum in the $k_{//}$- plane, the higher energy state will be a barrier for the lower energy state. The lower energy state may be confined in its sublattice, forming a subband

$$E(k_{//}, k_\perp) = (m_o/m*_{//})k_{//}^2 + \epsilon_s(k_\perp) \qquad (s = 1, 2..)$$

in this new quantum regime. The lowest subband state $\epsilon_1(k_\perp = 0)$ is a function of V_o, the width of the sublattice n , and the effective mass ($m_\perp*/m_o$). The smaller are n and

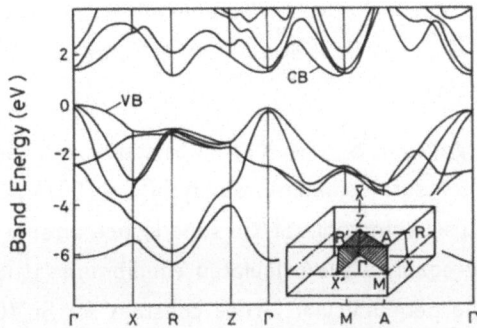

Fig. 1. Energy band structure of $(GaAs)_1/(AlAs)_1$ superstructure oriented in the [001] direction. The zero of energy is set to the maximum of the valence band. The inset shows the tetragonal superlattice Brillouin zone. The symmetry points of the fcc Brillouin zone is marked by bars.

($m_\perp*/m_o$), the higher is $\epsilon_1(k_\perp)$. This oversimplified situation is a direct consequence of the uncertainty principle, and is called the size effect. In $(GaAs)_n/(AlAs)_n$ the lowest Γ-conduction band state of AlAs is a barrier for the corresponding state of GaAs if n is large. However, owing to the quantum size effect (i.e. small n and small m*/m_o of GaAs) the Γ_c-state of GaAs in $(GaAs)_4/(AlAs)_4$ raises relative to that of AlAs, and becomes a barrier for the folded X_z-state of AlAs. The important features of the above discussion are: (i) A superlattice with very thin layer of alternation can support the confined states. (ii) Since the size effect is insignificant for large n, the location of the confined states are consistent with the predicted band alignment obtained in terms of the bulk states. (iii) For small n (for which simple effective mass approximation is conjectured to break down) the quantum size effect is significant, and relative alignment of sublattice states become important, and has to be taken into account individually.

4. STRAINED Si$_n$/Ge$_n$ SUPERLATTICES

4.1. Energetics

Si and Ge both being a covalent semiconductor with zero polarity, the lattice mismatch becomes the most crucial aspect of the pseudomorphic Si$_n$/Ge$_n$ superlattice[5,6]. Earlier experimental data by Abstreiter et al.[29] and calculations by Van de Walle and Martin[30] pointed to the fact that the band edge alignment depends strongly on the lattice strain of the sublattice. Not only the electronic structure, but also the stability of the superstructure is strongly dependent on the lattice mismatch. In a Si$_n$/Ge$_n$ superlattice restricted to the (001) surface of Si the lattice misfit of $\approx$4% is completely accommodated by the uniform lattice strain in the commensurate or pseudomorphic Ge layers. While the grown layers are in registry with the epilayer, the lattice constant in the perpendicular direction expands, leading to a tetragonal distortion. In contrast to polar/polar, but lattice matched (GaAs)$_n$/(AlAs)$_n$, the strain energy becomes dominant in ΔE^f of Si$_n$/Ge$_n$, and increases with n. However, $\Delta E^f(n)$ cannot exceed a threshold value, *i.e.* the activation energy of a defect relieving the strain.

The analysis on the energetics of Si$_n$/Ge$_n$ (1$\leq$n$\leq$6) starts with the determination of the equilibrium structure. To assure commensurability in the (001) atomic plane the lateral lattice constants of the pseudomorphic Si/Ge superlattice arranged in a tetragonal unit cell, a$_{//}$ are taken to be equal to the calculated equilibrium lattice constant of Si, a$_{si}^o$. The determination of the perpendicular lattice constant for Si$_n$/Ge$_n$ n$\geq$2 has required the optimization of the total energy with respect to two structural degrees of freedom (*i.e.* optimization with respect to interlayer spacings d$_2$(Si-Ge) and d$_3$(Ge-Ge). The interlayer spacing d$_1$(Si-Si) is taken equal to that of the bulk. Then the values of d$_2$ and d$_3$ are obtained by the optimization of total energy through their simultaneous variation in Si$_2$/Ge$_2$. Eventually these calculated values for d$_2$ and d$_3$ are used to find the perpendicular lattice constant of Si$_n$/Ge$_n$ with n$\geq$3. The appropriateness of this approximation was tested by the interlayer force calculations[10]. For example, the calculated atomic forces vary in the range of 10^{-2}-10^{-3} mdyn. The calculated equilibrium values for the interlayer spacings are d$_1$=2.56 a.u., d$_2$=2.61 a.u, and d$_3$=2.70 a.u. According to these values the tetragonal distortion $\epsilon_1 = (a_\perp - a_{//})/a_{Ge}^o$ is $\approx$5% in the Ge sublattice.

In order to provide a consistent comparison of the energetics for the stability analysis, the formation energy of the pseudomorphic Si$_n$/Ge$_n$ superlattice is calculated

$$\Delta E^f(Si_n/Ge_n) = E_T^s(Si_n/Ge_n) - [E_T(Si_{2n}) + E_T^o(Ge_{2n})]/2$$

where the total energies of the constituent strain-free crystal E$_T$(Si$_{2n}$) and E$_T$(Ge$_{2n}$) are calculated in a tetragonal unit cell corresponding to that of Si$_n$/Ge$_n$. All superlattice formation energies are found to be positive[10,31]. The decomposition into constituent crystals (*i.e.* segregation) is favored, as long as permitted by the kinetics of the reaction. Alternatively, the strain energy accumulated in the Ge sublattice can be relieved by the creation of the misfit dislocation or by other type of defects). The value of ΔE^f increases with increasing n, because the Ge sublattice has more strained layers. Also for large n

$\Delta E^f/n$ saturate to a value. As discussed in Sec.3., this is a different behaviour seen in $(GaAs)_n/(AlAs)_n$.

The formation energy of Si_n/Ge_n consists of the interfacial and strain energy, which can be dealt with separately. The interfacial energy E_i is related to the heteropolar (Si-Ge) bond formation, which is estimated from the formation energy of the strain-free SiGe in the zinc-blende structure. This is $E_i \approx 0.5$ mRy per atom-pair. The strain energy per atom in the Ge sublattice is defined as

$$E_s = [E_T^s(Ge_n) - E_T^o(Ge_n)]/n$$

in terms of the energy, $E_T^s(Ge_n)$ of the epitaxial Ge calculated in a tetragonal unit cell

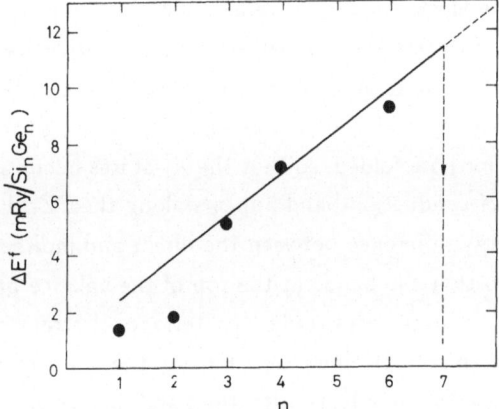

Fig. 2. Comparison of the calculated formation energies, ΔE^f (filled circles) of Si_n/Ge_n with those estimated from a simple relation developed in the text.

containing n atoms. It is found to be E_s=1.46 mRy per epitaxial Ge atom. Then the formation energy is defined as

$$E(Si_n/Ge_n) = 2E_i + nE_s$$

In Fig. 2 the formation energies is estimated from this simple expression are compared with the ab-inito results, displaying an almost perfect match for n≥3. In consideration of the fact that only six Ge layers can be grown, and beyond that thickness the misfit dislocations are formed to relieve the accumulated strain energy, the formation energy barrier Q, to form this defect is estimated to be $\Delta E^f(n=6) < Q < E^f(n=7)$. In summary the pseudomorphic Si_n/Ge_n, and also $Si_n/(Si_{1-x}Ge_x)_n$ superlattices and $Si_{1-x}Ge_x$ alloy itself are metastable structures corresponding to a local minimum in the Born- Oppenheimer surface. The lowest energy structure is, however the separation (or segregation) into unstrained constitient crystals, Si and Ge. The observation[32] that the $Si_{1-x}Ge_x$ ($x \approx 0.5$) alloy restricted to the Si(001) surface undergoes a neostructural phase transitions is the only experimental evidence confirming the above prediction[8,31].

4.2. Electronic Structure

Having discussed the energetics and the stability of Si_n/Ge_n we next discuss the electronic structures based on the results of SCF-pseudopotential[8,11] and empirical tight binding method[13]. It is known, however, that the SCF-pseudopotential method within the local density approach[14] is suitable for ground state properties, but underestimates the conduction band energies, and thus yields smaller band gap. As pointed out earlier[10,30], the average error in the conduction band energies relevant to our discussion (L, X, and Δ_{min} points of the fcc BZ) is ≈ 0.5 eV. Therefore, the local density functional method can be used safely to explore the electronic structure by applying a constant upward shift of 0.5 eV to the conduction band energies. The superlattice formation has three major effects on the electronic structure. These are zone folding along the superlattice direction [001], the strain in the Ge sublattice, and the band lineup. Bands along the Δ–direction (which is parallel to [001] direction) are folded, so that the X_z-states occur at the center of SBZ. Similarly the minimum of conduction band appears along the ΓZ- direction. This results in a decrease of the energy difference between the direct and indirect gap, δE_g, in the Si sublattice. In contrast to that the bands at the top of the valance band are split and the lowest conduction band at k=0 raises under the tetragonal strain of Ge, the net effect being an increase of δE_g in the strained Ge. Finally, the third effect, i.e. superlattice formation, combines the previous effects with the band alignment. The minizones with the flat conduction bands form along the superlattice direction (ΓZ).

After the above general remarks we now examine the calculated electronic structure and present a summary of the results. In comply with the discussion for $(GaAs)_1/(AlAs)_1$, Si_1/Ge_1 being only a strained SiGe in the zinc-blende structure, does not yield any quantum well structure with localized states. The lowest conduction band state of Si_2/Ge_2 is an interface state, which is localized between the adjacent Si and Ge layers. The features of Si_2/Ge_2 go beyond the trends obtained in the others. For example among the structures, $1 \leq n \leq 6$, it has the lowest indirect gap, and largest δE_g. By going to n=3, the charge distribution of the lowest conduction band state shifts towards the Si-sublattice, indicating some confinment character. As seen in Fig.3 the band structure of Si_4/Ge_4 clearly shows that the quantum well structure has set in. Two lowest folded band with k ($//[001]$) are flattened out, and form a subband structure for $k_{//}$ in the (001) plane. At the center of the SBZ, and in the plane parallel to the epilayer, these bands have a parabolic despersion ($E \propto k_\perp^2$). This is the well known behaviour of the confined states with 2-D character. In the valence band the top three states are relatively high weight in the Ge sublattice. The localization of the quantum well states increases with the increasing sublattice thickness.

The character of the quantum well states are analysed for Si_6/Ge_6 in the charge density contour plots presented in Fig.4. Since the calculated band offset for the valence band was reported[30] to be 0.84 eV (the average being 0.54 eV), the lack of confinment in the hole states can be explained by the quantum size effect. Since the effective masses of the states at the maximum of the Ge valence band are 0.34 m^o, 0.043 m^o, and 0.08 m^o, they

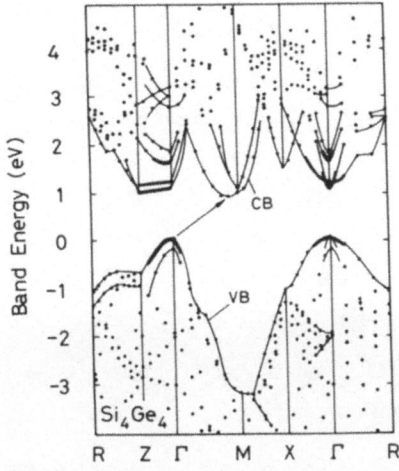

Fig. 3. Energy band structure of the pseudomorphic Si_4/Ge_4 superlattice. The zero of energy is set to the average energy of the top-most three valence band states. Confined states are indicated by thick lines.

lead to significant size effect. Therefore, in view of the calculated value of the valence band offset, the perfect confinement of the states near the maximum of the valence band is not expected for $n \leq \sim 12$. The calculated band offset for the conduction band is ≈ 0.28 eV for the Δ_{c1}-state (*i.e.* the lowest conduction band state of Si occuring along the Δ–direction). At this band minimum the transverse effective mass, m_t^* is only 0.19 m_o. Given a small sublattice thickness n=6, the small effective mass imposes a large size effect. Consequently, the states around this minimum can not be confined in spite of the fact that they have the lowest energy in the conduction band. According to the results of emprical tight-binding method, this particular state makes a sharp local-nonlocal transition beyond a value of the conduction band discontinuity. The lowest two conduction band states at Γ (but above Δ_{c1}) are strongly localized (confined) in the Si-sublattice. These

41

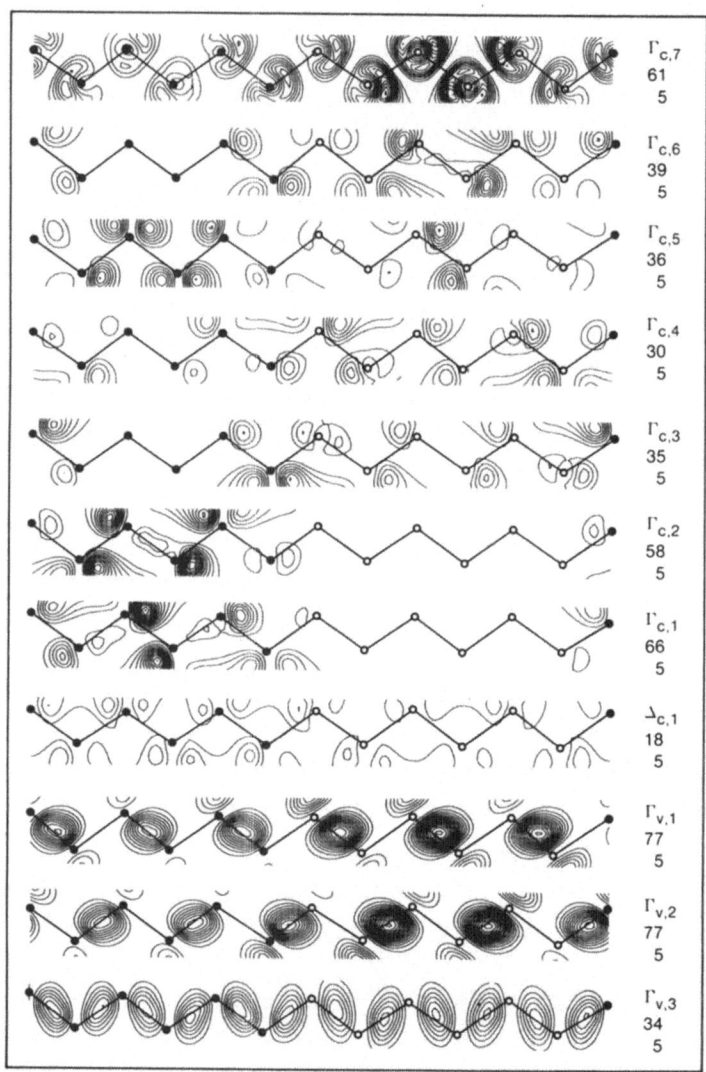

Fig. 4 . Charge density contour plots of the valence and conduction band states of Si_6/Ge_6 at Γ Numerals indicate the value of the maximum charge density (upper) and contour spacings (lower).

are the states producing the flat bands along ΓZ-shown in Fig.3 for Si_4/Ge_4. As noted, upon the zone folding the longitudial effective mass of the $\Delta_{c,1}$-state of Si, which close to m_o, enters in the effective mass hamiltonian, leading to a much smaller size effect than the lowest, extended conduction band state. Depending on the alignment of particular bands, the localization of the high energy states alternate in two sublattices.

The important question to be addressed now is the value of δE_g, or the degree of indirectness of Si_n/Ge_n. Both SCF-pseudopotential and ETB calculations show that all Si_n/Ge_n ($1 \leq n \leq 6$) have indirect band gap. ($\delta E_g > 0$). A more significant finding is that δE_g decreases from 2.1 eV to 0.07 eV by going from the bulk Si to Si_6/Ge_6. The values of the direct, indirect gaps, and δEg are plotted as a function of n in Fig.5.

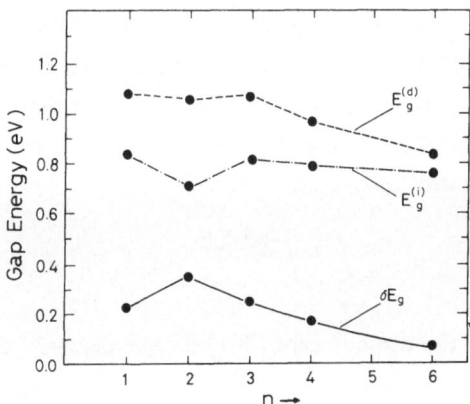

Fig. 5 Variation of direct (dashed-line), indirect (dash- dotted line), and the difference between the direct-and indirect band gap (continuous line) as a function of the sublattice periodicity, n for Si_n/Ge_n .

4.3. δ- doping in Si_n/Ge_n superlattice.

An alternative way to control the band alignment, and thus to modify the confined states in a Si/Ge quantum well structures is to implement an extremely sharp and high density doping profile, which is called a δ-doping[34-35]. Upon the growth of a high-density impurity sheet the localized states change into a 2-D impurity band. The delocalization of impurity states and the dispersion of the metallic bands produced therefrom is dependent on the concentration of dopant and thus the overlap of nearest neighbor impurity orbitals. While the impurity band modifies the band gap, the charge distribution and the potential at the δ-layer may affect the band diagram of the heterostructure. The form of the band diagram, and the stability of the δ-layer against the exchange-place reaction have to vary according to its position. Therefore, one expects significant differences whether the δ-layer is located at the interface, or in one of the sublattices.

The δ-doping is investigated by replacing one Si-layer near the center of the Si-sublattice in Si_6/Ge_6 by a Sb-layer laterally resricted to the Si-sublattice. The Sb-Si-Sb interlayer distances are determined by assuming that the sum of the covalent radii is conserved. Earlier calculations demonstrate that this is a valid approximation[20]. Important results of this study are summarized as follows: (i) The average potential of the Ge-sublattice raises relative to that of Si, which increases the depth of both quantum wells ($i.e.$ the electron quantum well in the Si-sublattice, and the hole quantum well in Ge). As a result the localization of the quantum well states, especially that of the holes increases. (ii) A-quasi 2-D band produced by the δ-layer causes the band gap to decrease, and the indirectness of the gap to increase. Owing to this band of the δ-layer the lowest conduction band states became localized in the Si-sublattice. The dispersion of this band has to depend on the impurity concentration, and thus provides a means to control the band gap. The data by Zeindl et al.[35] showed that the δ-doping involving a few layers with relatively lower impurity concentrations gives rise to the quantum well with the confined election states. The band structure of Si_3SbSi_2/Ge_6 is presented in Fig. 6. (iii) Defining the planarly averaged total charge density as

$$\bar{\rho}(z) = \frac{2}{a_o^2} \int_0^{\frac{a_o}{\sqrt{2}}} \int_0^{\frac{a_o}{\sqrt{2}}} \sum_{n,k}^{E \leq E_F} |\psi_n(\vec{k}, \vec{r})|^2 \, dx \, dy$$

the layer charge between the two adjacent (001) atomic planes, ℓ and $\ell+1$, is calculated to be

$$q(\ell+1\,,\,\ell) = \frac{a_o^2}{2} \int_\ell^{\ell+1} \bar{\rho}(z) \, dz$$

The amount of electronic charge between the adjacent Si_3-Sb, and Sb-Si_4 (001) atomic planes are found to be 4.51 and 4.56 electrons per ($a_o^2/2$) area, respectively. These self-consistent values of $q(Si_3, Sb)$ and $q(Sb, Si_4)$ indicate that simple bond picture predicting the excess charge of Q= -e(Z- 4)/2 (Z being the valency of the impurity atom) is justified. On the other hand we calculate that a small amount of charge (0.02 electrons per cell) is transferred from one side, Si_3SbSi_2 to the adjacent Ge-side of the superlattice. The important aspect revealed from the present study, and from the earlier experimental data[35] is that the electronic properties of superlattices are strongly influenced depending upon the type, concentration, location, and thickness of the impurity layer. Therefore, the δ-doping can develop as a new method to modify the electronic properties to obtain unusual device characteristics

5. COVALENT/POLAR SUPERLATTICES

In an effort to incorporate the photonics into microelectronics, the growth of a polar semiconductors on a Si substrate have been recently achieved[36,37]. The character and the operation of the devices fabricated through heteroepitaxy are know to depend mainly on the quality of the interface. Different electronegativities of the constituent atoms, and the lattice mismatch is known to lead to defects. The anion atoms (A) of the polar semiconductor form stronger bonds with the covalent semiconductor (E) than the cations (C) do. Earlier it was speculated[36] that the excess charge of the (E-A) bond is $Q^A = -e(Z_A-4)/4$ (Z_A, being the valency of A), while the charge of the (E-C) bond has been depleted by $Q_c = -e(Z_c-4)/4$. This certainly gives rise to the interface charging, and thus high electric field, which makes the heterostructure highly unstable. For the same reason, cross-dopings near the interface or antiphase boundaries at the steps of the odd number of layers on the (001) surface of the substrate are produced. If the lattice

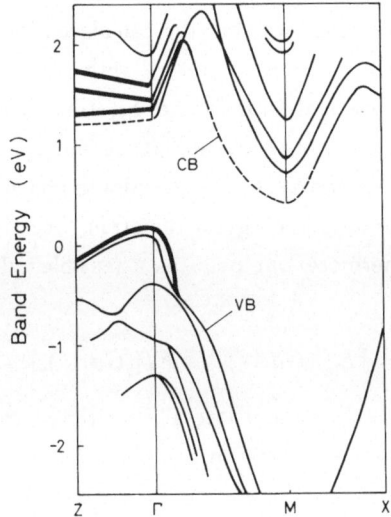

Fig. 6 . Energy band structure of the δ-doped Si_3SbSi_2/Ge_6. Bands indicated by the dashed and thick lines are the δ-layer (Si-Sb-Si), and confined superlattice states, respectively.

constants of the constituent crystals (for example Si/GaAs) are different, the lattice strain in the grown layer becomes another source of defect affecting the quality of heterostructure. Ge/GaAs and Si/GaAs superlattices grown on the (001) surface are two systems of particular interest. In the former the lattice misfit, and thus the lattice strain is negligibly small. The interface charging has to be dully responsible for the instability of the epitaxially grown GaAs. On the other hand the lattice constants of Si and GaAs differ by 4%. Therefore, in addition to the interface charging the strain energy is expected to play a crucial role in the instability.

5.1. Energetics

A thorough analysis of the covalent/polar interface is presented by providing ab-initio values for the superlattice energy, and interlayer charge density. This way one can contrast the effect of different factors leading to instability. To this end $Ge_4/(GaAs)_2$ and strained $Si_4/(GaAs)_2$ superlattices are taken as prototypes. Because of the lattice strain the structure of $Si_4/(GaAs)_2$ laterally restricted to the Si(001) surface is determined by the minimization of total energy and atomic forces.

In the minimum total energy structure, the layer charge q(Si,Ga) at the cation interface is found to be 3.59 electrons. Then the charge depletion (or possitive charging) is slightly smaller than $Q_c = -e(Z_c-4)/2$. The excess charge in the anion interface is only 0.08 electrons larger than $Q_A = -e(Z_A-4)/2$.

The formation energy ΔE_f of $Si_4/(GaAs)_2$ and $Ge_4/(GaAs)_2$ are found to be 51 mRy/cell, and 47 mRy/cell, respectively. It is seen that the calculated formation energies are rather high implying the instability of the heteroepotaxy with uniform atomic planes. ΔE^f of a typical, strained covalent/covalent superlattice, Si_4/Ge_4, is 7 mRy/cell, and 6 mRy of which arises from the strain energy of the Ge sublattice undergoing a tetragonal distortion. The contribution of the strain energy in the pseudomorphically grown GaAs has to be rather small, because the formation energy of the strain free $Ge_4/(GaAs)_2$ is comparable to $\Delta E^f[Si_4/(GaAs)_2]$. Furthermore, one deduces a sensible value for the contribution of the strain energy from

$$E_s = \{E_T^s[(GaAs)_4] - E_T^o[(GaAs)_4]\}/2,$$

Where E_T^s is calculated for GaAs crystal laterally restricted to the Si(001) surface, but undergone the same tetragonal distortion in $Si_4/(GaAs)_2$. The calculated value is $E_s \approx 4$ mRy per $(GaAs)_2$

A simple electrostatic model, in which $Si_4/(GaAs)_2$ is represented by a continuous media of two types of dielectric slab with +0.5e and -0.5e charge regions uniformly distributed in the spacing equivalent to Si-Ga and As-Si interfaces, is used to show the origin of the large positive formation energy. The calculated electrostatic energy (48 mRy/cell) is close to that obtained from the SCF-total energy calculations. Clearly, the large value for ΔE^f arises from the interface charging, and becomes the major cause of the instability, which has to increase with increasing sublattice thickness. The instability has to be fallowed by a large atomic rearrangement yielding electrically neutral interface.

In Fig.7, we present the variation of the 1-D potential energy, V(z), along the superlattice axis of $Si_4/(GaAs)_2$. This potential energy curve is obtained by the planarly (or xy) averaged SCF-pseudopotential. Owing to the interface charging the mean value of V(z) displays a triangular corrugation with a significant bowing of $\approx$1 eV. The bowing is expected to yield dramatic changes in the electronic structure. Even the metal-insulator transition can occur (provided that thick, epitaxial GaAs would be grown).

5.2. Electronic Structure

Further to the three effects (*i.e.* zone folding, strain and superlattice formation) discussed previously, the interface charging plays the dominant role in the electronic structure of the unstable $Si_4/(GaAs)_2$ superlattice. The topmost valence band state is localized near the interface region on the Si-Ga-Si bonds. This is an interface band which is split from the valence-band continua, and localized in the hole quantum well consisting of bowed Si and GaAs valence band edges. Because of the size effects the second and third valence bands do not display any confined character. The lowest conduction band is derived from As-Si, and thus is localized in the corresponding interface region. It is flat along ΓZ-direction, and has a minimum at the Γ-point. Second and third conduction band states are primarily confined in the Si-sublattice. They are flat along the ΓZ-direction, but have parabolic dispersion for k lying in the epilayer.

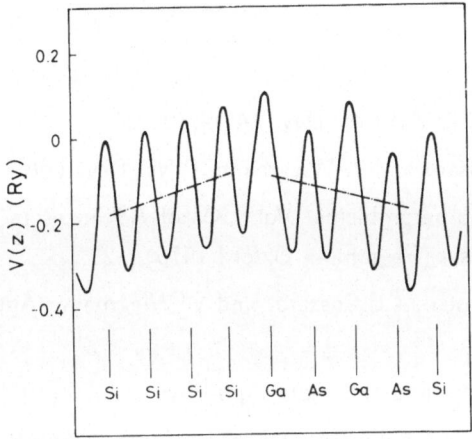

Fig. 7 . Planarly averaged SCF- pseudopotential, V(z), for $Si_4/(GaAs)_2$. An estimate of the mean value is shown by dash-dotted lines.

6. CONCLUSIONS

It becomes clear from the above discussion that all superlattices has positive formation energy, favoring the separation (or segregation) into constituent crystals. For a lattice mached and polar/polar superlattice, such as $(GaAs)_n/(AlAs)_n$, the formation energy is not large, and decreases with increasing sublattice periodicity. Small formation energies is consistent with small polarity difference of compound crystals[38]. For a covalent/covalent, strained Si_n/Ge_n superlattice the strain energy is dominant in the formation energy. The formation energy of Si-Ge heterobonds is also positive. The transfer of charge from one sublattice to the adjacent one is negligibly small. In the covalent/polar superlattices the interface charging yields very large formation energy, which is the major cause of the instability requiring massive atomic rearrangements. The electronic structure is found to depend strongly on the sublattice periodicity n, when n is small. For example, while $(GaAs)_n/(AlAs)_n$ has a type-I band discontinuity, the confined electron and hole states are separated in direct and momentum space for $n \approx 4$, as if it has a staggered type-II band offset. In Si_n/Ge_n (≥ 3) the direct, indirect band gap decreases with increasing n. The difference between the direct and indirect gap is postive, but receeds to 0.07 eV for n=6. Upon a Sb-monolayer doping the band gap is further decreased, but indirectness increased. The interface charging of an epitaxial, covalent/polar superlattice (if it can be produced) may lead to a metal-insulator transition.

Acknowledgement: I would like to thank Dr. I.P. Batra, Dr. Ş. Ellialtıoğlu, Dr. J.N. Nelson, O. Gülseren, E. Tekman, and E. Özbay for their contribution to several works, on which this review is based. I also thank to Dr. M. Akgül for his valuable help during the preparation of the manuscript.

REFERENCES

1 L.Esaki and R. Tsu, IBM J.Res. Dev. 14, 6l(1970).

2 R. Dingle, A.C. Gossard and W.Wiegmann, Phys. Rev. Lett. 21, 1327 (1975).

3 R. Dingle, in Festkörperprobleme, Vol. XV of Advances in Solid State Physics, edited by H.J.Queisser (Pergamon, Oxford 1975) p.21.

4 R. Dingle, H.L.Störmer, A.G.Gossard, and W.Wiegmann, App.Phys.Lett. 33, 665 (1978).

5 E.Kasper, H.J.Herzog, and H.Kimble, App.Phys. 8, 199 (1975).

6 A.T.Fiory, J.C.Bean, L.C.Feldman, and I.K.Robinson, J.App.Phys. 56, 1227 (1984).

7 T.P. Pearsall, J.Bevk, L.C.Feldman, J.M.Bonar, J.P.Mannaerts, and A.Ourmazd, Phys. Rev. Lett. 58, 729 (1987).

8 S.Ciraci and I.P.Batra, Phys. Rev. Lett. 58, 2114 (1987).

9 S.Ciraci and I.P.Batra, Phys. Rev. B36, 1225 (1987).

10 S.Ciraci and I.P. Batra, Phys. Rev. B (in press).

11 I.P.Batra, S.Ciraci and E.Özbay, Phys. Rev. B.

12 S.Ciraci, I.P.Batra and E.Tekman, Phys. Rev. B.

13 S.Ciraci, O.Gülseren and S.Ellialtıoğlu, Solid State Commun. 65, 1285 (1988).

14 P.Hohenberg and W.Kohn, Phys. Rev. 136, B864 (1964); W.Kohn and L.J. Sham, Phys. Rev. A1133 (1965).

15 J.Ihm, A.Zunger, and M.L.Cohen, J.Phys. C12, 4409 (1979).

16 G.B.Bachelet, D.R.Hamann, and M.Schlüter, Phys. Rev. B26, 4199 (1982).

17 D.M. Ceperley and B.J. Alder, Phys. Rev. Lett. 45, 566 (1980), J.Perdew and A.Zunger, Phys. Rev. B23, 5048 (1981).

18 I.P.Batra, S.Ciraci, G.P.Srivastava, J.S.Nelson and C.Y.Fong, Phys.Rev. B34, 8246 (1986).

19 T.S.Kuan, T.F.Kuech, W.I.Wang and E.L.Wilki, Phys. Rev. Lett. 54, 201 (1985).

20 G.P.Srivastava, J.L.Martin and A.Zunger, Phys. Rev. B 31, 2561 (1985).

21 D.M.Bylander and L.Kleinman, Phys.Rev.B34, 5680 (1986).

22 D.M.Wood, S.-H.Wei and A.Zunger, Phys.Rev.Lett. 58, 1123 (1987).

23 K.Kunc and I.P.Batra (to be published).

24 I.P.Batra, S.Ciraci and J.S.Nelson, J.Vac.Sci.Technol. B5, 1300 (1987).

25 M.Jaros and K.B.Wong, J.Phys. C17, L765 (1984); M.Jaros, K.B.Wong and M.A.Gell, Phys.Rev. B31, 1205 (1985).

26 E.Finkman, M.D.Sturge,, and M.C.Tamargo, App.Phys.Lett. 49, 1299 (1986).

27 M.A. Gell, D. Ninno, M. Jaros, and D.C. Herbert, Phys. Rev. B34, 2416 (1986).

28 W.A.Harrison, J.Vac.Sci.Technol. 14, 1016 (1977);ibid B3, 1231 (1985).

29 G.Abstreiter, H.Brugger, T.Wolf, H.Jorke, and H.Herzog, Phys. Rev.Lett. 54, 2441 (1985).

30 C.H.Van de Walle and R.M.Martin, J.Vac.Sci.Technol. B3, 1256(1985);ibid. Phys. Rev. B34, 5261 (1986).

31 S.Froyen, D.M.Wood and A.Zunger, Phys Rev. B36,4547(1987).

32 A.Ourmazd and J.C.Bean, Phys. Rev.Lett. 55, 765 (1985).

33 J.L.Martins and A.Zunger, Phys. Rev. Lett. 56, 1400 (1986).

34 D.Streit, R.A.Metzger and F.G.Allan, App.Phys. Lett., 44, 234 (1984).

35 H.P.Zeindl, T.Wegehaupt, I.Eisele, H.Oppolzer, H.Reisinger, G.Temple and F.Koch, App. Phys. Lett. 50, 1164 (1987).

36 H.Kroemer, Journal of Crystal Growth 81, 193 (1987).

37 P.R.Pukite and P.I.Cohen, Journal of Crystal Growth 81, 214 (1987).

38 W.A.Harrison and S.Ciraci, Phys. Rev. B10, 1516 (1974).

BAND OFFSETS AT SEMICONDUCTOR HETEROJUNCTIONS:

BULK OR INTERFACE PROPERTIES?

Stefano Baroni and Raffaele Resta

Scuola Internazionale Superiore di Studi Avanzati
Strada Costiera 11, I-34014 Trieste - Italy

Alfonso Baldereschi

Dipartimento di Fisica Teorica, Università di Trieste
Strada Costiera 11, I-34014 Trieste - Italy
and
Institut de Physique Appliquée, Ecole Polytechnique Fédérale de Lausanne
CH-1015 Lausanne - Switzerland

The problem of whether band offsets at semiconductor interfaces are determined by bulk properties of the constituents or substantially affected by interface phenomena is critically readdressed. In particular, the conditions under which band offsets do not depend on the interface orientation are examined. State-of-the-art ab-initio pseudopotential calculations are performed for $(GaAs)_3(AlAs)_3$ grown in the (001), (110), and (111) directions. Our results are analysed through a novel definition of the interface charge distribution which does not make any use of ideal *reference* interfaces: the dipole corresponding to such a distribution directly yields the potential drop across the interface. Our calculations give for the (001), (110), and (111) interfaces a band offset of 0.49, 0.51, and 0.49 eV respectively, thus indicating that orientation independence holds in this case. However, in the case of the (111) orientation, two inequivalent interfaces exist whose offsets slightly differ (0.07 eV); associated with this difference we also found a net interfacial charge accumulation at the two inequivalent interfaces ($\pm 2.8 \times 10^{-4}$ electrons per unit surface cell). Our results are finally interpreted through a new model based on crystal symmetry and whose only ingredients are the bulk charge densities of the the two constituents. The model – though not reproducing the fine details of the (111) superlattice – is in excellent agreement with our first-principles results and with available experimental data.

I. Introduction

One of the main goals of band structure engineering is the tailoring of transport properties at semiconductor heterojunctions. The key parameters which determine such transport properties are the band offsets at the interface. The band offset can be split into two contributions: the first is the difference between the band edges calculated when the average electrostatic potentials in the two bulk materials are aligned; the second is the so called *potential line-up*, i.e. the difference between the average electrostatic potential on the two sides of the junction. The former contribution is a quantum property of the two bulk materials and can be calculated from the band-structures of the individual periodic systems. As for the latter, due to the long-range nature of the Coulomb interaction, it depends in principle on the details of the electronic charge distribution at the interface, which is only accessible to elaborate calculations for the full semi-infinite system. For this reason, much interest has been devoted to establishing whether the potential line-up can also be predicted – though approximately – using information from the two bulk materials only. Experimentally two facts suggest that, for a large class of materials, this is indeed the case: the *orientation independence* and the *transitivity* of potential line-ups[1]; i.e. line-ups are found to be roughly independent of the crystallographic orientation of the interface, and seem to obey the rule that the line-up Δ between two materials A and B, nearly equals the sum of the line-ups between A, B, and a third material C: $\Delta_{AB} = \Delta_{AC} + \Delta_{CB}$. Many of the existing theories and models[2-4] share this point of view, though they start from seemingly very distant physical assumptions.

The determination of electrostatic potential line-ups at semiconductor interfaces is far from being straightforward: the reason is that the average of the electrostatic potential in an infinite solid is ill-defined. Of course, the difference of such averages across the junction between two semiinfinite solids is well defined, although it depends on the details of the electronic charge distribution at the interface. From a physical point of view, this is due to the fact that a dipole layer at the interface modifies the potential line-up across it, leaving any other *genuine* bulk property (such as the energy per cell, the optical gap, etc.) unchanged. From a mathematical point of view, we observe that, if the solid is thought to be made out of periodically repeated localized units:

$$n_{cryst}(\mathbf{r}) = \sum_{\mathbf{R}} n_{loc}(\mathbf{r} - \mathbf{R}), \tag{1}$$

and if these units do not carry any net charge, dipole, nor quadrupole, then the average electrostatic potential is given by:

$$< V > = \frac{4\pi e^2}{\Omega} \int n_{loc}(\mathbf{r}) r^2 d\mathbf{r}, \tag{2}$$

where Ω is the volume of the unit cell. Throughout this paper, the word *potential* will indicate the potential energy felt by a negative test charge, while the density n is the number density of negative charges (i.e. it is positive for electrons). Of course, in a periodic crystal, only the total charge density is well defined, and the partition of the density into localized contributions is equivalent to a prescription for interpolating the *discrete* Fourier coefficients of the electron density, $\tilde{n}_{cryst}(\mathbf{G})$, with a *continuous* function $\tilde{n}_{loc}(\mathbf{q})$. The average electrostatic potential given above can indeed be expressed in reciprocal space as: $< V > = \frac{4\pi e^2}{\Omega} \lim_{q \to 0} \tilde{n}_{loc}(\mathbf{q})/q^2$. In the case of a semiconductor heterojunction, the partitioning of the two bulk materials into elementary building blocks gives a prescription for the shape of the interface and fixes the potential line-up across it. Of course, starting from a given prescription, charge readjustment occurs at the interface, and the actual line-up Δ depends on both the reference prescription (giving the *bare* line-up Δ_{ref}) and the charge readjustment at the interface (which gives the *relaxation* contribution to the line-up, which is referred to as Δ_{dip} in Ref. 5). From this point of view, the definition of a reference interface is unnecessary and may lead to erroneously attribute separate physical meaning to Δ_{ref} and Δ_{dip}. However, as noted above, both experiments and many of the existing models and theories suggest that Δ is mainly determined by the properties of the two bulk materials. This indicates that a reference interface can be found (i.e. a partition of bulks into elementary building blocks) such that charge readjustment with respect to it is negligible, and Δ_{ref} is a very good approximation to Δ. In Sec IV we provide such an optimum reference.

II. Potential line-ups from linear-response theory

As a simple theoretical approach to the potential line-up at the interface between two lattice-matched semiconductors A and B, we consider the actual interface as a perturbation with respect to an infinite periodic virtual crystal $A_{\frac{1}{2}} B_{\frac{1}{2}}$. The description of the electronic distribution within linear-response theory is expected to be particularly good for those common-anion (or common-cation) heterojunctions such as GaAs/AlAs, where the chemical differences between the constituents on the two sides are rather small. In the following, we will focus our attention on such a simple case.

As a starting point, we take the virtual crystal $Ga_{\frac{1}{2}} Al_{\frac{1}{2}} As$ and we investigate the linear response to a *single* isovalent substitution, where a mixed cation is replaced with a Ga (or Al) ion. The change in the electrostatic potential due to electronic rearrangement is:

$$\Delta \tilde{V}_{Ga}^{H}(\mathbf{q}) = \frac{4\pi e^2}{q^2} \Delta \tilde{n}_{Ga}(\mathbf{q}), \qquad (3)$$

where $\Delta \tilde{n}_{Ga}$ is the induced charge density which has the full point symmetry of the crystal (T_d in the present case). Because of this, $\Delta \tilde{n}_{Ga}$ does not carry dipole nor quadrupole moments, while – within linear-response theory – isovalent substitutions carry no net charge; therefore $\Delta \tilde{V}_{Ga}^{H}(\mathbf{q})$ is non singular as $\mathbf{q} \to 0$, the lowest order term in Eq. (3) being a constant which we call $\Delta/2$ and can be recast as:

$$\frac{\Delta}{2} = \lim_{q \to 0} \frac{1}{\Omega} \Delta \tilde{V}_{Ga}^{H}(\mathbf{q}) = \frac{4\pi e^2}{\Omega} \int \Delta n_{Ga}(\mathbf{r}) r^2 d\mathbf{r}. \qquad (4)$$

The bare perturbation due to an Al substitution is by construction equal in magnitude and opposite to the Ga case; thus – within linear response – one has: $\Delta_{Al} = -\Delta_{Ga}$.

Going now from single substitutions to the actual interface, we simply have to replace any mixed cation with either a Ga or Al ion; using linearity, the electrostatic potential line-up is easily proved to be equal to Δ. We stress the importance and generality of this result: *for all those lattice-matched isovalent interfaces where the electronic response to relevant ion substitutions can be described within linear-response theory, the potential line-up is determined by bulk properties only.* This means that the line-up is independent not only of the growth *direction*, but even more generally of the *abruptness* of the interface. For non-isovalent interfaces linear-response theory maintains its power, but care must be taken in handling net transfers of charge.

The linear response to any given perturbation can be efficiently calculated along the lines indicated in Ref. 7. For the sake of simplicity, we follow here a *direct* approach where the response is calculated from the difference between two independent calculations for the perturbed and unperturbed systems. In the present case, the unperturbed system is the virtual crystal $Ga_{\frac{1}{2}} Al_{\frac{1}{2}} As$. We study the response to a single cation substitution using a simple cubic (SC) supercell, containing four zincblende (FCC) unit cells; in the perturbed system, one of the four cations in the SC cell is replaced by either Ga or Al, and Δn is simply obtained by difference.

All the calculations in this work have been performed using state-of-the-art density-functional (DFT) local-density (LDA) techniques: norm-conserving pseudopotentials[8], large plane-wave basis sets (up to a kinetic-energy cutoff of 14 Ryd), Ceperley-Alder exchange-correlation data[9]; the special points used for Brillouin-zone integrations are those of the (444) Monkhorst-Pack cubic mesh[10], appropriately folded for the various geometries considered here and below.

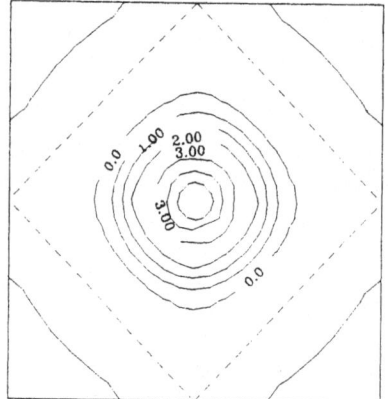 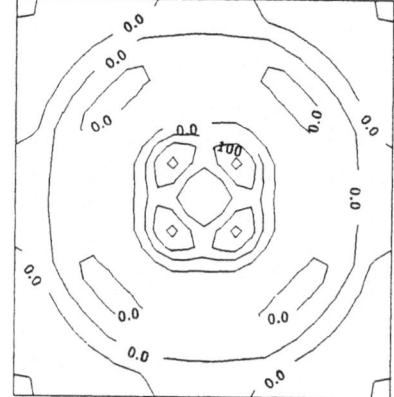

Figure 1. Contour plots in the (001) plane of the linear ($n^{(1)}$, left) and quadratic ($n^{(2)}$, right) density response of the $Ga_{\frac{1}{2}}Al_{\frac{1}{2}}As$ virtual crystal to a single cation substitution. Units are electron per FCC cell. The intersection of a cation-centered FCC Wigner-Seitz cell with the plane is indicated with a dashed line.

In Fig. 1 we report the contour plot of $\Delta n^{(1)} = \frac{1}{2}(\Delta n_{Ga} - \Delta n_{Al})$ and $\Delta n^{(2)} = \frac{1}{2}(\Delta n_{Ga} + \Delta n_{Al})$ on the (001) plane, as obtained from our calculations. $\Delta n^{(1)}$ and $\Delta n^{(2)}$ approximately represent the linear and quadratic response to a single substitution, respectively. The intersection of a cation-centered FCC Wigner-Seitz cell (WSC) with the plane is also indicated. Inspection of Fig. 1 shows that linear-response is an adequate approximation to treat potential line-ups in this case, and that the response to a single substitution is sufficiently short-range to be practically confined within one WSC. Using the calculated $\Delta n^{(1)}$ in Eq. (4) we get an electrostatic potential line-up $\Delta = 0.43$ eV.

The quantities calculated so far are electronic ground-state properties, therefore within the reach of our DFT calculation, the only essential approximation being LDA. As mentioned in the introduction, band offsets are the sum of the potential line-ups discussed above plus the difference between band edges of the two bulk materials calculated when the arbitrary average values of the electrostatic potentials are aligned. The latter quantity is not directly accessible to DFT, and one has in principle to resort to more elaborate many-body techniques[11,12]. It has long been assumed that many-body corrections to LDA are small for valence bands. According to our calculations for the individual infinite systems, the top of the valence band lies 0.05 eV higher in $GaAs$ than in $AlAs$. We notice at this point that we include in the potential line-up only the electrostatic effect of the valence electrons: exchange-correlation effects, as well as the electrostatic effects from the core pseudopotential are accounted for in the difference between bulk band edges. Of course, the calculated band offsets do not depend on this (arbitrary) partition: the procedure is in fact equivalent to calculating the offset in the valence edges of the local density of states across the interface. Adding the band structure term of 0.05 eV and spin-orbit effects (0.03 eV) to the potential line-up calculated above, we obtain a (geometry-independent) valence-band offset of 0.51 eV. This value compares quite well with available experimental data which range from 0.40 to 0.55 eV[1]. Recent investigations have shown that many-body contributions to valence-band offsets are not quite negligible[11], and amount (in the present case of $GaAs/AlAs$) to ~ 0.1 eV. We stress that the above many-body correction is by definition a property of the two bulk materials which does not affect our conclusions on potential line-ups. A proper account of many-body effects would result in a total offset of ~ 0.6 eV.

III. First-principles supercell calculations

We have performed state-of-the-art density functional theory (DFT) calculations for $(GaAs)_3(AlAs)_3$ superlattices oriented in the (111), (110) and (001) directions. The physical quantities $f(\mathbf{r})$ we are interested in (such as the electron density $n(\mathbf{r})$ or the electrostatic potential $V(\mathbf{r})$) are periodic in the planes perpendicular to the growth direction (z axis). As we are mainly interested in the z dependence of such quantities, it is convenient to define $\bar{f}(z)$ as the xy planar average of $f(\mathbf{r})$. The function $\bar{f}(z)$ is nonperiodic in the interface region, and goes asymptotically into two different periodic functions (having the same periods for lattice-matched heterojunctions) far from the interface: a typical result for the self-consistent charge density $\bar{n}(z)$ and potential $\bar{V}(z)$ of $(GaAs)_3(AlAs)_3$ (001), (110), and (111) is shown in Fig. 2. Note that the three figures are on the same scale and that the charge distribution looks very different in the three cases: yet, our previous considerations on linear response as well as the present first-principles calculations indicate that band offsets are effectively independent of interface orientation. The strong atomic oscillations are bulk-like and hide interface effects such as the barely visible potential shift across the interface. In order to get rid of bulk effects and to blow up interface features, it has been proposed[5,6] to define a function $\Delta \bar{f}(z)$ subtracting, on each side of the interface, the appropriate bulk function from $\bar{f}(z)$. The relevant features of this construction are: i) The interface region corresponds to $\Delta \bar{n}(z)$ significantly different from zero or, equivalently, to non-constant $\Delta \bar{V}(z)$. ii) $\Delta \bar{n}(z)$ and $\Delta \bar{V}(z)$ are discontinuous at the interface and they are not physically linked to each other by Poisson equation. iii) the potential drop generated by $\Delta \bar{n}(z)$ (i.e. Δ_{dip} in the notation of Refs. 4-5) is only part of the total potential line-up Δ, the latter being recovered by adding the difference Δ_{ref} between the average potentials of slabs of bulk materials. While Δ is in principle a physically measurable property of the interface, its decomposition $\Delta = \Delta_{ref} + \Delta_{dip}$ bears no physical meaning since it depends on the arbitrary shape of the reference interface[13]. iv) On the contrary, $\Delta \bar{V}(z)$ tends to different constant values on the two sides far from the interface whose difference is the potential line-up Δ if the arbitrary average values of the bulk Hartree potentials are aligned.

We propose a new procedure to subtract bulk effects from $\bar{f}(z)$, which avoids the definition of an ideal interface and its use as a reference. In fact such a definition is unnecessary, arbitrary, and might lead one to erroneously attribute physical meaning to Δ_{dip}. We define the *macroscopic average* $\bar{\bar{f}}(z)$ as the one-dimensional average of $\bar{f}$ over a period centered at z: $\bar{\bar{f}}(z) = \frac{1}{a} \int_{z-\frac{a}{2}}^{z+\frac{a}{2}} \bar{f}(\zeta) d\zeta$. This is equivalent to performing the three-dimensional average of $f(\mathbf{r})$ over a slab-adapted unit cell[13] centered at point $\mathbf{r}$ and therefore corresponds to the usual definition of macroscopic quantities in electrostatics. When applied to $\bar{n}(z)$ and to $\bar{V}(z)$, the construction gives functions $\bar{\bar{n}}(z)$ and $\bar{\bar{V}}(z)$ which have the following features: i) They are continuous functions. ii) They are derived from a single ground-state calculation and not from differences. iii) In the two bulk regions, $\bar{\bar{n}}(z)$ tends to a constant value n_o (8 electrons per cell in our case), while $\bar{\bar{V}}(z)$ tends to constants differing by Δ. iv) The interface region is unambiguously

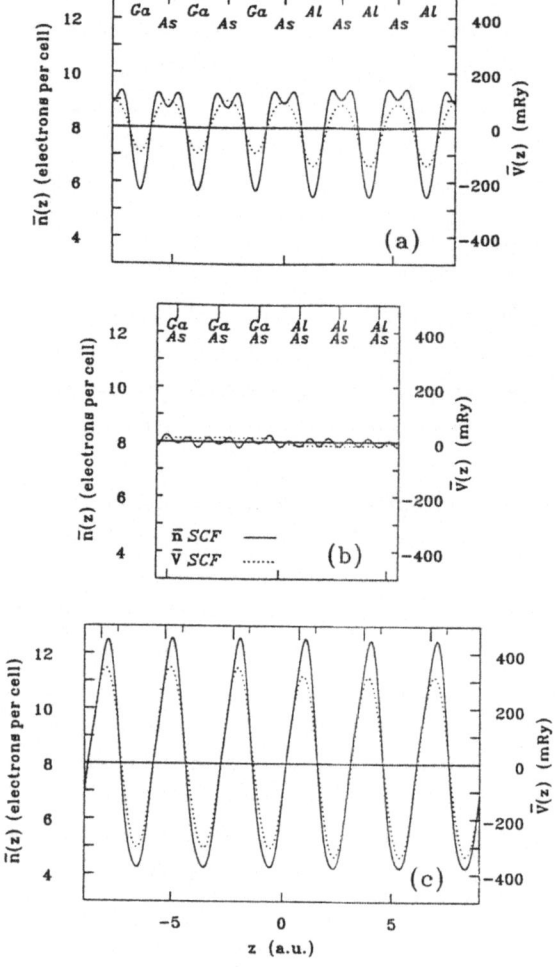

Figure 2 Planar average of the first-principles electron density $\bar{\bar{n}}(z)$ and electrostatic potential $\bar{\bar{V}}(z)$ of $(GaAs)_3(AlAs)_3$ oriented along (001) (a), (110) (b), and (111) (c).

defined as the region where both $\bar{\bar{n}}(z)$ and $\bar{\bar{V}}(z)$ significantly deviate from constancy. *v)* The density $\bar{\bar{n}}(z)$ is related to $\bar{\bar{V}}(z)$ by the one-dimensional Poisson equation and is a physically meaningful interface electron distribution, since $\bar{\bar{n}}(z) - n_o$ is the finite-range charge distribution which generates the interface macroscopic dipole.

In Fig. 3 we display $\bar{\bar{n}}(z)$ and $\bar{\bar{V}}(z)$ as obtained from first-principles calculations for the interfaces studied here. An inspection of the figure shows that our lattice is thick enough to satisfactorily reproduce bulk features midway between the two interfaces and that computational noise is small even on this expanded scale. Note that $\bar{\bar{n}}(z)$ has a typical dipolar shape around n_o across the interface. The difference between the values of $\bar{\bar{V}}(z)$ calculated in the two bulk-like regions gives a potential line-up of 0.41, 0.43, and 0.41 eV for the (001), (110), and (111) geometries respectively. These values, compared to the (geometry-independent) linear-response value of 0.43 eV, prove the soundness of the previous approximate viewpoint. Taking into account bulk-band and spin-orbit effects the resulting band offsets are 0.49, 0.51, and 0.49 eV. Many-body effects would further raise these values to ~ 0.6 eV.

The (111) interface deserves some further comments: while the two interfaces in our supercells are equivalent by symmetry for the (001) and (110) superlattices, they are not for (111). In fact the two interfaces differ according to whether the interface bond which is parallel to the z direction is $Ga\text{-}As$ or $Al\text{-}As$. The interface in the middle of Fig. 3(c) is of

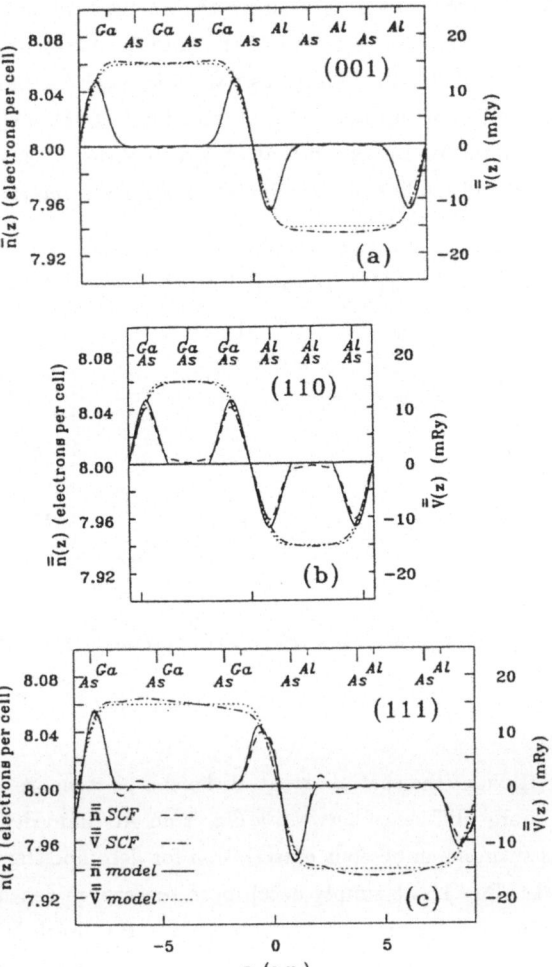

Figure 3 Macroscopic averages of the first-principles (SCF) and model electron density ($\bar{\bar{n}}(z)$) and electrostatic potential ($\bar{\bar{V}}(z)$) of $(GaAs)_3$ $(AlAs)_3$ oriented along: (a): (001); (b) (110); (c) (111). The predictions of the model described in Sec IV are also displayed.

the former type (type A), while those at the figure borders are of the latter (type B). Type A and type B line-ups can in principle be different ($\Delta_A \neq \Delta_B$), and net interfacial charges σ_A and σ_B can pile up, overall charge neutrality requiring $\sigma_A = -\sigma_B$. The average electric field in the superlattice is determined by the overall boundary conditions and *not* by the local charge distribution; the use of periodic boundary conditions (as it is done here) amounts to assuming a zero average field, which physically corresponds to short-circuiting the two free surfaces of a finite sample. The combined effect of $\Delta_A \neq \Delta_B$, $\sigma_A = -\sigma_B \neq 0$, and of the periodic boundary conditions may result in nonvanishing slopes of the average electrostatic potentials in each of the bulk-like regions: such slopes are indeed visible in Fig. 3 (c). Although we have verified that the superlattice charge densities in the middle of the bulk regions do coincide with their bulk counterparts, the differences in the macroscopic averages extend more deeply far from the interface, due to the averaging procedure. In order to blow up the effects due to the difference between the two line-ups and to net interfacial charges, we have considered a $(GaAs)_6(AlAs)_6$ (111) superlattice whose charge density is obtained from our $(GaAs)_3(AlAs)_3$ calculations cutting the $3+3$ superlattice in the middle of the two bulk regions and inserting a slab of 6 planes of pure material therein. The resulting macroscopic average of the potential and charge density is displayed in Fig. 4, where the potential slopes in the middle of the bulk regions are clearly visible. Simple electrostatic considerations show that the average between the slopes in the two bulk regions is related to the difference between the line-ups, while their difference is related to the net charges at the interfaces. Analysis of the data presented in Fig. 4 gives $\Delta_B - \Delta_A = 0.07$ eV, and $\sigma_B = -\sigma_A = 2.8 \times 10^{-4}$ electrons per surface cell. A similar value (0.06 eV) for the difference in the two (111) line-ups has been obtained in Ref. 14 aligning suitably defined midgap levels; our results differ from theirs in that they also obtain a sensible difference between the *average* (111) line-up and the line-ups for the (001) and (110) interfaces.

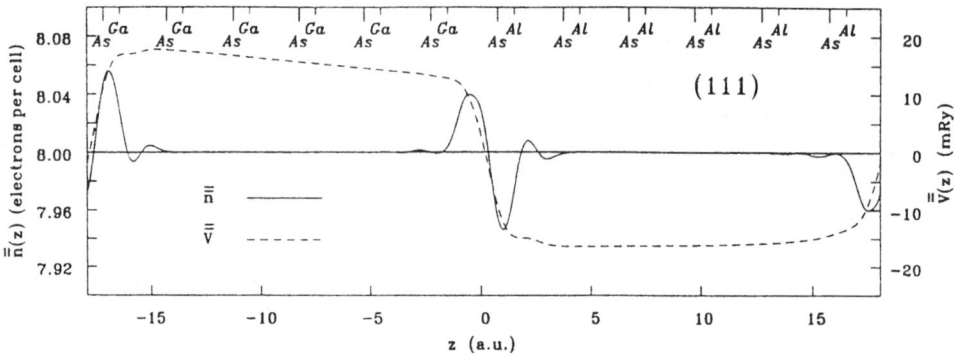

Figure 4. Macroscopic average of the charge density $\bar{n}(z)$ and electrostatic potential $\bar{V}(z)$ of a $(GaAs)_6(AlAs)_6$ (111) superlattice (see text).

IV. A new model

The considerations developed in Sec II about the spatial extent of the charge response to a single cation substitution show that the charge difference between a Ga or an Al "impurity" is confined within a single FCC WSC. This suggests an obvious prescription for decomposing the infinite solid into localized building blocks, Eq. 1: we simply decompose the charge densities of the two bulk regions into nonoverlapping WSC's: $n_{cryst} = \sum_{\mathbf{R}} n_{WSC}(\mathbf{r} - \mathbf{R})$. Each WSC is centered at a cation site and has fractions of anions at four of the corners: it is neutral and by symmetry does not have dipole nor quadrupole moments. Therefore the potential line-up is simply the difference between the integrals given by Eq. 2, calculated for $n_{loc} = n_{WSC}^{GaAs}$, and $n_{loc} = n_{WSC}^{AlAs}$. This gives a potential line-up of 0.41 eV, in excellent agreement with both our first-principles calculations and linear response results. Besides this single figure, the soundness of the physical picture underlying our model is best judged from an analysis of the predicted interface charge. Our model interface electronic density is obtained by rigid juxtaposition of WSC's. In Fig. 3, the macroscopic averages of the first-principles electronic charge density and the electrostatic potential are compared with the predictions of the model for $(GaAs)_3(AlAs)_3$ oriented along (001), (110), and (111). The agreement is extremely good. In particular the spatial extent of the interface regions and the shape of the dipolar charge distributions are well reproduced for all the interfaces. Of course, starting from the model, a small electronic rearrangement must occur in order to ensure at least charge continuity. Our results show that this rearrangement does not affect the total line-up for the (001) and (110) geometries, while it is responsible for the small line-up difference and interfacial charges in the case of the (111) orientation.

A model where the elementary blocks are spherical atoms has been recently proposed by Van de Walle and Martin (VWM)[3]; it approximates the two bulk densities with a reasonable accuracy and gives a valence-band offset of 0.60 eV. Our procedure has several advantages over that of VWM. It *exactly* reproduces the electronic density in the bulk, it provides a more accurate description of density profiles at the interface, and it yields a 0.49 eV valence-band offset which is one order of magnitude more accurate than the VWM model and practically coincides with first-principles calculations. Of course, our model is less general than that of VWM one in that in its present form it only applies to common-anion (or common-cation) lattice-matched heterojunctions. Extensions of this simple model to more general interfaces are possible and presently under way[15].

Both the model presented in this section and the considerations on linear response developed in Sec II are in agreement with a recent experiment by Shih and Spicer on $Cd_xHg_{1-x}Te$ alloys[16]. These authors studied the Hg $5d$ and Cd $4d$ core-level binding energy relative to the valence-band maximum, E_c^v, as a function of the alloy composition x. Their results can be summarized as follows: i) The difference between Cd $4d$ and Hg $5d$ binding energies is roughly independent of the composition x: $E_c^v(Hg, x) - E_c^v(Cd, x) \sim const$; ii) The differences $\Delta E_c^v(Hg, x) = E_c^v(Hg, x) - E_c^v(Hg, 0)$ and $\Delta E_c^v(Cd, x) = E_c^v(Cd, x) - E_c^v(Cd, 1)$ are such that $\Delta E_c^v(Hg, x) - \Delta E_c^v(Cd, x) = const$, the constant being to a good approximation equal to the heterojunction valence-band offset measured independently[17]. Using the above results it is simple to deduce that the position of the individual core levels are independent of the composition x, when referred to some appropriate *absolute* energy scale. This is in agreement with the so called *new common-anion rule* by Zunger, according to which *deep core levels of cations in common-anion pairs are nearly unchanged relative to a common reference energy ... in going from a binary to a ternary (including alloy) system*[18]. The considerations developed in the present section and in Sec II provide a simple and natural explanation of the above findings. In fact the dependence of core levels on composition essentially comes from the dependence of the electrostatic potential at the cationic nuclei. If we assume that the alloy is made up of cation-centered bulk-like WSC's (or alternatively that its difference from a virtual crystal is linear and localized around each cation), we can separate the electrostatic potential at cation nuclei into two contributions: the first is from the on-site WSC which is by construction composition independent; the second comes from the other WSC's and depends in principle on x. However – as WSC's are neutral, without dipole nor quadrupole, it is not surprising that it is small. These considerations show that – within the range of validity of the present assumptions – the potential line-up across the interface between a pure material and an alloy A_xB_{1-x}/A is a linear function of the composition x.

V. Conclusions

We conclude summarizing the main goals achieved in the present work. A proper use of perturbation theory allows us to draw important and general conclusions about the orientation independence of band offsets on interface orientation. In particular, for all those isovalent interfaces where linear response with respect to the virtual crystal is a good approximation, the band offset is expected to be independent of interface orientation and even abruptness. Inspection of the spatial extent of the response of the virtual crystal to a single ion substitution allows one to introduce a model interface which uses crystal symmetry and bulk charge densities of the constituents as its only ingredients. We have shown that an appropriate use of basic concepts of electrostatics allows one to define an interface dipole at semiconductor interfaces avoiding any unnecessary reference to arbitrary ideal interfaces. Contrary to previous definitions, this dipole is directly related to the electrostatic potential line-up. We suggest that the above arbitrariness can be exploited to define a model interface such that electronic relaxation with respect to it is small in most cases. Last but not least, evidence has been given that a small difference exists between the line-ups of the two inequivalent (111) interfaces which is associated with net interface charges and internal electric fields.

Acknowledgements

The present work has been partially supported by the Italian Ministry of Education through the Centro Interuniversitario di Struttura della Materia and is part of the collaborative project between Scuola Internazionale Superiore di Studi Avanzati and the CINECA computing center.

References

[1] R.S. Bauer and G. Margaritondo, Phys. Today, **40**, 27 (1987) and ref. quoted therein. See also the issues: J. Vac Sci. and Technol. B 4 No. 4 (1986), and B **5**, No. 4 (1987).

[2] For a review of models of semiconductor heterjunction interfaces, see R.S. Bauer and G. Margaritondo, Ref. 1.

[3] C. Van de Walle and R.M. Martin, in *Computer-Based Microscopic Description of the Structure and Properties of Materials*, edited by J. Broughton, W. Krakow and S.T. Pantelides, (Materials Research Society, Pittsburg, 1986), p. 21; J. Vac. Sci. Technol. B **4**, 1056; Phys. Rev. B **35**, 8154 (1987).

[4] M. Cardona and N.E. Christensen, Phys. Rev. B **35**, 6182 (1987); N.E. Christensen, preprint

[5] D.M. Bylander and L. Kleinman, Phys. Rev. B **34**, 5280 (1986); *ibid.* **36**, 3229 (1987); Phys. Rev. Lett. **59**, 2091 (1987).

[6] S. Massidda, B.I. Min and A.J. Freeman, Phys. Rev. B **35**, 9871 (1987).

[7] S. Baroni, P. Giannozzi, and A. Testa, Phys. Rev. Lett. **58**, 1861 (1987).

[8] G.B. Bachelet, D.R. Hamann and M. Schlüter, Phys. Rev. B **26**, 4199 (1982).

[9] D.M. Ceperley and B.J. Alder, Phys. Rev. Lett. **45**, 566 (1980); J. Perdew and A. Zunger, Phys. Rev. B **23**, 5048 (1981).

[10] H.J. Monkhorst and J.P. Pack, Phys. Rev. B **13**, 5188 (1976).

[11] R.W. Godby, M. Schlüter and L.J. Sham, Phys. Rev. B **35**, 4170 (1987); *ibid.* **36**, 6497 (1987).

[12] S.B. Zhang, D. Tománek, and S.G. Louie, preprint

[13] L. Kleinman, Phys. Rev. B **24**, 7412 (1981).

[13] A. Muñoz, J. Sánchez-Dehesa and F. Flores, Phys. Rev. B **35**, 6468 (1987).

[15] M. Peressi, A. Baldereschi, S. Baroni, and R. Resta, to be published.

[16] C.K. Shih and W.E. Spicer, Phys. Rev. Lett. **58**, 2594 (1987).

[17] S.P. Kowalczyk, J.T. Cheung, E.A. Kraut, and R.W. Grant, Phys. Rev. Lett. **56**, 1605, (1986). T.M. Duc, H. Hsu, and J.P. Faurie, Phys. Rev. Lett. **58** 1127, (1987).

[18] S.H. Wei and A. Zunger, Phys. Rev. Lett. **59**, 144 (1987).

THE PHYSICS OF Hg-BASED HETEROSTRUCTURES

M. Voos

Groupe de Physique des Solides de l'Ens

24 rue Lhomond, 75005 Paris, France

We describe some properties of Hg-based superlattices and heterojunctions obtained from far infrared magneto-absorption experiments. A brief discussion of the valence band offset between HgTe and CdTe is also presented.

INTRODUCTION

Until recently two types of superlattices[1] (SL), I and II, have been investigated. GaAs-Al_xGa_{1-x}As structures correspond[2] to type I, and the situation is schematically represented in Fig. 1. The GaAs conduction band edge is at lower energy than that of Al_xGa_{1-x}As, while its valence band edge is at higher energy than that of Al_xGa_{1-x}As. As a result, the GaAs layers are potential wells for both electrons and holes which are thus confined in these layers. This situation is very frequent, and corresponds, in fact, to many of the structures which have been studied up to now. In InAs-GaSb systems[3], which belong to type II structures, the InAs conduction band edge lies at lower energy than the GaSb valence band edge, as illustrated in Fig. 2. It follows that the InAs and GaSb layers serve as potential wells for electrons and holes, respectively. Electrons and holes are thus spatially separated, but photon absorption and emission are nevertheless possible because there is a slight overlap of the electron and hole wave functions. In addition to the spatial separation of electrons and holes, there is another important difference between InAs-GaSb and GaAs-Al_xGa_{1-x}As superlattices. Indeed, the electron effective mass in InAs ($0.023\ m_0$[4]) is much lighter than in GaAs ($0.066\ m_0$[4]), leading to stronger tunneling interactions between InAs layers than between GaAs layers. As a result, in InAs-GaSb structures even for rather large values of the layer thicknesses (≈ 200 Å), the E_1 subband has an appreciable width ΔE_1 in the z direction[5], while ΔE_1 is often negligible in GaAs-Al_xGa_{1-x}As systems. Besides, due to the large heavy-hole effective mass in III-V compounds ($\approx 0.38\ m_0$[4]), the H_1 subband is essentially flat in the z direction.

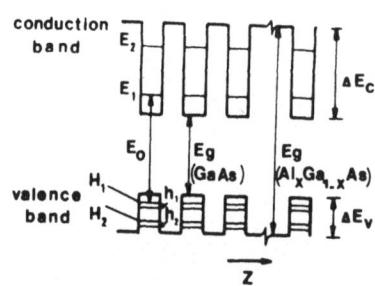

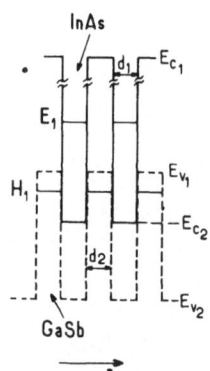

Fig. 1 Conduction and valence bands in a GaAs-AlGaAs superlattice in the z direction perpendicular to the plane of the layers. Heavy-hole (H_1, H_2), light-hole (h_1, h_2) and conduction (E_1, E_2) subbands are shown. E_O is the band gap, namely E_1-H_1.

Fig.2 Valence (E_{V1}, E_{V2}) and conduction (E_{c1}, E_{c2}) subbands in an InAs-GaSb superlattice in the z direction the perpendicular to the plane of the layers.

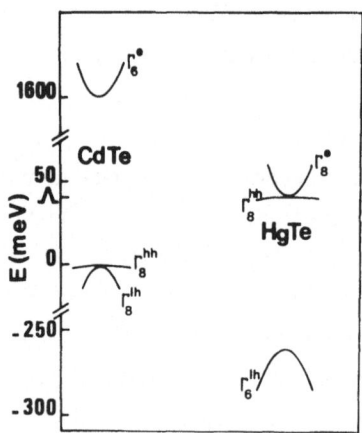

Fig. 3 Band structure of bulk HgTe and CdTe.

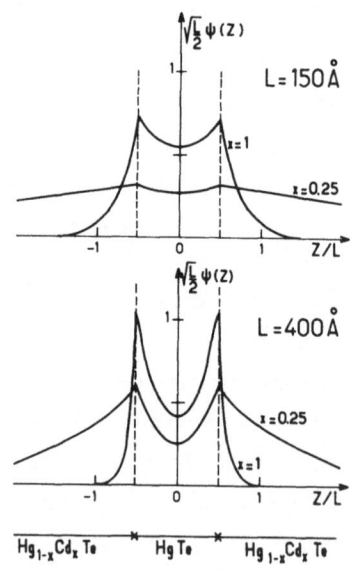

Fig.4 Interface wavefunctions in HgTe-Hg$_{1-x}$Cd$_x$Te double heterostructures. The thickness of the HgTe layer is L, and z is perpendicular to the layer plane.

Most of the superlattices investigated up to now involve III-V semiconductor compounds but, recently, II-VI compounds have been used to grow such heterostructures, namely HgTe-CdTe[6,7], $Hg_{1-x}Mn_xTe$-CdTe[7], HgTe-ZnTe[7], $Hg_{1-x}Cd_xTe$-CdTe[7], $CdTe-Cd_{1-x}Mn_xTe$[8,9], CdTe-ZnTe[7] and $ZnSe-Zn_{1-x}Mn_xSe$[10] SL's. It is thought that the $Hg_{1-x}Cd_xTe$-CdTe (for x > 0.16), $Hg_{1-x}Mn_xTe$-CdTe (for x > 0.07), $CdTe-Cd_{1-x}Mn_xTe$, CdTe-ZnTe and $ZnSe-Zn_{1-x}Mn_xSe$ SL's correspond to type I structures, while the HgTe-CdTe, HgTe-ZnTe, $Hg_{1-x}Cd_xTe$-CdTe (for x < 0.16) $Hg_{1-x}Mn_xTe$-CdTe (for x < 0.07) ones belong to a new category, i.e. type III superlattices whose physical aspects will be described below and result from the unusual band structure of bulk HgTe.

We will describe briefly some properties of Hg-based heterostructures obtained from far infrared magneto-optical experiments performed at low temperature. The heterostructures which will be considered here are HgTe-CdTe SL's, HgMnTe-CdTe SL's and HgCdTe-CdTe heterojunctions. We will also discuss the value obtained from some of these investigations for a very important parameter, namely the valence band offset $\wedge$ at each interface.

BAND STRUCTURE OF BULK HgTe AND CdTe

HgTe-CdTe heterostructures are particularly interesting from the point of view of basic physics because of the specific band structure of bulk HgTe and CdTe and of the band alignment of the two host materials (see Fig. 3). CdTe is a wide gap semiconductor (≈ 1.6 eV at low temperature) with a direct gap at the Brillouin zone center (Γ point). At k = 0, the conduction band edge has a s-type symmetry (Γ_6) and the upper valence band edge (Γ_8) is fourfold degenerate (p type symmetry, J = 3/2). The spin-orbit split-off Γ_7 band (p type symmetry, J = 1/2) lies below the Γ_8 states with $\Delta = E_{\Gamma 8} - E_{\Gamma 7} \approx 1$ eV. On the other hand, HgTe is a zero gap semiconductor as a result of the inversion of the relative positions of the Γ_6 and Γ_8 edges. The Γ_8 light hole band in CdTe becomes a conduction band in HgTe, and the Γ_6 conduction band in CdTe forms a light-hole band in HgTe. The ground valence band is the Γ_8 heavy-hole one, and the Γ_8 states correspond to the top of the valence band and to the bottom of the conduction band, resulting in a zero gap configuration with a spin orbit separation Δ which is ≈ 1 eV. The superlattice band structure depends of course on the offset between the conduction and valence band edges at the HgTe-CdTe interface. From Harrison's common anion argument[11], one can think that the offset $\wedge$ between the Γ_8 band edges of HgTe and CdTe is small. As shown later from the analysis of the experimental data, we believe that $\wedge$ is positive and smaller than ≈ 100 meV, this parameter being measured from the top of the CdTe valence band (Fig. 3). This value of $\wedge$ implies that the HgTe layers are potential wells for heavy holes, but the situation for light particles (electrons or light holes) is more complicated because the bands which contribute most significantly to the light-particle SL states are the Γ_8 conduction band in HgTe and Γ_8 light valence band in CdTe. These two bands have <u>opposite</u> curvatures and the <u>same</u> Γ_8 symmetry. This mass-reversal for the light particles at each interface is a unique property of the HgTe-CdTe system. An important consequence of these very unusual features is the existence of interface states[12] which are evanescent in both the HgTe and CdTe layers with a wavefunction peaking at the interfaces, as shown, for example, in Fig. 4 in the case[13] of a HgTe layer sandwiched between two CdTe layers. This special situation, which corresponds to type III structures, is thus very different from the more common one met, for instance, in GaAs-Al$_x$Ga$_{1-x}$As SL's (type I) where the SL states arise mainly from bands in GaAs and AlGaAs displaying the same curvature. Note also, for example, that for bulk Hg$_{1-x}$Cd$_x$Te the band structure is similar to that of HgTe for x < 0,16 at low temperature and corresponds therefore to a band gap equal to zero. For x > 0.16, the band structure of these alloys is similar to that of CdTe.

HgTe-CdTe SUPERLATTICES

The infrared magneto-absorption experiments described here were performed at liquid helium temperature using a grating monochromator (3 μm ≤ λ ≤ 5 μm), a CO2 laser (9 μm ≤ λ ≤ 11 μm), a far-infrared laser (41 μm ≤ λ ≤ 255 μm), and carcinotrons (600 μm ≤ λ ≤ 1 μm). The transmission signals, observed at fixed photon energies (in the Faraday configuration), were detected by a carbon bolometer. The magnetic field B, provided by a superconducting coil, could be varied from 0 to 10 T and was perpendicular to the plane of the layers.

Fig. 5(a) presents typical transmission spectra obtained[14] in a HgTe-CdTe SL consisting of one hundred periods of (100 Å) HgTe-(36 Å) CdTe grown[7] by molecular beam epitaxy on a (111) CdTe substrate. Figs.5(b) and 6 give the energy positions of the transmission minima (i.e. absorption maxima) as a function of B from data such as those which are shown in Fig. 5 (a). It can be seen from these results that no transmission spectra have been observed around 20 meV which corresponds to the LO phonon energy in CdTe and to the restrahlen band of the substrate.

To interpret the data, the SL band structure should be calculated, taking into account the band structure of bulk HgTe and CdTe. This has been done, for instance, in the envelope function formalism[14], and the results are presented in Fig. 7 (for ∧ = 40 meV) which shows, along k$_x$ and k$_z$ (where z is perpendicular to the plane of the layers), the lowest conduction subband E$_1$, the ground light-particle subband I (interface state), and heavy-hole subbands HH$_1$, HH$_2$ and HH$_3$. The SL gap (E$_1$ - HH$_1$ at k = 0) is found to be 17 meV at 4 K. One can see that the valence band structure is rather complicated for k$_x$ ≠ 0 as a result of the strong hydibrization occuring between the heavy-hole and I states. Then, to obtain the corresponding Landau levels, and thus the energy of the observed transitions, one has to calculate the band structure when a magnetic field is applied along the z direction. Such calculations[14] are formally the same as those performed at B = 0, replacing **k** by **k** - eA/c in the Kane Hamiltonian and taking into account the direct coupling of the electron and hole spins by

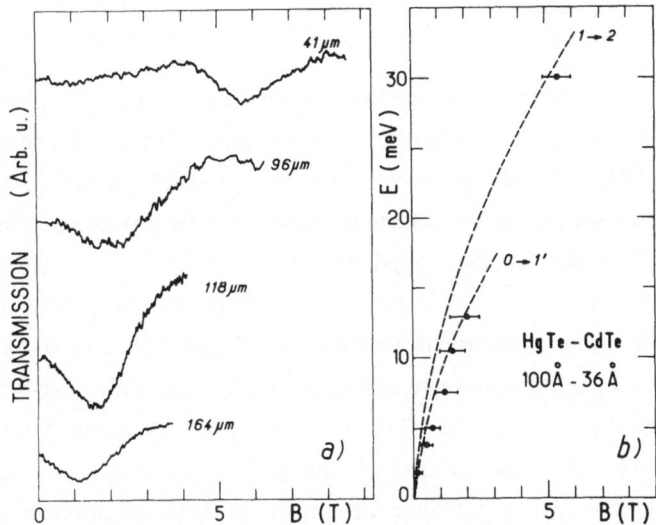

Fig. 5 (a) Typical transmission spectra at 1.6 K in a HgTe-CdTe SL as a function of B for several infrared wavelengths. (b) Energy position of the observed transmission minima as a function of B (solid dots) ; the dashed lines correspond to theoretical fits to the data.

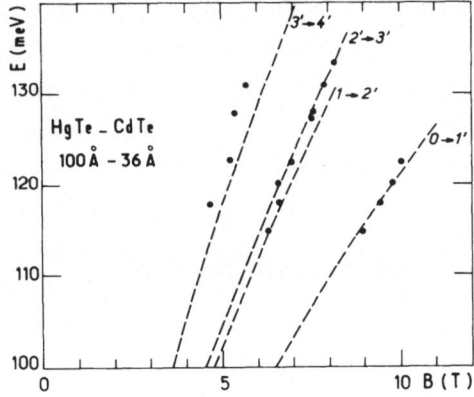

Fig. 6 Energy position of transmission minima detected in a HgTe-CdTe SL at 1.6 K (solid dots). The dashed lines are theoretical fits to the data.

introducing the additional valence band parameter κ[15]. The motion parallel to the layers is thus given[14] by a six-component vector :

$$\varphi_n = (C_1 \psi_{n-1},\ C_2 \psi_{n-2},\ C_3 \psi_n,\ C_4 \psi_n,\ C_5 \psi_{n-1},\ C_6 \psi_{n+1})$$

where ψ_n is the nth harmonic oscillator funciton and $n = -1, 0, 1, 2...$ For $n \leq 1$, the C_1 coefficients corresponding to the negative oscillator index vanish. The calculated Landau levels[14] of E_1, I, HH_1 and HH_2 are given, also for $\wedge = 40$ meV, in Fig. 8. The situation is obviously very complicated, and the Landau levels are strongly mixed, again as a result of the coupling between the I state and the heavy-hole ones.

The magneto-optical transitions shown in Fig. 5 extrapolate to an energy $E \approx 0$ at $B = 0$, and are attributed to cyclotron resonance in the E_1 band, which is consistent with the n-type nature of this superlattice. The first intraband magneto-optical transitions corresponding to the selection rule $\Delta n = +1$ are[14] the $1 \to 2$ and $0 \to 1'$ transitions (see Fig. 8). The dashed lines in Fig. 5(b) give the energies of these transitions obtained from the calculations of the E_1 Landau levels shown in Fig. 8 for $\wedge = 40$ meV. At low photon energies (< 15 meV), the theoretical results correspond fairly well to the observed broad absorption lines (Fig. 5(a)), evidencing that the $n = 1$ and $n = 0$ levels are populated. For $E = 30$ meV, the calculated magnetic field separation between the corresponding lines is broader than the observed absorption line. Only one type of transition, namely $1 \to 2$, is detected, which indicates that the $n = 1$ Landau level is populated for the corresponding value of the magnetic field, namely $B \approx 5$ T. It is worth noting that interband transitions from valence up to conduction Landau levels are not detectable in the investigated infrared region (0-30 meV) because of the population of the ground conduction levels and of the value of the superlattice band gap. Such transitions have been studied in the CO_2 laser energy region, as shown in Fig. 6. They are interpreted as being due to interband transitions from HH_1 to E_1 Landau levels with the selection rule $\Delta n = \pm 1$. The dashed lines in Fig. 6 correspond to such transitions calculated using $\Delta n = +1$ and $\wedge = 40$ meV (Fig. 8), showing that the agreement between experimental and theorical results is satisfying. In fact, the experimental data could also be interpreted with the selection rule $\Delta n = -1$ because of the width of the observed absorption lines but, for simplicity, only one type of transition is shown in Fig. 6. These investigations give essentially information on the E_1 Landau levels because most of the transition energies arise from the conduction band. Besides, the transitions shown in Fig. 6 extrapolate to an energy $E \approx 20$ meV at $B = 0$, which is consistent with the theoretical value of the SL band gap (17 meV). It can thus be concluded that these investigations support the value of $\wedge$ (40 meV) determined from the first magneto-optical experiments performed[16] on HgTe-CdTe superlattices. We would also like to emphasize that optical and magneto-optical transitions between HH_1 and E_1 can be intense because the HH_1 subband wavefunction, which is localized in the HgTe layers, is p-like while the part of the E_1 subband wavefunction confined in the HgTe layers has[17] a non-negligible s-character.

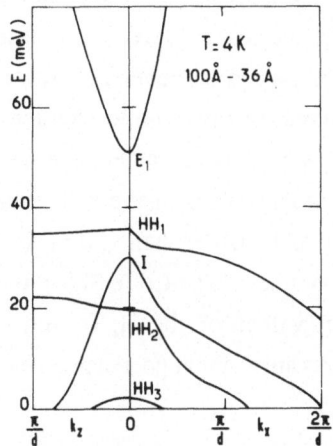

Fig. 7 Calculated band structure along k_x and k_z of a (100 Å) HgTe- (36 Å) CdTe superlattice ; d is the superlattice periodicity.

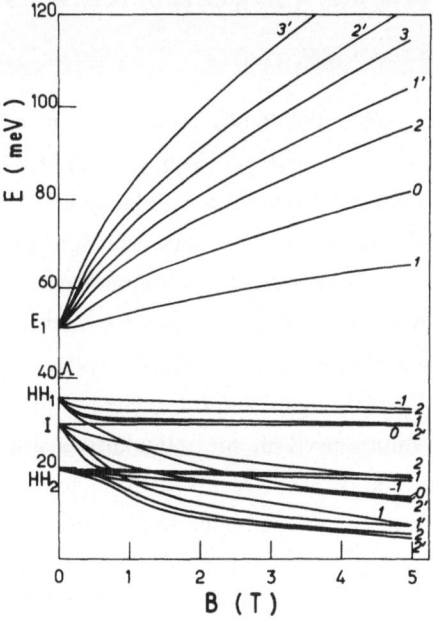

Fig. 8 Calculated Landau levels for the same sample as in Fig. 7 for the HH_2, HH_1, I and E_1 subbands.

We have investigated the sensitivity of the fitting procedure to the value of ∧. At first, one should notice that SL band gap is smaller than 5 meV for ∧ > 100 meV and that, in addition to cyclotron resonance, interband transitions should be detected in the studied infrared region (0-30 meV). We have also checked that a reasonable agreement between the experimental and theoretical results can be obtained for $0 < ∧ < 120$ meV by taking into account the possible uncertainties on the sample characteristics, the HgTe and CdTe parameters and the data themselves (namely the rather large width of the absorption lines). We are therefore led to consider from these investigations that the valence band offset ∧ at the HgTe-CdTe interface is positive and smaller than 120 meV. In addition, we have recently performed similar experiments on other HgTe-CdTe SL's grown on CdTe, $Cd_{0.96}Zn_{0.04}Te$ or GaAs substrates with different values of the HgTe and CdTe layer thicknesses, and experimental and theoretical results are again in good agreement for ∧ smaller than 120 meV.

HgMnTe-CdTe SUPERLATTICES

Here we report temperature-dependent magneto-absorption experiments on a $Hg_{0.96}Mn_{0.04}Te$-CdTe superlattice grown by molecular beam epitaxy. The superlattice was grown on a (100) GaAs substrate with a buffer layer of ≈ 2 μm thick (111) CdTe[7]. The superlattice, which is n-type, consists of 100 periods of $d_1 = 168$ Å thick $Hg_{0.96}Mn_{0.04}Te$ quantum wells interspaced by $d_2 = 22$ Å thick CdTe barriers, with a SL electron concentration $n = 6 \times 10^{16}$ cm^{-3}.

The band structure for such a superlattice has been calculated[18] in the envelope function approximation. The band structures of $Hg_{0.96}Mn_{0.04}Te$ and CdTe near the Γ point under strong magnetic field are described by the 6 x 6 Pidgeon and Brown Hamiltonian[19]. This includes the Γ_6 and Γ_8 bands and neglects the spin-orbit split-off Γ_7 band, due to the large spin-orbit splitting, $\Delta \approx 1$ eV, in the Te-based II-VI semiconductors. The interaction with the higher bands is included up to second order through the modified Luttinger parameters γ_1, $\gamma_2 = \gamma_3 = \gamma$ (spherical approximation), and κ. The two other parameters which enter the calculation are E_g, the Γ_6-Γ_8 energy gap, and E_p, related to the square of the Kane matrix element. The parameter values used in the calculations are given[18] in Table 1. For $Hg_{0.96}Mn_{0.04}Te$, the Pidgeon and Brown model is modified to take into account the exchange interactions between localized d electrons bound to the Mn^{2+} ions and the Γ_6 and Γ_8 s- and p-band electrons. The s-d and p-d interactions in the molecular field approximation are introduced through two additional parameters, r and A[20,21]. The parameter $r = \alpha/\beta$ is the ratio between the Γ_6 and Γ_8 exchange integrals, α and β. The parameter A is the normalized magnetization defined by $A = (1/6) \beta \times N_0 < S_z >$, where N_0 is the number of unit cells per unit volume of the crystal, x is the Mn composition (x = 0.04 here) and $< S_z >$ is the thermal average of the spin operator along the direction of the applied magnetic field.

Table 1 : Band parameter values used in the calculations (see text).

	γ_1	γ	κ	E_g(meV)	E_p(eV)
$Hg_{0.96}Mn_{0.04}Te$	4.50	1.1	- 0.85	- 125	18
CdTe	1.54	0.015	0.6	1600	18

Previous magneto-optical data[20,21] and direct magnetization measurements[22,23], obtained on alloys with dilute Mn concentration, have shown that $< S_z >$ can be well described by a modified spin 5/2 Brillouin function, replacing the temperature T with an empirical temperature parameter $(T + T_0)$. T_0 is found to be ≈ 5 K for x = 0.04[23], while[21] the exchange integral $N_0\beta \approx 1$ eV and $r \approx -1$. For a HgMnTe-CdTe superlattice, a system of six differential equations for the six component envelope function is established from the Pidgeon and Brown Hamiltonian. The boundary conditions at the interfaces are satisfied by the continuity of the wave functions and integration of the six coupled differential equations across an interface. Taking into account the superlattice periodicity through the Bloch theorem, a numerical solution for the Landau level energies is obtained. Figure 9 shows these results[18] at $k_z = 0$ for the first conduction band E_1 and the first heavy hole band HH_1. We have set the CdTe valence band edge at zero energy and the $Hg_{0.96}Mn_{0.04}Te$ valence band edge at 40 meV, as in HgTe-CdTe superlattices[14]. The energy levels are enumerated as in Ref. 14. Note that this sample is a very narrow gap superlattice (≈ 12 meV), due to the small confinement energies with such thick quantum wells and thin barriers. The effect of the magnetization, A, on the Landau level energies is illustrated by the three different values of A given in the figure. Finally, the magnitude of A appropriate for x = 0.04 at various magnetic fields and temperatures is given[23,24] in Table 2.

The magneto-absorption experiments[18] reported here used as infrared sources a CO_2 laser ($\lambda = 9 . 3$ um - $10 . 8$ µm) and a molecular laser ($\lambda = 41$ µm - 255 µm) pumped by the CO_2 laser. The transmission signal, detected by a carbon bolometer, was measured at fixed wavelengths while the magnetic field, provided by a superconducting coil, was varied continuously to 12 T. As usual, the cyclotron resonance or interband transitions correspond to minima in the transmission spectrum. The magnetic field positions of the minima for several different far-infrared wavelengths are given in Figure 10 as open circles. These data are interpreted as electron cyclotron resonance transitions since the sample is n-type and the data extrapolate to $E \approx 0$ at B = 0. The solid lines in Fig. 10 correspond to all calculated cyclotron resonance transitions observable with a selection rule $\Delta n = + 1$ and a Fermi energy $E_F \approx$ 30 meV above the conduction band edge, consistent with the large electron concentration for this sample. Here we have taken the magnetization A = - 1 meV/T at all magnetic fields for simplicity.

Figure 11 shows the transmission minima observed at several infrared wavelengths of the CO_2 laser. The calculated interband transitions corresponding to $\Delta n = \pm 1$ (solid lines) are

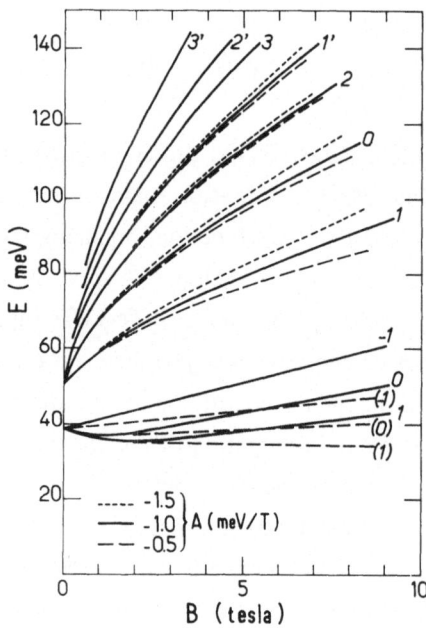

Fig. 9 Influence of the magnetization on the E_1 and HH_1 Landau levels.

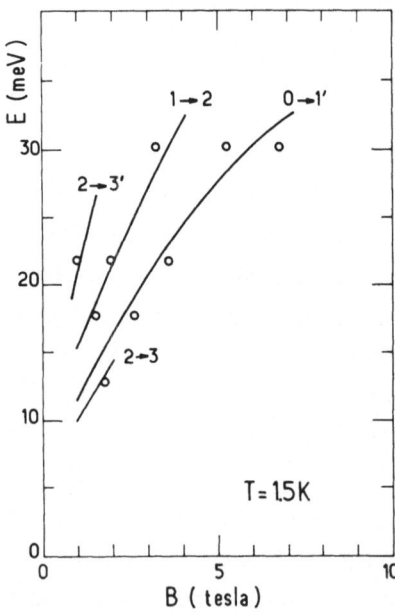

Fig. 10 Experimental magneto-transmission minima (open circles) and calculated cyclotron resonance transitions (solid lines).

in agreement with the observed transitions. Each observed minimum (denoted → 2' and →1') most probably results from the two unresolved interband transitions. Calculated transitions to conduction band levels n = 1, 0, and 2 lie at magnetic fields above our experimental range and are therefore not observed. The data and calculated transitions contained in Figs. 10 and 11 show that this superlattice is well-behaved and adequatly modelled by the envelope function approximation.

Let us consider now the temperature dependence of the observed transitions as evidence of the magnetic nature of the Mn impurities. Figure 12 shows four transmission spectra at $\lambda = 41$ μm (E = 30 . 2 meV) for temperatures from 1.5 K to 10 K. The two identified minima shift to lower magnetic field : the 1 → 2 transition by $\Delta B_r = -0.5$ T and the 0 → 1' transition by ≤ - 0.2 T. There are two contributions to these temperature dependent shifts : the $Hg_{0.96}Mn_{0.04}Te$ energy gap and the Mn magnetization. From the behavior of HgTe[25], as T increases up to 10 K, the magnitude of the energy gap for bulk $Hg_{0.96}Mn_{0.04}Te$ decreases by less than 3 meV. The resulting decrease of the cyclotron mass gives calculated shifts of at most $\Delta B_r \approx -0.05$ T and - 0.15 T, respectively. This may account for the shift of the 0 → 1' transition, but the shift of the 1 → 2 transition is about ten times larger.

As temperature increases over our experimental range, the Mn magnetization decreases dramatically at lower magnetic fields (Table 2).

Table 2 : Magnetization, A (meV/T) at different magnetic fields and temperatures for bulk $Hg_{0.96}Mn_{0.04}Te$ (see text).

	B = 0.5 T	3 T	5 T	7 T	10 T
T = 1.5 K	- 2.3	- 1.3	- 1.0	- 0.8	- 0.6
4.2 K	- 1.4	- 1.1	- 0.9	- 0.7	- 0.6
10.0 K	- 0.5	- 0.5	- 0.45	- 0.45	- 0.4

This has a direct effect on the Landau levels in the superlattice (Fig. 9). Note that the n = 1 level is depressed more than the n = 2 level, increasing the cyclotron resonance energy. This effect is enhanced in superlattices as compared to bulk semi-magnetic semiconductors, in which the exchange interaction is primarily observed in the spin and combined resonances[26]. For magnetization A = - 1.3 meV/T decreasing to - 0.5 meV/T, we calculate a shift $\Delta B_r \approx -0.5$ T for the 1 → 2 transition, which accounts perfectly for the enhanced experimental shift. The calculated magnitude of the temperature dependent shift, ΔB_r, is shown in Fig. 12. The 0 → 1' transition does not exhibit this magnetic effect, because the n = 0 and n = 1' levels shift the same for - 0.5 ≤ A ≤ - 0.9 meV/T (Table 2). As might be expected, the experimental interband transitions (Fig. 11) also exhibit temperature dependences. As with the cyclotron resonance transitions, the → 1' transition shifts to lower field with increasing temperature ($\Delta B_r = -0.4$ T for T = 1.5 K → 10 K). This matches our

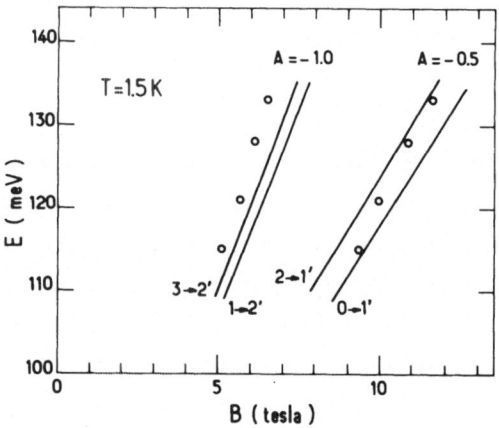

Fig. 11 Observed data (open circles) and calculated energies (solid lines) corresponding to interband transitions (A is given in meV/T).

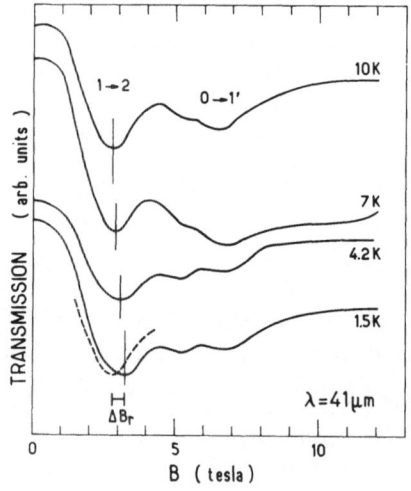

Fig. 12 Far infrared magneto-transmission spectra between 1.5 K and 10 K. The bracket ΔB_r indicates the magnitude of the calculated shift in the $1 \rightarrow 2$ cyclotron resonance transition.

calculations ($\Delta B_r \leq -0.5$ T) for the temperature dependence of the $Hg_{0.96}Mn_{0.04}Te$ energy gap. No temperature dependent magnetization effect exists here due to saturation of the Mn magnetization at high magnetic field. The remaining interband transition, the $\rightarrow 2'$ transition, shifts in the direction contrary to all other transitions ($\Delta B_r \approx +1$ T for T = 1.5 K $\rightarrow$ 10 K).

The calculated shift due to the band gap ($\Delta B_r \leq -0.05$ T) is dominated by a larger shift in the proper direction ($\Delta B_r \approx +0.5$ T) due to the change in Mn magnetization. The net calculated shift remains smaller than the one which is observed. The data and calculations suggest an even larger effect due to the presence of Mn.

In conclusion, we have observed cyclotron resonance transitions and interband transitions in a $Hg_{0.96}Mn_{0.04}Te$-CdTe superlattice which are in good agreement with calculations in the envelope function approximation. These transitions exhibit temperature dependent energy shifts (1.5 K $\leq$ T $\leq$ 10 K) which require the inclusion of both magnetization and superlattice effects.

HgCdTe-CdTe HETEROJUNCTIONS

We describe now magneto-optical experiments at low temperatures performed on a two-dimensional electron gas in $Hg_{1-x}Cd_xTe$-CdTe heterojunctions grown by molecular beam epitaxy (MBE). With this growth technique, the interfaces are better defined than in the $Hg_{1-x}Cd_xTe$ metal-oxide-semiconductor (MOS) structures which have been used until now to study two-dimensional electron gases in II-VI compounds[27]. The samples (S1, S2) used here grown[7] by MBE on GaAs substrates with the following growth sequence : for sample S1, a (100) CdTe buffer layer with thickness $d_1 = 2.02$ μm, a $Hg_{0.8}Cd_{0.2}Te$ layer with thickness $d_2 = 1.02$ μm, and a CdTe cap layer with thickness $d_3 = 300$ Å ; for sample S2, a (100) CdTe buffer layer ($d_1 = 2.06$ μm), a HgTe sandwich layer with thickness $d_{SW} = 100$ Å, a $Hg_{0.8}Cd_{0.2}Te$ layer ($d_2 = 0.89$ μm), and a CdTe cap layer ($d_3 = 300$ Å). The far-infrared magneto-absorption experiments reported here were done at 1.6 K using as an infrared source a molecular laser pumped by a CO_2 laser.

Typical transmission spectra obtained at different infrared wavelengths are given in Figure 13 for sample S1 and $\theta = 0°$, where θ is the angle between the direction of the magnetic field and the normal to the layer plane. Two transmission minima can be clearly seen. As shown in Figure 14, the magnetic field positions of the minima depend on θ, and varies roughly as $(\cos \theta)^{-1}$, indicating that we are dealing with a two-dimensional system. Figure 15 gives as a function of B the infrared photon energy, E, corresponding to the transmission minima detected for $\theta = 0°$ (Fig. 13). The observed optical transitions extrapolate to an energy E $\approx$ 0 at B = 0. Taking into account the n-type nature of the investigated structures, as obtained from Hall experiments, these transitions are attributed to two-dimensional electron cyclotron resonance transitions. The weak feature appearing in

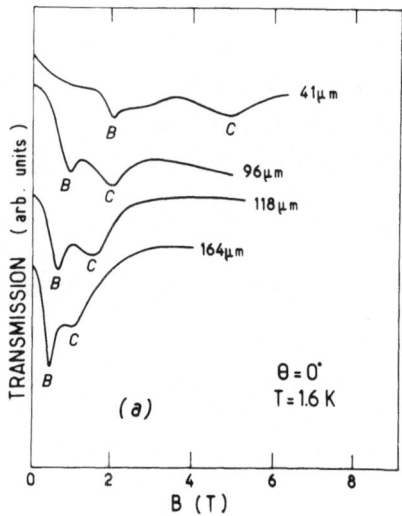

Fig. 13 Transmission spectra as a function of the magnetic field, for different infrared wavelengths.

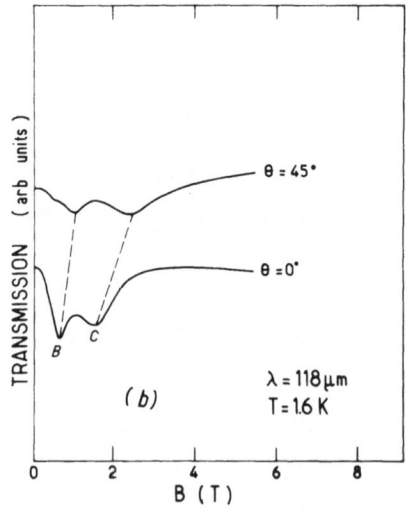

Fig. 14 Transmission spectra versus B for an infrared wavelength $\lambda = 118\ \mu$m.

Fig. 14 at B ≈ 0.5 T for θ = 45° may correspond to bulk cyclotron resonance in the HgCdTe layer, with an electron effective mass $m^* = 0.0055\ m_0$. Quantum Hall effect measurements at 4.2 K yield a two-dimensional electron density, n_S, a few times $10^{11}\ cm^{-2}$, and a simple analysis of the width of these cyclotron lines yields an electron mobility $\mu \approx 5 \times 10^4\ cm^2\ V^{-1}s^{-1}$. Thus, it is quite reasonable to consider that a two-dimensional electron gas occurs in this HgCdTe-CdTe heterojunction. We now seek to establish that : (i) the charge transfer occurs at least in part from the wide gap CdTe deep traps to the narrow gap HgCdTe layer, and (ii) the observed two-dimensional electron gas occurs at the heterojunction interface and not at the surface of the sample.

Figure 16 gives schematically the variation of the Γ_6 conduction and Γ_8 valence band-edges in sample S1 in the z direction perpendicular to the plane of the layers, taking into account electron transfer from the CdTe to the $Hg_{0.8}Cd_{0.2}Te$ layer. The band gaps of CdTe and $Hg_{0.8}Cd_{0.2}Te$ at 1.6 K are 1.6 eV and 60 meV, respectively. The valence band offset between these two materials is not known, but it should be roughly similar to the HgTe-CdTe offset. The offset in HgTe-CdTe heterostructures is not yet firmly established, since reported values range from 40 to 350 meV[16,28,29] but, in any case, it is substantially smaller than 0.8 eV, half the band gap of CdTe. In addition, the Fermi level, E_F, in the CdTe layer is approximately located at the CdTe mid-gap, as shown by XPS measurements and transport measurements on CdTe epilayers grown under similar MBE conditions. This leads to the situation shown in Fig. 16 and, as a result, we believe that the two-dimensional electron gas is at least partly due to electron transfer from CdTe mid-gap deep levels to the HgCdTe layer, even if bulk HgCdTe electrons contribute to the two-dimensional layer. Turning to point (ii), the data shown in Fig. 15 imply that two electron subbands, with energies E_1 and E_2, are populated. The corresponding electron effective masses are $m_1 = 0.016\ m_0$ and $m_2 = 0.007\ m_0$ for B ≈ 1 T. The value of m_1 is comparable to that obtained recently (≈ 0.015 m_0) in an inversion layer occuring in a (p-type) $Hg_{0.8}Cd_{0.2}Te$ MOS structure with a comparable electron density[30]. Detailed agreement is not expected because the shape of the quantum well in the MOS structure is necessarily different from that in sample S1, where one is dealing with an accumulation layer.

In sample S2, the situation is somewhat more complicated because of the inclusion of a thin HgTe sandwich layer. The experimental data[31] from S2 suggest the existence of a two-dimensional electron gas with two populated electron subbands, as in S1. However, in S2, due to the thinness of the HgTe sandwich layer, the two-dimensional electron gas is partially located in both the HgTe and $Hg_{0.8}Cd_{0.2}Te$ layers, a new situation in semiconductor heterojunctions. In S2, the experimental electron effective masses are $m_1 = 0.023\ m_0$ and $m_2 = 0.013\ m_0$ at B ≈ 1 T with value for n_S similar to that in S1. As a consequence of the HgTe conduction band mass (≈ 0.03 m_0), m_1 and m_2 are larger in S2 than in S1 but, nevertheless, smaller than the HgTe bulk mass. This is consistent with the two-dimensional electron gas being located at the heterojunction interface in these samples. Finally, the

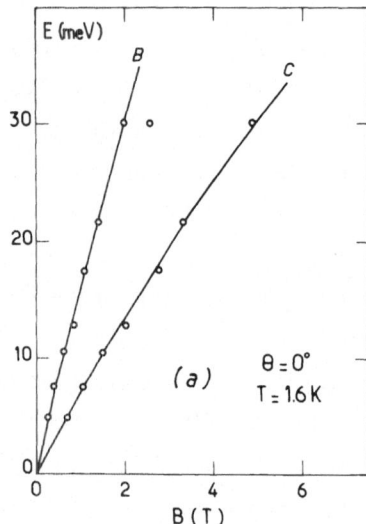

Fig. 15 Position of the transmission minima as a function of far-infrared photon energy, E, and magnetic field B (open circles). The solid lines are guides to the eye.

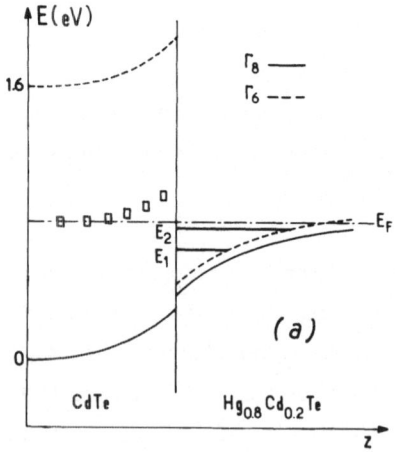

Fig. 16 Schematic variation of the conduction and valence bands.

experimental results are unaffected by chemical etching (using bromine methanol) of the entire CdTe cap layer and up to one half of the $Hg_{0.8}Cd_{0.2}Te$ epilayer. Thus, we believe that the observed two-dimensional electron gas is located at the heterojunction interface. Unfortunately, due to large differences in n_S, it is not possible to make a significant comparison between our results and those previously reported[32,34]. Also, previous theoretical results correspond to larger values of[35] n_S. However, unpublished calculations[36] for $n_S \approx 3 \times 10^{11}$ cm^{-2} yield two occupied subbands with $m_1 = 0.015\ m_0$ and $m_2 = 0.010\ m_0$ in the case of MOS $Hg_{0.8}Cd_{0.2}Te$ structures with $N_A - N_D = 10^{14}$ cm^{-3}. These values are consistent with the results obtained in sample S1. In any case, precise calculations will be complicated by the resonance between the interface wavefunction and the bulk $Hg_{0.8}Cd_{0.2}Te$ valence band[27,35]. In the case of sample S2, further complication results from the presence of the HgTe sandwich layer.

In summary, it has been demonstrated that a two-dimensional electron gas can occur in $Hg_{1-x}Cd_xTe$-CdTe heterojunctions. The electron transfer involves certainly deep centers in the gap of the CdTe layer, corresponding to a new charge transfer mechanism.

THE VALENCE BAND OFFSET

The valence band offset $\wedge$ at the HgTe-CdTe interface is a very important parameter, at least because it governs the SL band structure and thus many properties of such heterostructures. It should be first emphasized that $\wedge$ is necessarily positive because, in the opposite situation ($\wedge < 0$), charge transfer would occur between HgTe and CdTe layers, and this is not compatible with the p-type conduction observed[37] at low temperatures in most of the HgTe-CdTe superlattices investigated up to now. From the magneto-optical investigations presented here, we are led to think that $\wedge$ lies in the range 0-120 meV, which is a rather small value. This is corroborated by different experiments performed in Hg-based heterostructures. For instance, resonant Raman scattering measurements[38] done at 12 K in HgTe-CdTe SL's in the backscattering geometry give data which are consistent with $0 < \wedge < 120$ meV. Another example is provided by resonant tunneling experiments[39] in double-barrier HgCdTe-CdTe heterostructures which yield $\wedge \approx 130$ meV. On the other hand, XPS experiments[28,29] give $\wedge = 0.35$ eV, and Schottky barrier measurements[40] provide an upper bound for $\wedge$ (0.5 - 0.8 eV) but, in this case, this result is not reliable because the HgTe-CdTe interface was not steep since its extension was 300 Å about.

From a theoretical point of view, some investigations yield[41] $\wedge = (0.5 \pm 0.2)$ eV, in contradiction with the common anion rule argument[11,42] and with recent calculations[43].

At this stage, it should also be noticed that all the HgTe-CdTe SL's used in the far infrared magneto-optical experiments are calculated (in the envelope function scheme) to be semimetallic at 4 K for $\wedge$ = 0.35 eV, which does not seem to be consistent[13] with the magneto-absorption data. It can thus be only concluded that more experimental and theoretical investigations are required to reconcile the different values obtained up to now for the valence band offset in the HgTe-CdTe system.

CONCLUSION

We believe that the investigations described here show unambiguously that the band structure of Hg-based heterostructures is more complicated and subtle than that of usual III-V compound structures such as GaAs-AlGaAs superlattices for example. We are led to think that the band offset $\wedge$ is small and positive, but it is really necessary to reconcile the different values reported for this parameter because it is very important from the point of view of basic and applied physics. Indeed the band structure, and thus many properties of these heterostructures, depend on the value of $\wedge$.

It is worth noting that II-VI compound heterostructures broaden the field of quasi-two-dimensional systems, at least from the point of view of basic physics. Type I and type III situations can be achieved, depending on the Hg content, and type II configurations can certainly be obtained, for instance under magnetic field in heterostructures involving manganese[44]. Another interesting system is the HgTe-ZnTe one[7] because it is highly strained, and we have recently obtained promising preliminary data on such superlattices.

ACKNOWLEDGEMENTS

The author would like to emphasize that all the investigations described here involved G. Bastard, J. M. Berroir, Y. Guldner and J. P. Vieren from the Ecole Normale Supérieure, and J. P. Faurie from the University of Illinois at Chicago.

REFERENCES

1. L. Esaki and R. Tsu, IBM J. of Res. and Develop. 14, 61 (1970).
2. See, for example, R. Dingle in : "Festkörperprobleme (Advances in Physics)", H. J. Queisser, ed., Pergamon-Vieweg, Braunschweig (1975).
3. See, for example, L. Esaki and L. L. Chang, J. Magn. Magn. Mater. 11 208 (1979).
4. "Handbook of Electronic Materials", M. Neuberger, ed., Plenum, New York (1971).
5. Y. Guldner, J. P. Vieren, P. Voisin, M. Voos, L. L. Chang and L. Esaki, Phys. Rev. Lett. 45 1719 (1980).

6. J. P. Faurie, A. Million and J. Piaguet, Appl. Phys. Lett. $\underline{41}$ 713 (1982).

7. J. P. Faurie, IEEE J. Quantum Electron, QE-$\underline{22}$, 1656 (1986) ; X. Chu, S. Sivananthan and J. P. Faurie, Appl. Phys. Lett. $\underline{50}$, 597 (1987).

8. L. A. Kolodziejski, T. C. Bonsett, R. L. Gunshor, S. Datta, R. B. Bylma, W. M. Becker and N. Otsuka, Appl. Phys. Lett. $\underline{45}$ 440 (1984).

9. R. N. Bicknell, R. W. Yanka, N. C. Giles-Taylor, D. K. Blanks, E. L. Buckland and J. F. Schetzina, Appl. Phys. Lett. $\underline{45}$ 92 (1984).

10. L. A. Kolodziejski, R. L. Gunshor, T. C. Bonsett, R. Venkatasubramanian, S. Datta, R. B. Bylsma, W. M. Becker and N. Otsuka, Appl. Phys. Lett. $\underline{47}$ 169 (1985).

11. W. A. Harrison, J. Vac. Sci. Technol. $\underline{14}$ 1016 (1977).

12. Y. C. Chang, J. N. Schulman, G. Bastard, Y. Guldner and M. Voos, Phys. Rev. B $\underline{31}$ 2557 (1985).

13. M. Voos, Surf. Sci. Reports $\underline{7}$, 189 (1987).

14. J. M. Berroir, Y. Guldner, J. P. Vieren, M. Voos and J. P. Faurie, Phys. Rev. B $\underline{34,}$ 891 (1986).

15. J. M. Luttinger, Phys. Rev. $\underline{102}$, 1030 (1956).

16. Y. Guldner, G. Bastard, J. P. Vieren, M. Voos, J. P. Faurie and A. Million, Phys. Rev. Lett. $\underline{59}$, 907 (1983).

17. J. N. Schulman and Y. C. Chang, Phys. Rev. B $\underline{33}$, 2594 (1986).

18. G. Boebinger, Y. Guldner, J. M. Berroir, M. Voos, J. P. Vieren and J.P. Faurie, Phys. Rev. B $\underline{36}$, 7930 (1987).

19. C. R. Pidgeon and R. N. Brown, Phys. Rev. $\underline{146}$, 575 (1966).

20. G. Bastard, C. Rigaux, Y. Guldner, J. Mycielski and A. Mycielski, J. Phys. $\underline{39}$, 87 (1978).

21. G. Bastard, C. Rigaux, Y. Guldner, A. Mycielski, J. K. Furdyna and D. P. Mullin, Phys. Rev. B $\underline{24}$, 1961 (1981).

22. W. Dobrowolski, M. von Ortenberg, A. M. Sandaner, R. R. Galazka, A. Mycielski and R. Pauthenet, Physics of Narrow Gap Semiconductors, Lecture Notes in Physics, edited by E. Gornik (Springer, Heidelberg, 1982) $\underline{152}$, 302 (1982).

23. J. R. Anderson, M. Gorska, L. J. Azevedo and E. L. Venturini, Phys. Rev. B $\underline{33}$, 4706 (1986).

24. G. Barilero and C. Rigaux, private communication.

25. See M. H. Weiler in Semiconductors and Semimetals, Vol. 16, edited by R. K. Willardson and A. C. Beer (Academic, New York, 1981).

26. R. E. Kremer, A. M. Witowski, M. Jaczynski and J. K. Furdyna, Physics of Narrow Gap Semiconductors, Lecture Notes in Physics, edited by E. Gornik (Springer, Heidelberg, 1982) $\underline{152}$, 307 (1982).

27. See, for instance, F. Koch, NATO Workshop on Optical Properties of Narrow-Gap Low-Dimensional Structures, St. Andrews, 1986, edited by C. M. Sotomayor Torres et al. (Plenum, New York, 1987), p 187.

28. S. P. Kowalczyk, J.T. Cheung, E. A. Kraut and R. W. Grant, Phys. Rev. Lett. 56, 1605 (1986).

29. Tran Minh Duc, C. Hsu and J. P. Faurie, Phys. Rev. Lett. 58, 1127 (1987).

30. R. Wollrab and F. Koch, private communication.

31. Y. Guldner, G. S. Boebinger, J. P. Vieren, M. Voos and J. P. Faurie, Phys. Rev. B 36, 2958 (1987).

32. J. Scholz, F. Koch, J. Ziegler and H. Maier, Sol. St. Commun. 46, 665 (1983).

33. J. Singleton, F. Nasir and R. J. Nicholas, Surf. Sci. 170, 409 (1986).

34. F. Koch, in Springer Series in Solid State Sciences, vol. 53, edited by G. Bauer, p. 20, (1984).

35. Y. Takada, K. Arai and Y. Uemura, Physics of Narrow Gap Semiconductors, Lecture Notes in Physics, edited by E. Gornik (Springer, Heidelberg, 1982) 152, 101 (1982).

36. Y. Takada, private communication to J. C. Thuillier.

37. J. P. Faurie, M. Boukerche, S. Sivananthan, J. Reno and C. Hsu, Superlattices and Microstructures 1, 237 (1985).

38. D. Olego, J. P. Faurie and P. M. Raccah, Phys. Rev. Lett. 55, 328 (1985).

39. M. A. Reed, private communication to Y. Guldner.

40. T. F. Kuech and J. O. McCaldin, J. Appl. Phys. 53, 312 (1982).

41. J. F. Tersoff, Phys. Rev. Lett. 56, 2755 (1986) ; M. Jaros, A. Zoryk and D. Ninno, Phys. Rev. B 35, 8277 (1987).

42. J. O. McCaldin, T. C. McGill and C. A. Mead, Phys. Rev. Lett. 36, 56 (1976).

43. N. F. Johnson, H. Ehrenreich, K.C. Hass and T. C. McGill, Phys. Rev. Lett. 59, 2352 (1987).

44. J. A. Brum, G. Bastard and M. Voos, Sol. St. Comm. 59, 561 (1986).

*Laboratoire associé au CNRS

VALENCE BAND DISCONTINUITIES IN H_gT_e-CdTe-ZnTe HETEROJUNCTION SYSTEMS

Dr. Jean-Pierre Faurie

Department of Physics
University of Illinois at Chicago
Chicago, Illinois 60680

ABSTRACT

II-VI HgTe-based semiconductor microstructures have recently attracted much interest because of their unique fundamental properties and their great technological interest for novel infrared devices. An important parameter, which determines most of the heterostructure properties is the valence-band discontinuity ΔE_v. The value of ΔE_v is presently disputed in HgTe-CdTe heterojunctions. Although large discrepancies exist between optical or magneto-optical and x-ray photoemission experiments it turns out that ΔE_v can be classified into two groups: small ΔE_v (0-120 meV) and large ΔE_v (300-400 meV). This paper presents an overview of these experimental data.

INTRODUCTION

Band edge discontinuities at semiconductor heterojunctions are the fundamental parameters involved in the designing of devices and in the prediction of their characteristics. In other words, bandgap engineering requires a precise knowledge of the band discontinuities between the semiconductors involved in the proposed microstructure. In addition to this, the understanding of the band discontinuities represents a real challenge from both a theoretical point of view for their calculation and an experimental point of view for their determination.

Despite the considerable effort spent on the GaAs-AlAs system there is still much controversy on the value of the band discontinuities.

II-VI Te-based semiconductor microstructures such as superlattices (SLs) and heterojunctions (HJs) have recently attracted much interest because of their unique fundamental properties and their great technological potential for novel infrared devices.[1] The specific characters of the systems where HgTe is involved arise mainly from the zero-gap configuration of HgTe. More generally, II-VI structures involving a zero-gap mercury compound such as HgTe and an open-gap semiconductor such as CdTe, form a new class of heterostructures which are called Type III. Their band structure can be calculated by using the LCAO[2] or the envelope function[3] models which give very similar results. An important parameter, which determines most of the heterostructure properties, is the valence-band discontinuity ΔE_v. The value of ΔE_v is presently disputed in Te-based semiconductor heterojunctions.

If ΔE_v has only been investigated by XPS between the three binary compounds HgTe, CdTe, and ZnTe, the HgTe-CdTe system has been the object of numerous optical studies and theoretical consideration. Therefore, in the scope of this paper, more attention will be given to the HgTe-CdTe system.

From the phenomenological common-anion rule[4] and from the LCAO approach of Harrison,[5] one can deduce that ΔE_v is small ($\Lambda \lesssim 0.1$ eV), because the valence-band energy depends essentially on the anion and because HgTe and CdTe are closely matched in lattice constant (within 0.3%). Nevertheless, recent theoretical results, based on the role of interface dipoles and the analogy of an heterojunction with a metal-metal junction, do not support the common-anion rule and predict a much larger value for ΔE_v.

The main point of the theoretical debate on the understanding of ΔE_v is the role played by dipoles at the interface. According to Tejedor et al.,[6] Tersoff[7] and Harrison et al.[8] the existence of interface dipoles in their models explain the failure of the common anion rule. Recently Van de Walle et al.[9] and Wei et al.[10] using different approaches without any dipole contribution have also calculated large values for ΔE_v.

This situation is slightly confusing since both types of theory, with and without dipole, give results which are in qualitative agreement as it can be seen in Table 1.

It is important to point out that these theoretical values for ΔE_v are obtained from an energy difference between two large quantities.

Table 1. Theoretical predictions of the valence band discontinuities ΔE_v at the HgTe-CdTe-ZnTe interfaces

Theoretical model	ΔE_v HgTe-CdTe	ΔE_v ZnTe-CdTe	ΔE_v HgTe-ZnTe	Reference(s)
Common anion rule	0	0	0	4
Harrison's LCAO approach	<0.1 eV			5
Induced density of interface states	0.51 eV	0.01 eV	0.05 eV	7, 8
Self consistent density-function calculation	0.27 eV			9
Self consistent tight-binding method	0.39 eV (average)			11,12
Tight-binding (role of cation d orbitals)	0.36 eV	0.13 eV	0.22 eV	10
Linear muffin tin orbital (LMTO)	0.61 eV			13

This means that ΔE_v cannot be known from theoretical calculations to better than a fraction of an eV, which is clearly insufficient to predict most of the electronic properties of these materials. However, it is important to note that all these recent theoretical calculations including self consistent tight binding and linear-muffin-tin orbital methods predict a large ΔE_v of 0.3-0.6 eV.[11,12,13]

A more accurate determination can be obtained experimentally, for example from optical measurements. Experimental data of magneto-optical,[14] resonant Raman scattering,[15] infrared transmission[16,17] and infrared photoluminescence[18] experiments have been carried out on superlattices and have been mostly performed at low temperature. All these experiments agree with a small ΔE_v of less than 0.12 eV. None of these experiments give a direct determination of ΔE_v .

On the one hand, the band structure is theoretically calculated. The energy differences between bands are mostly determined by ΔE_v. On the other hand, transitions are observed experimentally and are attributed to specific transitions in the calculated band structure. The proposed ΔE_v represents the best fit between the experimental and the theoretical energy difference.

Therefore, two sources of error are possible in such an approach. The first one lies in the transition attribution, because of the numerous transitions involved. The second is related to the reliability of the theoretical calculation.

The details of some of these experiments have been published in a recent review article[14] and will not be presented extensively again here. Only a summary of magneto-optical measurements will be given in the present paper, because they have been so far the major support for the small valence band discontinuity.

A larger valence band offset of 0.36 eV has been found using X-ray photoelectron spectroscopy (XPS) at 300 K on HgTe-CdTe heterojunctions.[19,20] These experiments are presented hereafter. This large value is supported by recent measurements carried out on a single barrier heterostructure.[21]

II. GROWTH OF HgTe-CdTe SUPERLATTICES AND HgTe-CdTe-ZnTe HETEROJUNCTIONS

The investigated superlattices (SLs) and heterojunctions (HJs) have been grown in situ by molecular beam epitaxy (MBE) in RIBER MBE machines. All the structures were grown in the (111)B orientation. The growth of the superlattices is currently carried out using three different effusion cells containing CdTe for the growth of CdTe, Te and Hg for the growth of HgTe. In order to grow high quality SL crystals the temperature of the substrate must be 180°C or above. In this range of temperature the condensation coefficient for mercury is very small[22].

This means that a large mercury overpressure is needed to grow HgTe. It also implies that mercury will easily and noncongruently evaporate from HgTe. Due to this problem, the common growth technique for HgTe-CdTe superlattices and other microstructures such as single and double barrier tunneling structures involves leaving the Hg source open at all times.[23,24] Thus, there is a mercury flux on the sample during the growth of the CdTe layers. A competition then occurs between the Hg and Cd atoms for lattice sites. As a result, the CdTe layers may not be pure CdTe but instead be $Hg_{1-x}Cd_xTe$ with some percentage of mercury.

This problem was recognized by the first people to grow HgTe-CdTe superlattices on CdTe(111)B substrates. They grew thick layers of CdTe under the same conditions as in the superlattice, including the presence of the Hg flux. The Hg content was then measured by energy dispersive spectroscopy (EDS). It was found that the CdTe contained less than 5% mercury.[25]

We have recently measured the amount of mercury incorporated in ultra thin CdTe layers grown with a mercury flux.[26] The amount of mercury was found to be between 3 and 9% for CdTe($\bar{1}\bar{1}\bar{1}$)B, depending on the growth conditions. Incidently it was found that under the same conditions much more mercury (~15%) was incorporated in the (100) orientation.

Therefore it is confirmed that for HgTe-CdTe SLs grown in the ($\bar{1}\bar{1}\bar{1}$)B orientation the amount of Hg incorporated in CdTe is low enough not to influence significantly the band structure calculations when this incorporation is not taken into consideration. However in the (100) orientation the effect cannot be neglected.

More detail on the growth of HgTe-CdTe SLs can be found in previous articles.[14,23] Concerning heterojunctions, ($\bar{1}\bar{1}\bar{1}$)B and (100) orientations have been investigated. The linearity test as well as meaningful comparison with theory requires interfaces to be abrupt at the atomic scale. Thus the growth temperature was maintained at 190°C, a temperature known to give no interdiffusion across the interface when the growth time is short. Moreover, the interface abruptness is inferred from the exponential attenuation with overlayer thickness of the substrate XPS peak.

The samples were directly transferred to the attached spectrometer, at a pressure of 10^{-10} Torr, without transition through the air. No contamination occurs, avoiding any cleaning procedure.

III. MAGNETO-OPTICAL MEASUREMENTS

Interesting information on the SL band structure can be obtained from far-infrared (FIR) magnetoabsorption experiments. When a strong magnetic field B is applied perpendicular to the layers, the SL bands are split into Landau levels. At low temperature, the FIR transmission signal being recorded at fixed photon energies as a function of B, presents pronounced minima which correspond to resonant optical transitions between the different Landau levels. Intraband (namely, cyclotron resonance) and interband magneto-optical transitions can be observed, depending on the Fermi-level position at low temperature. From the theoretical analysis of the data, the SL band structure and, in particular, the valence-band offset ΔE_v can be deduced.[27-29]

The first experimental determination of ΔE_v was in fact obtained from far-infrared magneto-optical experiments at T = 1.6 K on superlattice S_1 consisting of 100 periods of HgTe(180 Å)-CdTe(44 Å) (Table II). The best agreement between experiment and theory (done in the envelope function approximation) was obtained for $\Delta E_v = \Gamma_{8HgTe} - \Gamma_{8CdTe} = 40$ meV[27].
For this particular value of ΔE_v, S_1 presents a zero-gap configuration because E_1 and HH_1 are degenerate at $k_z = 0$. Note that similar results were also obtained from LCAO calculations[30].

Quite different results in terms of band structure were expected for sample S_2 which is an open gap superlattice with n-type conduction at low temperature.

Again the results for sample S_2 were consistent with a valence-band offset $\Delta E_v = 40$ meV. The sensitivity of the fitting procedure to the value of ΔE_v was studied and it turned out that an acceptable agreement between experiment and the calculated transitions could be obtained for ΔE_v within the limits (0-100 meV), if one takes into account the uncertainties of the sample characteristics on the data (broad absorption minima) and of the

Table 2. Characteristics of HgTe-CdTe superlattices used in the magneto-optical investigations. (d_1 = HgTe layer thickness and d_2 = CdTe layer thickness.) The samples are grown in the (111)B orientation.

	d_1 (Å)	d_2 (Å)	n	Substrate
S_1	180	44	100	CdTe
S_2	100	36	100	CdTe
S_3	77	38	70	GaAs
S_4	38	20	250	$Cd_{0.95}Zn_{0.05}Te$

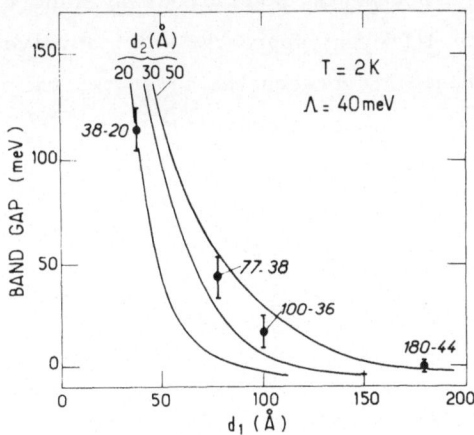

Fig. 1. Variation of the superlattice bandgap E_g as a function of the HgTe layer thickness d_1. The experimental data for samples S_1, S_2, S_3 and S_4 are given by the solid dots; for each sample, the first number corresponds to d_1 and the second one to d_2 (in Å). The solid lines are the theoretical variations $E_g(d_1)$ for three values of d_2.

band parameters of HgTe and CdTe used in the model. In addition, the S_2 band gap becomes nearly zero for $\Delta E_v > 100$ meV, and interband transitions should have been observed, in addition to cyclotron resonance, in the 0-30 meV FIR region, but we did not observe any.[28]

A small positive valence-band offset within the limits (0-100 meV) also provided the best fit for the two other SL's.[14]

Finally, for each sample, a precise determination of the SL bandgap E_g at low temperature can be obtained from magneto-absorption experiments,[14,29] and fig. 1 shows the value of E_g deduced from such experiments for the samples S_1, S_2, S_3 and S_4. The solid lines in fig. 1 are the theoretical variation of E_g as a function of d_1 for $d_2 = 20$, 30 and 50 Å using $\Delta E_v = 40$ meV and $T = 4$ K. Experiments and theory show a very satisfying agreement. An acceptable agreement can in fact be obtained for ΔE_v within the range 0-100 meV by taking into account the uncertainties of the sample characteristics, of the experimental data and of the band parameters of HgTe and CdTe.

Magneto-optical experiments carried out by Yang et al. on HgTe-CdTe superlattices using a different approach, which involves the determination of the energy difference ΔE between the first two valence subbands at the center of the Brillouin zone, have concluded that a ΔE_v of 63 meV represents the best fit between experiment and theoretical calculation.[31]

IV. X-RAY PHOTOEMISSION SPECTROSCOPY

Core Level XPS

Figure 2 illustrates schematically the principle for measuring ΔE_v at the interface between two semiconductors A and B with XPS. If ΔE_v is small ($\lesssim 0.5$ eV), as for the Te based heterojunctions, a direct investigation of the valence-band edges, E_v^A and E_v^B, is difficult and inaccurate due to the large uncertainty on the measurement, but not impossible as it will be discussed later. Indirect measurement involving core levels have to be used. By selecting two core levels E_{c1}^A and E_{c1}^B, well resolved in energy and by measuring their energy difference ΔE_{c1} across the interface, ΔE_v can be directly deduced according to the following relation [see Fig. 1(b)].[32]

$$\Delta E_v(A-B) = \Delta E_{c1}(A-B) + (E_{c1}^A - E_v^A) - (E_{c1}^B - E_v^B). \tag{1}$$

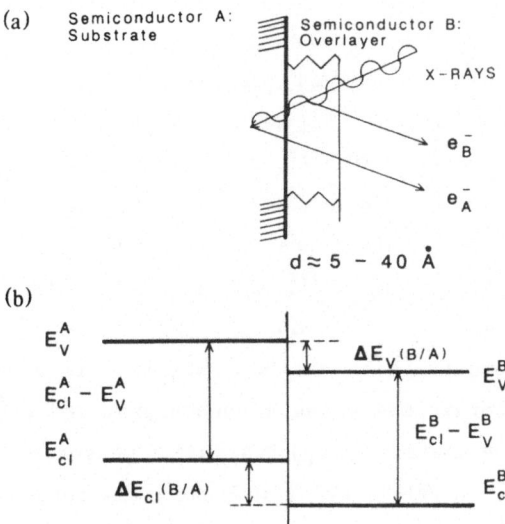

(a) Semiconductor A: Semiconductor B:
     Substrate   Overlayer

                                    X - R A Y S

                                         e_B^-

                                         e_A^-

                    d ≈ 5 - 40 Å

(b)

E_V^A _____

$E_{cl}^A - E_V^A$ $\Delta E_V (B/A)$ E_V^B

E_{cl}^A _____ $E_{cl}^B - E_V^B$

          $\Delta E_{cl}(B/A)$ E_{cl}^B

Fig. 2. Principle of determining ΔE_V with XPS. (a) By irradiating with
         x-rays semiconductor A covered by overlayer of semiconductor B,
         XPS spectra of both semiconductors are recorded if overlayer
         thickness is smaller than electron escape depth. (b) Schematic
         flat-band energy diagram illustrating relation (1).

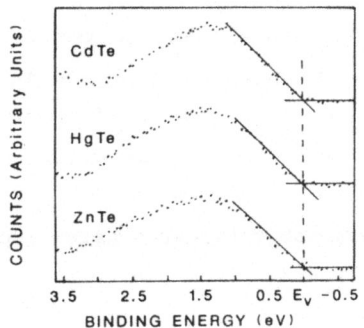

Fig. 3. Linear extrapolation of valence-band leading edge locates the
         same characteristic feature of the bands taken as E_V. The
         spectra are shifted to align E_V.

$E_{cl}-E_v$, the binding-energy (BE) differences between the core level and the top of the valence band for each semiconductor, are determined independently on the bulk semiconductors. All information pertinent to the interface in relation (1) are clearly contained in $\Delta E_{cl}(A-B)$.

Commutativity holds in $\Delta E_v(A/B)=\Delta E_v(B/A)$ and transitivity between A, B and C interfaces does so if $\Delta E_v(A-C) = \Delta E_v(A-B) = \Delta E_v(B-C)$. However, it is not necessary to determine ΔE_v for verifying their linearity. From relation (1), clearly all the parameters of the form $(E_{cl}-E_v)$ cancel in the above expressions. Therefore commutability and transitivity are verified when $\Delta E_{cl}(A/B) = \Delta E_{cl}(B/A)$ and $\Delta E_{cl}(A-C) = \Delta E_{cl}(A-B) + \Delta E_{cl}(B-C)$

The XPS spectrometer which has been utilized is a SSX-100 model from Surface Science Laboratories using a monochromatized and focussed Al, Kα excitation line. The overall energy resolution measured on the $Au_{4f7/2}$ line at a binding energy of 83.93 eV is 0.7 eV. The core levels selected in this work are the resolved spin-orbit components $Cd_{4d4/2}$, $Hg_{5d5/2}$. The splitting for the Zn3d doublet is small and unresolved here.

The nearly lattice-matched HgTe-CdTe ($\bar{1}\bar{1}\bar{1}$)B heterojunctions have been investigated here in greatest detail.

It has been found that $\Delta E_{cl}(\Delta E_v)$ does not depend of overlayer coverage and of the growth order. A value of 2.696 ± 0.030 eV has been found as it is shown in Table 3.[20]

The two remaining systems present larger lattice mismatch which is 6.5% for ZnTe-CdTe($\bar{1}\bar{1}\bar{1}$), heterojunction. The measurements of ΔE_{cl} as a function of growth order are given in Table 3. For ZnTe-CdTe($\bar{1}\bar{1}\bar{1}$), noncommutativity may be found: a 50 meV difference in ΔE_{cl} between the two growth orders is observed, but at the very limit of the experimental uncertainties, and so was not considered here. However, this point which has attracted some attention[33,34] deserves further investigation. Nevertheless, we concluded that for ZnTe-CdTe and HgTe-ZnTe, ΔE_{cl} are coverage independent and commutative. They are, respectively, 0.629 ± 0.050 eV and 2.076 ± 0.040 eV. It follows immediately that transivity is very accurately obeyed: $\Delta E_{cl}(HgTe-CdTe)=\Delta E_{cl}(HgTe-ZnTe)+\Delta E_{cl}(ZnTe-CdTe)$ (i.e., 2.696eV = 2.067eV + 0.629eV). See Table 3.

For obtaining the $(E_{cl} - E_v)$ used in relation (1), E_v is simply located by linear extrapolation of the valence-band leading edge. This

Table 3. Values measured for ΔE_{c1}

	$E_{Cd4d_{5/2}} - E_{Hg5d_{5/2}}$ (eV)	ΔE_v (eV)
CdTe/HgTe	2.694 ± 0.020	
HgTe/CdTe	2.697 ± 0.025	
Average Value	2.696 ± 0.030	0.36 ± 0.05
	$E_{Cd4d_{5/2}} - E_{Zn3d}$	
CdTe/ZnTe	0.652 ± 0.040	
ZnTe/CdTe	0.605 ± 0.030	
Average value	0.629 ± 0.050	0.10 ± 0.06
	$E_{Zn3d} - E_{Hg5d_{5/2}}$	
ZnTe/HgTe	2.074 ± 0.025	
HgTe/ZnTe	2.060 ± 0.031	
Average value	2.067 ± 0.040	0.25 ± 0.05

Table 4. Values measured for ($E_{c1} - E_v$ of ZnTe, CdTe, and HgTe semiconductors along the (100) and ($\bar{1}\bar{1}1$)B direction.

II-VI Compound	$E_{c1} - E_v$ (eV) ($\bar{1}\bar{1}\bar{1}$)B Orientation		$E_{c1} - E_v$ (eV) (100) Orientation
ZnTe	9.62	$E_{c1} = E_{Zn_{3d}}$	9.597
CdTe	10.145	$E_{c1} = E_{Cd_{4d_{5/2}}}$	10.175
HgTe	7.805	$E_{c1} = E_{Hg_{5d_{5/2}}}$	7.833

Our value ΔE_v for the CdTe-HgTe HJ compares very closely with Ref.

procedure is well justified by the close similarity of the band structure of the tellurides near E_v (see Fig. 3). It is quite accurate as shown by the standard deviation σ over numerous measurements on independent samples. Table 4 displays $\Delta(E_{c1}-E_v)$ measured values for ZnTe, CdTe and HgTe in both (111)B and (100) orientations.

15. An excellent agreement is also observed between ΔE_v values and those calculated by S. H. Wei and A. Zunger for the three heterojunctions[10] (see Table 3).

Direct measurement of ΔE_v by XPS

The direct measurement of ΔE_v by XPS is difficult when ΔE_v is small, i.e. when the leading edges of the valence band structures of the two semiconductors are very close in energy. The principle of this process is to convolute the normalized valence band spectra of HgTe and CdTe in order to fit the interface spectrum. This procedure can be applied to this system since the strain can be neglected due to the very small lattice mismatch between the two components. The only fitting parameter is ΔE_v. These measurements have shown that ΔE_v is not only insensitive to the overlayer thickness but also independent of the growth sequence. The best fit was obtained for a ΔE_v = 0.30 eV, a result in agreement with the core-level XPS determination.[35] More details of this will be published elsewhere.

Orientation Dependence of ΔE_v

The influence of the crystallographic orientation on the offsets is currently an open question in III-V heterojunctions. Such a dependence has been observed in the Ge-GaAs system (see ref. 36 p. 181). For heterojunctions with a common anion, such as HgTe-CdTe, the theoretical models have not considered the effect on the band offset's value.

ΔE_v has been measured for HgTe-CdTe-ZnTe (HJ) in the (100) orientation.[37] The growth conditions related to the (100) orientation have been described in detail previously. The test procedures are the same as those used for the HJs grown in the (111)B orientation.

ΔE_{c1} are measured from two growth sequences and are shown in Table 5. The values of ΔE_{c1} for the two growth sequences of the HgTe-CdTe HJ are identical confirming the property of commutativity for this system.

Table 5. Average values measured for ΔE_{c1} and ΔE_v of HgTe-CdTe, ZnTe-CdTe and HgTe-ZnTe heterojunctions in the (100) orientation.

	$E_{Cd_{4d_{5/2}}} - E_{Hg_{5d_{5/2}}}$ (eV)	ΔE_v eV
CdTe//HgTe	2.63	0.288
HgTe//CdTe	2.63	0.288
Average value	2.63	0.288

	$E_{Cd_{4d_{5/2}}} - E_{Zn_{3d}}$	ΔE_v
CdTe//ZnTe	0.73	0.152
ZnTe//CdTe	0.56	- 0.018
Average value	0.64	0.067

	$E_{Zn_{3d}} - E_{Hg_{5d_{5/2}}}$	ΔE_v
ZnTe//HgTe	2.165	0.401
HgTe//ZnTe	2.075	0.311
Average Value	2.120	0.356

The value of ΔE_v is calculated according to relation 1 and it is found to be equal to 0.29 eV, i.e. 70 meV less than ΔE_v in the (111)B orientation. This difference is larger than the uncertainty of the measurement demonstrating that ΔE_v is orientation dependent. Whether it is an intrinsic property of the heterojunction or due to the approach itself has not been yet elucidated.

For the two other HJ systems, HgTe-ZnTe and ZnTe-CdTe, a large non-commutativity is observed implying that the HgTe-CdTe-ZnTe HJ system does not obey a linear model in the (100) orientation. It is not understood why a energy difference of 0.170 eV between core levels of has been observed in the CdTe/ZnTe system depending on the growth order. Such a non-commutability due to the strain is not expected to be as large on the core levels.

V. ELECTRICAL DETERMINATION OF ΔE_v

Direct electrical measurements of ΔE_v at the HgTe-CdTe interface have been reported recently[21]. The active region consisted of a CdTe barrier layer of 180Å sandwiched between two $Hg_{0.78}Cd_{0.22}Te$ electrodes doped n-type with indium, to a carrier concentration of 3.6×10^{16} cm^{-3} at 30K. The detail of the experiment has been given in ref. 21. At temperatures above 160K, energy band diagrams suggest that the dominant low-bias current is thermoionic hole emission across the CdTe barrier layer. This interpretation yields a direct determination of 390 ± 75meV for the HgTe-CdTe valence band discontinuity at 300K. Similar analysis of current-voltage data taken at 190 - 300K seems to indicate that the valence band offset decreases at low temperature in this heterojunction. It should be pointed out that a transport mechanism which has not been considered may be contributing to the observed currents. This could lead to a false determinations of the low-temperature band offsets.

CONCLUSION

In Table 6 is presented a summary of the experimental data reported so far concerning the valence band discontinuity ΔE_v in HgTe-CdTe superlattices and heterojunctions. Although large discrepancies exist between optical and XPS measurements it is interesting to note that ΔE_v values can be classified into two groups: small (0-120 meV) and large (300-400 meV) ΔE_vs. In fact it turns out that small ΔE_vs have been obtained at low temperature whereas large ΔE_v (300-400 meV) have been measured at 300K. It has been suggested that these two sets of measurements could be consistent with each other if ΔE_v is temperature dependent.[38] Such a large change in temperature is theoretically very difficult to explain and the electrical determination of ΔE_v presented here, is not a formal proof of this hypothesis.

Recently XPS experiments have been performed on HgTe-CdTe (111)B HJs from 300 K down to 120 K.[39] At 120 K a ΔE_v of 300 ± 50 meV has been measured i.e. 60 meV smaller than the value found at 300 K. It is not possible to conclude that such a difference, which is within the uncertainty of the measurement, is significant. It is usually admitted that XPS measurements are more direct than optical measurements since no calculation from theory is involved. It has been shown that XPS measurements carried out on HgTe-CdTe HJs give a ΔE_v of 0.35 eV. In addition, XPS experiments performed on $Hg_{1-x}Cd_xTe$ alloys have also concluded that the "natural' ΔE_v between HgTe and CdTe is equal to 0.35 eV.[40,41]

Table 6. Summary of the experimental data reported for the valence band discontinuity ΔE at the HgTe-CdTe interface. ^{1}Temperature, ^{2}Superlattice, ^{3}Heterojunction, ^{4}Measurement not very sensitive to ΔE_v.

EXPERIMENTAL TECHNIQUE	T[1]	SL[2] or HJ[3]	ΔE_v 0-120meV	ΔE_v 300-400meV	REF.
Magnetooptics	2K	SL	X		14,27,28,29,31
Resonant Raman Scattering	12K	SL	X		15
Single Barrier Heterostructure (I-V)	4K	HJ	X		43
Double Barrier Heterostructure (I-V)	15K	HJ	X		44
Infrared Photoluminescence	25K	SL	X		18
IR Transmission (Optical absorption coefficient)	300K	SL	X[4]		16,17
Photoemission (XPS) (Core-Levels)	300K	HJ		X	19,20,40,41
Photoemission (XPS) Direct Measurement	300K	HJ		X	35
Single Barrier Heterostructure (I-V)	300K	HJ		X	21
Intrinsic Carrier Density n_i vs T and Eg	300-200K	SL		X	45

Therefore if the discrepancy observed between optical and XPS measurements cannot be explained by a temperature dependence of ΔE_v, the current trend is to believe that more work is required on the theories, since all the experiments which have suggested a small ΔE_v involved a fitting procedure between theory and experiment. The two theories used to fit the optical and magneto-optical data i.e. envelope function formalism and tight binding approach lead to very similar results. It has been postulated that they might not be fully satisfactory in the case of a semiconductor-semimetal interface where a mass-reversal for light particle is observed. Jaros et al.[42] are currently proposing a new approach based on pseudopotential calculations which is supposed to reconcile optical and XPS experiments. Unfortunately if the magnitude of the SL bandgaps are accounted for in this microscopic theory an appreciable difference still exists between experimental and calculated bandgaps. Therefore it is premature to conclude to a break down of the effective mass model used in the envelope function and tight binding formalisms.

ACKNOWLEDGEMENTS

I would like to acknowledge many participants in the Microphysics Laboratory at the University of Illinois at Chicago with a special mention to Dr. C. Hsu who has performed most of the XPS measurements. All the magneto-optical experiments have been performed at Ecole Normale Superieure in the Groupe de Physique des Solides by Y. Guldner and J.M. Berroir.

This work has been supported by the Defense Advanced Research Project Agency (DARPA) under contracts No. 903-85K-0300, F49620-87-C-0021 and also by SDIO/IST monitored by Naval Research Laboratory under contract No. N0001A-86-K-2023.

REFERENCES

1. D.L. Smith, T.C. McGill and J.N. Schulman, Appl. Phys. Lett. 43, 180 (1983).
2. J.N. Schulman and T.C. McGill, Appl. Phys. Lett. 34, 663 (1979), and Phys. Rev. B23, 4149 (1981).
3. G. Bastard, Phys. Rev. B25, 7584 (1982).
4. J.O. McCaldin, T.C. McGill and C.A. Mead, Phys. Rev. Lett. 36, 56 (1976)
5. W. Harrison, J. Vac. Sci. Technol. 14, 1016 (1977).
6. C. Tejedor and F. Flores, J. Phys. C11, L19 (1978); F. Flores and

C. Tejedor, J. Phys. C12, 731 (1979).

7. J. Tersoff, Phys. Rev. Lett. 56, 2755, (1986).

8. W.A. Harrison and J. Tersoff, J. Vac. Sci. Technol. B4, 1068 (1986).

9. C.G. Van de Walle and R.M. Martin, J. Vac. Sci. Technol. B5, 1225 (1987).

10. S.H. Wei and A. Zunger, J. Vac. Sci. Technol. B5, 1239 (1987), and Phys. Rev. Lett. 59, 144 (1987).

11. C. Priester, G. Allan and M. Lannoo, Journal de Physique C5, 48, 203 (1987).

12. A. Munoz, J.S. Dehesa and F. Flores, Phys. Rev. B35, 6468 (1987); A. Munoz, J.C. Duran and F. Flores, Surf. Sci. 181, L200 (1987).

13. M. Cardona, N.E. Christensen, Phys. Rev. B35, 6182 (1987).

14. J.P. Faurie and Y. Guldner. Heterojunction band discontinuities: Physics and Device Applications. Edited by F. Capasso and G. Margaritondo, Elsevier Science Publishers B.V. 1987 - Chapter 7, p. 283.

15. D.J. Olego, J.P. Faurie and P.M. Raccah, Phys. Rev. Lett. 55, 328 (1985). and D.J. Olego and J.P. Faurie, Phys. Rev. B33, 7357 (1986).

16. J. Reno, I.K. Sou, J.P. Faurie, J.M. Berroir, Y. Guldner, J.P. Vieren, Appl. Phys. Lett. 49, 106 (1986).

17. L.S. Kim, Y. Lansari, J.W. Han, J.W. Cook, J.F. Schetzina and J.N. Schulman, March meeting of the American Physical Society (1988) (unpublished).

18. J.P. Baukus, A.T. Hunter, O.J. Marsh, C.E. Jones, G.Y. Wu, S.R. Hetzler, T.C. McGill and J.P. Faurie, J. Vac. Sci. Technol., A4, 2110 (1986).

19. S.P. Kowalczyk, J.T. Cheung, E.A. Kraut and R.W. Grant, Phys. Rev. Lett. 56 (1986).

20. T.M. Duc, C. Hsu and J.P. Faurie, Phys. Rev. Lett. 58 (1987) 1127.

21. D.A. Chow, J.O. McCaldin, A.R. Bonnefoi, T.C. McGill, J.P. Faurie and I.K. Sou, Appl. Phys. Lett. 51 (1987).

22. J.P. Faurie, A. Million, R. Boch and J.L. Tissot, J. Vac. Sci. Technol. A1, 1593 (1983).

23. J.P. Faurie, IEEE J. Quantum Electron. QE-22, 1656 (1986).

24. K.A. Harris, S. Hwang, D.K. Blanks, J.W. Cook, J.F. Schetzina, N. Otsuka, J.P. Baukus, and A.P. Hunter, Appl. Phys. Lett. 48, 396 (1986).

25. J.P. Faurie, A. Million, and J. Piaguet (unpublished results, 1982).

26. J. Reno, R. Sporken, Y.J. Kim, C. Hsu, and J.P. Faurie, Appl. Phys. Lett. 51, 1545 (1987).

27. Y. Guldner, G. Bastard, J.P. Vieren, M. Voos, J.P. Faurie, and A. Million, Phys. Rev. Lett. 51, 907 (1983).

28. J.M. Berroir, Y. Guldner, J.P. Vieren, M. Voos, and J.P. Faurie, Phys. Rev. B34, 891 (1986).

29. J.M. Berroir, Y. Guldner, and M. Voos, IEEE J. Quantum Electron. 22, 1793 (1986).

30. J.N. Schulman and Y.-C. Chang, Phys. Rev. Lett. B33, 2594 (1986).

31. Z. Yang and J.K. Furdyna, Appl. Phys. Lett. 52, 498 (1988).

32. The notation (B/A) specifies the growth order as B deposited on A and the notation (A-B) is for whatever the growth order is. In addition, the convention adopted regarding the sign of ΔE_v is the following: $\Delta E_v(A-B) > 0$ corresponds to $E_v^B > E_v^A$ in binding-energy scale.

33. J. Tersoff and C.G. Van de Walle, Phys Rev. Lett. 59 (1987).

34. T.M. Duc, C. Hsu and J.P. Faurie, Phys. Rev. Lett. 59, 947 (1987).

35. C. Hsu and J.P. Faurie, American Physical Society Meeting, 1988 (unpublished results).

36. R.W. Grant, E.A. Kraut, J.R. Waldrop and S.P. Kowalczyk in Heterojunction band discontinuities, Physics and device applications edited by F. Capasso and G. Margaritondo, Elsevier Science Publishers B.V., 181 (1987).

37. C. Hsu and J.P. Faurie, J. Vac. Sci. Technol. (to be published).

38. J.P. Faurie, C. Hsu and T.M. Duc, J. Vac. Sci. Technol. A5, 3074 (1987).

39. R. Sporken, C. Hsu and J.P. Faurie, 1988 (unpublished results).

40. C.K. Shih and W.E. Spier, Phys. Rev. Lett. 58, 2594 (1987).

41. C. Hsu, T.M. Duc and J.P. Faurie, Journal de Physique C5-48,307 (1987).

42. M. Jaros, A. Zoryk and D. Ninno, Phys. Rev. B35, 6182 (1987).

43. D.H. Chow, T.C. McGill, I.K. Sou, J.P. Faurie and C.W. Nieh, Appl. Phys. Lett. 52, 54 (1988).

44. M.A. Reed, R.J. Koestner, M.W. Goodwin and H.F. Schaake, J. Vac. Sci. Technol. (to be published).

45. C.A. Hoffman, J.R. Meyer, E.R. Youngdale, J.R. Lindle, F.J. Bartoli, K.A. Harris, J.W. Cook Jr. and J.F. Schetzina, Phys. Rev. (in press).

EXACT ENVELOPE FUNCTION EQUATIONS FOR MICROSTRUCTURES AND THE PARTICLE IN A BOX MODEL

M. G. Burt

British Telecom Research Labs, Martlesham Heath, Ipswich IP5 7RE, UK

ABSTRACT: An exact formulation of the envelope function method recently published by the author is extended to include spin orbit interaction, nonlocal microscopic potentials and nonlatticematched structures. A systematic derivation of the particle in a box model is outlined. It is also shown how the basic ideas behind the envelope function method can lead to an intuitive derivation of the particle in a box model without ambiguities concerning boundary conditions.

INTRODUCTION

In a previous paper the author (Burt 1987) has given an exact formulation of the envelope function method and shown (Burt 1988a) how it can be used to give a systematic derivation of the particle in a box model (effective mass equation for multilayer structures). The purpose of this paper is three fold; firstly to extend the author's previous work to include spin orbit interaction and nonlatticematched multilayer structures; secondly to show how the systematic derivation of the particle in a box model equation can lead to more understanding concerning the correct boundary conditions to be used with the model; thirdly, to show how basic ideas of the exact formulation of the envelope function method can lead to an attractive intuitive derivation of the particle in a box model without any ambiguities concerning boundary conditions. Previous work on the envelope function method by other authors will not be summarised here since this has been done elsewhere (Burt 1988a).

THE ENVELOPE FUNCTION EXPANSION

In their famous paper on the extension of effective mass theory to degenerate bands, Luttinger and Kohn (1955) showed how the standard expansion of the wavefunction ψ in Bloch waves could be transformed into the envelope function expansion

$$\psi(\underline{R}) = \sum_n F_n(\underline{R}) \, U_n(\underline{R}) \qquad (1)$$

where the $U_n(\underline{R})$ are the zone centre eigenstates of the semiconductor (ψ and $U_n(\underline{R})$ are spinors in general). The envelope functions, $F_n(\underline{R})$, have a plane wave expansion limited to the first Brillouin zone. It can be readily inferred from the work of Luttinger and Kohn (1955), that an expansion of the type (1) can be made using any complete set of linearly independent periodic functions $U_n(\underline{R})$ and that the envelope functions $F_n(\underline{R})$ are uniquely defined thereby (Burt 1988a).

If cyclic boundary conditions are applied to a large volume containing the microstructure of interest, then the plane wave expansion of F_n $(\underline{R})$ contains a large, but finite number of terms. If the wavefunction is finite everywhere, then the coefficients of this planewave expansion are all finite (Burt 1988a). Hence F_n $(\underline{R})$ and ∇F_n $(\underline{R})$ are continuous everywhere.

That the envelope function expansion (1) is unique leads to the important property that if two envelope functions expansions are equal for all $\underline{R}$, then the expansion coefficients must be equal (Burt 1988a). Hence, if

$$\sum_n F_n^{(1)}(\underline{R}) \ U_n(\underline{R}) = \sum_n F_n^{(2)} \ (\underline{R}) \ U_n(\underline{R}) \tag{2}$$

then $\quad F_n^{(1)} \ (\underline{R}) = F_n^{(2)} \ (\underline{R})$ \hfill (3)

provided the planewave expansions of the $F_n^{(1)}$ and $F_n^{(2)}$ are restricted to the first Brillouin zone. Envelope function expansions (1) need not be restricted to wave functions and hence results (2) and (3) apply to any expansions of the type (1).

EXACT ENVELOPE FUNCTION EQUATIONS

The basic method for deriving exact envelope function equations is to substitute for ψ in the Schroedinger equation using the envelope function expansion (1) and then after suitable manipulation to equate 'coefficients' according to (2) and (3). This procedure has been described in detail for scalar wavefunctions and hamiltonians with local potentials (Burt 1988a). Here a more general procedure that can deal with nonlocal operators and spin orbit interaction will be derived.

The state vector $|\psi>$ will have a spinor representative $<\underline{R}|\psi>$ and we wish to express the spinor $<\underline{R}|O|\psi>$, where O is an arbitary operator, in envelope expansion form. We introduce a complete set of plane wave states (including spin) where the $\underline{G}$'s are the reciprocal lattice vectors of a Bravais lattice to be specified more closely in due course and the $\underline{K}$ are wavevectors lying within the corresponding first Brillouin zone. For notational simplicity we do not display the spin variables explicitly in the states $|\underline{R}>$ and $|\underline{K}+\underline{G}>$. Introducing complete sets of states we have

$$<\underline{R}|O|\psi> = \sum_{\substack{KK' \\ GG'}} <\underline{R}|\underline{K}+\underline{G}><\underline{K}+\underline{G}|O| \ \underline{K}'+\underline{G}'><\underline{K}'+\underline{G}' \ |\psi> \tag{4}$$

In the spinor notation used $<\underline{R}|\underline{K}+\underline{G}>$ and $<\underline{K}+\underline{G}|O|\underline{K}'+\underline{G}'>$ are 2x2 matrices and in particular

$$<\underline{R}|\underline{K}+\underline{G}> = \Omega^{-\frac{1}{2}} \ \underline{I} \ \exp \ (i \ (\underline{K}+\underline{G}).\underline{R}) \tag{5}$$

where $\underline{I}$ is the 2x2 unit matrix and Ω is the normalising volume. Using the envelope expansion (1) we have

$$<\underline{K}'+\underline{G}' \ |\psi> = \Omega^{\frac{1}{2}} \sum_n F_n(\underline{K}') \ U_{n\underline{G}'} \tag{6}$$

where

$$F_n(\underline{R}) = \sum_{\underline{K}} F_n \ (\underline{K}) \ e^{i\underline{K}.\underline{R}} \tag{7}$$

and

$$U_n(\underline{R}) = \sum_{\underline{G}} \ U_{n\underline{G}} \ e^{i\underline{G}.\underline{R}} \tag{8}$$

The spinors $U_n(\underline{R})$ are chosen to be orthonormal so that

$$\int_{\substack{\text{unit}\\\text{cell}}} \frac{d^3R}{\Omega_c} \; U_n^{T*} \; U_{n'} = \delta_{nn'} \tag{9}$$

where the superfix T denotes the transpose of a spinor and Ω_c is the volume of a unit cell of the aforementioned Bravais lattice. Using completeness of the U_n we find

$$\underline{I} \; e^{i\underline{G}.\underline{R}} = \sum_n U_n(\underline{R}) \; U_{n\underline{G}}^{T*} \tag{10}$$

Using (5) (6) and (10) in (4) we obtain

$$\langle\underline{R}|O|\psi\rangle = \sum_{nn'} \sum_{\substack{\underline{K}\underline{K}'\\\underline{G}\underline{G}'}} U_n(\underline{R}) \; U_{n\underline{G}}^{T*} \; e^{i\underline{K}.\underline{R}} \; \langle\underline{K}+\underline{G}|O|\underline{K}'+\underline{G}'\rangle \; U_{n'\,\underline{G}'} \; F_{n'}(\underline{K}') \tag{11}$$

Using the inverse of (7) ie

$$F_{n'}(\underline{K}') = \int \frac{d^3R'}{\Omega} \; F_{n'}(\underline{R}') \; e^{-i\underline{k}'.\underline{R}'} \tag{12}$$

in (11) we find that the envelope function form for $\langle\underline{R}|O|\psi\rangle$ is obtained

$$\langle\underline{R}|O|\psi\rangle = \sum_{nn'} U_n(\underline{R}) \int d^3R' \; Q_{nn'}(\underline{R},\underline{R}') \; F_{n'}(\underline{R}') \tag{13}$$

with

$$Q_{nn'}(\underline{R},\underline{R}') = \frac{1}{\Omega} \sum_{\underline{K}\underline{K}'} e^{i\underline{K}.\underline{R}} \left[\sum_{\underline{G}\underline{G}'} U_{n\underline{G}}^{T*} \; \langle\underline{K}+\underline{G}|O|\underline{K}'+\underline{G}'\rangle U_{n'\,\underline{G}'} \right] e^{-i\underline{K}'.\underline{R}'} \tag{14}$$

The RHS of (13) is in envelope function expansion form because the 'coefficients' of the $U_n(\underline{R})$ have plane wave expansions restricted to the first Brillouin zone.

When O contains derivatives, such as occur in the momentum operator, $\langle\underline{K}+\underline{G}|O|\underline{K}'+\underline{G}'\rangle$ contains terms that depend on $\underline{K}'$. If this $\underline{K}'$ dependence is expressed as a gradient operation on $e^{i\underline{K}'.\underline{R}'}$ in (14), then integration by parts can bring the gradient operator to bear on the envelope function in (13). In this way we can derive the following exact set of envelope function equations

$$-\frac{\hbar^2}{2m} \nabla^2 F_n(\underline{R}) \; -i\frac{\hbar}{m} \sum_{n'} \underline{p}_{nn'} \cdot \nabla F_{n'}(\underline{R}) \; -i\hbar \sum_{n'} \int d^3R' \; \underline{v}_{nn'}(\underline{R},\underline{R}') \cdot \nabla F_{n'}(\underline{R}')$$

$$+ \sum_{n'} \int d^3R' \; H_{nn'}(\underline{R},\underline{R}') \; F_{n'}(\underline{R}') = E \; F_n(\underline{R}) \tag{15}$$

101

The $\underline{p}_{nn'}$ are momentum matrix elements with respect to the periodic basis spinors so that

$$\underline{p}_{nn'} = \int\limits_{\substack{\text{unit}\\\text{cell}}} \frac{d^3R}{\Omega_c}\, U_n^{T*}\, \underline{p}\, U_{n'} \tag{16}$$

The standard spin orbit interaction, H^{SO}, is written as $\underline{v}(\underline{R}).\underline{p}$, where $\underline{v}(\underline{R})$ is the 2x2 matrix proportional to the vector product of the Pauli spin vector and the local crystalline electric field. The function $\underline{v}_{nn'}(\underline{R},\underline{R}')$ is given by (14) with O taken as $\underline{v}(\underline{R})$. The function $H_{nn'}(\underline{R},\underline{R}')$ is given by

$$H_{nn'}(\underline{R},\underline{R}') = T_{nn'}\,\Delta(\underline{R}\text{-}\underline{R}') + H^{SO}_{nn'}(\underline{R},\underline{R}') + V_{nn'}(\underline{R},\underline{R}') \tag{17}$$

$T_{nn'}$ is the matrix element of T with respect to U_n and $U_{n'}$ in analogy with (16). $\Delta(\underline{R}\text{-}\underline{R}')$ is given by

$$\Delta(\underline{R}\text{-}\underline{R}') = \frac{1}{\Omega} \sum_{\underline{K}} \exp(i\underline{K}.(\underline{R}\text{-}\underline{R}')) \tag{18}$$

and acts as a delta function for functions that have their plane wave expansions restricted to the first Brillouin zone. $H^{SO}_{nn'}(\underline{R},\underline{R}')$ is given by

$$H^{SO}_{nn'}(\underline{R},\underline{R}') = \frac{1}{\Omega} \sum_{\underline{K}\underline{K}'} e^{i\underline{K}.\underline{R}} \left[\sum_{\underline{G}\underline{G}'} U_{n\underline{G}}^{T*} <\underline{K}+\underline{G}|v|\underline{K}'+\underline{G}'>.\hbar\underline{G}'\; U_{n'\underline{G}'} \right] e^{-i\underline{K}'.\underline{R}'} \tag{19}$$

and $V_{nn'}(\underline{R},\underline{R}')$ is given by (14) with O replaced by V, which can be nonlocal.

The equations (15) are completely equivalent to the original Schroedinger equation and are valid for any microstructure. However, they are only really useful in their present form for lattice matched systems. An appropriate set of equations for multilayers composed of nonlatticematched materials will be given in the next section.

In bulk material the nonlocal aspect of the equations (15) disappears if we take the basis spinors $U_n(\underline{R})$ to have the same periodicity as the crystal lattice. In such cases

$$\left[H_{hn'}(\underline{R},\underline{R}')\right]_{bulk} = H_{hn'}\,\Delta(\underline{R}\text{-}\underline{R}') \tag{20}$$

with

$$H_{hn'} = \sum_{\underline{G}\underline{G}'} U_{n\underline{G}}^{T*} <\underline{G}|H|\underline{G}'> U_{n'\underline{G}'}$$

$$= \int\limits_{\substack{\text{unit}\\\text{cell}}}\!\!\!\int \frac{d^3\underline{R}d^3\underline{R}'}{\Omega_c} U_n^{T*}(\underline{R})H(\underline{R},\underline{R}')U_{n'}(\underline{R}') \tag{21}$$

being the matrix elements of H with respect to U_n and $U_{n'}$. Similar equations hold for $\underline{v}_{nn'}(R,R')$. The envelope function equations (15) for bulk material are

$$- \frac{\hbar^2}{2m} \nabla^2 F_n(\underline{R}) - i \frac{\hbar}{m} \sum_{n'} (\underline{p}_{nn'} + m\underline{v}_{nn'}).\nabla F_{n'}(\underline{R})$$

$$+ \sum_{n'} H_{nn'} F_{n'}(\underline{R}) = E F_n(\underline{R}) \tag{22}$$

where $\underline{v}_{nn'}$ and $H_{nn'}$ (defined by (21)) are independent of position and are properties of the material and the basis set chosen alone. If the U_n are chosen to diagonalise $H_{nn'}$ ie to be zone centre eigenstates, then (22) becomes the real space version of the K.p equations (see eg Kane 1966) with the wavevector dependent spin orbit interaction terms included.

In multilayer structures the equations (22) are a good approximation to the true envelope function equations (15) except in the vicinity of interfaces. The deviations of $H_{nn'}(\underline{R},\underline{R}')$ from the bulk form $H_{nn'}\Delta(\underline{R}-\underline{R}')$ near an abrupt heterojunction have been discussed in detail elsewhere (Burt 1988a). The most striking point is that the major deviations do not contribute to the energy in first order perturbation theory if the envelope functions contain only wavelengths greater than four lattice periods ie their plane wave expansions terminate less than half way to the Brillouin zone edge. This may well explain the striking success of effective mass theory for small dimension superlattices (Herbert and Gell 1987).

ENVELOPE FUNCTION EQUATIONS FOR NONLATTICEMATCHED MULTILAYERS

The exact envelope function equations (15), while formally valid for nonlatticematched structures, (eg strained layer superlattices) are not suitable for application to such structures. While one may choose a set of U_n with the same periodicity as the lattice in one region, this set will be out of step with the lattice in other regions and even away from interfaces will not have the simple form (22) of the bulk equations for lattice matched structures. Fortunately, this problem can be tackled by a genealisation of the approach of Pikus and Bir (1959) to the effects of strain on bulk bandstructures. Their approach was to change position variable so that the distorted lattice in the new variable had the same periodicity as the undistorted lattice in the old variable. Our approach here is to change variable so that in the new position variable the lattice periodicity is the same throughout the structure. We can then apply the envelope function expansion to the transformed Schroedinger equation. While the procedure to be outlined below is completely general we will restrict our attention to the case of a multilayer taking the z axis perpendicular to the interfaces. It will be assumed that all the layers have the same translational symmetry in the x and y directions. Since it is only the lattice periodicity in the z direction that is a problem, we only need to introduce one new variable, $\zeta(z)$ such that in the coordinates (x,y,ζ) the unit cells of the multilayer lie on a Bravais lattice with a uniform period throughout. All the terms in the hamiltonian need to be expressed as functions of ζ. In particular for the kinetic energy for motion in the z direction

$$- \frac{\hbar^2}{2m} \frac{\partial^2}{\partial z^2} = - \frac{\hbar^2}{2m} (1 + D(\zeta)) \frac{\partial^2}{\partial \zeta^2} - \frac{\hbar^2}{2m} D^{(i)}(\zeta) \frac{\partial}{\partial \zeta} \tag{23}$$

where

$$D(\zeta) = \left[\frac{d\zeta}{dz}\right]^2 - 1 \tag{24}$$

and

$$D^{(i)}(\zeta) = d^2\zeta/dz^2 \tag{25}$$

The presence of the function $D^{(i)}(\zeta)$ is in explicit recognition that the transformation $\zeta(z)$ is inhomogeneous. We now choose, as in the previous section, a complete set of orthonormal functions with the periodicity of the above mentioned Bravais lattice and make an envelope function expansion. To avoid in essential complication we will consider the case in which the wavevector component parallel to the layers is zero so that the envelope functions $F_n(\zeta)$ are functions of ζ only. The envelope function equations in this case are

$$-\frac{\hbar^2}{2m} \frac{d^2 F_n(\zeta)}{d\zeta^2} - \frac{i\hbar}{m} \sum_{n'} p^{\zeta}_{nn'} \frac{dF_{n'}(\zeta)}{d\zeta}$$

$$+ \sum_{n'} \int d\zeta' \; D_{nn'}(\zeta,\zeta') \left[-\frac{\hbar^2}{2m} \frac{d^2 F_{n'}(\zeta)}{d\zeta'^2} - i\frac{\hbar}{2} \sum_{n''} p^{\zeta}_{n'n''} \frac{dF_{n''}(\zeta')}{d\zeta'} \right]$$

$$-\frac{\hbar^2}{2m} \sum_{n'} \int d\zeta' \; D^{(i)}_{nn'}(\zeta,\zeta') \frac{dF_{n'}(\zeta')}{d\zeta'}$$

$$-i\hbar \sum_{n'} \int d\zeta' \; \left[v_z \frac{d\zeta}{dz} \right]_{nn'}(\zeta,\zeta') \frac{dF_{n'}(\zeta')}{d\zeta'}$$

$$+ \sum_{n'} \int d\zeta' \; H_{nn'}(\zeta,\zeta') \; F_{n'}(\zeta') = E \; F_n(\zeta) \tag{26}$$

The quantities of the type $O_{nn'}(\zeta,\zeta')$ are obtained from the corresponding function or operator O using the prescription (14) where $\underline{R}$ is now (x,y,ζ) after integration by parts to allow any gradient operators in O to act on the envelope functions. They are independent of x and y because of translational symmetry parallel to the (x,y) plane. For the kinetic energy contributing to H we must use

$$-\frac{\hbar^2}{2m} \left[\frac{\partial^2}{dx^2} + \frac{\partial^2}{\partial y^2} + \frac{\partial^2}{\partial \zeta^2} \right] \tag{27}$$

and the ζ component, p^{ζ}, of momentum is

$$p^{\zeta} = -i\hbar \frac{\partial}{\partial \zeta} \tag{28}$$

$H_{nn'}(\zeta,\zeta')$ also contains a term

$$-i \frac{\hbar}{2m} \sum_{n''} D^{(i)}_{nn''}(\zeta,\zeta') \; p^{\zeta}_{n''n'} \tag{29}$$

arising from the spatial derivative of the strain. If one assumes a nonzero component of wavevector in the x and y directions, one just adds terms to the LHS of (26) familiar from K.p theory.

In bulk material, $D^{(i)}(\zeta)$ vanishes and the other quantities of the type $O_{nn'}(\zeta,\zeta')$ become diagonal in ζ and ζ'. One then recovers the real space equivalent of the Pikus and Bir bandstructure equations for strained crystals after a similarity transformation to diagonalise $H_{nn'}$ and ignoring quadratic and higher order terms in the strain.

104

The great advantage of equations (26) compared with (15) for nonlatticematched multilayer structures is that they approximate linear differential equations <u>with constant coefficients</u> except in the immediate vicinity of an interface. The nonlocality is restricted to a relatively small region and one may be optimistic that the effects of the deviations from the bulk equations may be estimated using perturbation theory.

Besides the nonlocal nature of the exact envelope function equations, the presence of the terms depending on the derivative of the strain is a totally new feature, which has been implicitly ignored in previous treatments (eg Poetz and Ferry 1986). The expression for $D_{nn}^{(i)}(\zeta,\zeta')$ for an abrupt interface (Burt 1988b) indicates that its omission produces fractional errors of the order of the fractional lattice mismatch in structures where the electron spends appreciable time in the interface region.

Finally, it should be pointed out that the envelope functions, $F_n(\zeta)$, and their derivatives, $\partial F_n/\partial \zeta$, are still continuous functions of ζ everywhere even for nonzero wavevector parallel to the interfaces. This can be seen by noting that the envelope function expansion can be defined and the cyclic boundary conditions applied to the large normalising volume <u>after</u> the change of variable $\zeta(z)$ has been applied to the original Schroedinger equation.

FORMAL DERIVATION OF THE PARTICLE IN A BOX MODEL EQUATION

This topic has been dealt with in considerable detail elsewhere (Burt 1988a) and we shall only concern ourselves with the salient points here. We take the simplest case of a latticematched multilayer, no spin orbit interaction and zero component of wavevector parallel to the interfaces. If the eigenvalues, E, of interest are close to one nondegenerate band extremum (say the sth) throughout the multilayer structure (ie the band edge discontinuities are small), then we may be reasonably confident the the solutions will be slowly varying and that F_s will be much larger than any of the other envelope functions. The slow variation of the envelope functions makes it reasonable to make what we might call the local bulk approximation to $H_{nn'}$ $(\underline{R},\underline{R}')$ ie at each point $\underline{R}$, $H_{nn'}$ $(\underline{R},\underline{R}')$ is replaced by the bulk value for the corresponding material. So we write (z axis taken perpendicular to the interfaces)

$$H_{nn'} \ (\underline{R},\underline{R}') \approx H_{nn'}^{bulk} \ (z) \ \Delta(\underline{R}\text{-}\underline{R}') \tag{30}$$

$H_{nn'}^{bulk}$ (z) takes on the bulk $H_{nn'}$ value for the material at z, and hence is a piecewise constant function of z with discontinuities at abrupt interfaces. One might also call (30) the abrupt step approximation. Detailed consideration (Burt 1988a) of the corrections to (30) show that the dominant ones vanish in first order perturbation theory for slowly varying envelope functions. The remaining correction is expected to produce only a small contribution in first order because its fourier transform has a large wavevector denominator.

Using (30) in the envelope function equations (15) we have

$$-\frac{\hbar^2}{2m} \frac{d^2F_n}{dz^2} - \frac{i\hbar}{m} \sum_{n'} p_{nn'} \frac{dF_{n'}}{dz} + \sum_{n'} H_{nn'}^{bulk} \ (z) \ F_{n'} = E \ F_n \tag{31}$$

which corresponds to the abrupt step approximation. If (31) has slowly varying solutions, then the corrections to E due to neglected terms will be small. It will also be assumed that the zone centre eigenfunctions of the materials of which the multilayer are composed are similar so that, if the U_n are chosen to be some suitable linear combination of the corresponding zone centre eigenfunctions, then

$$H_{nn'}^{bulk}(z) \approx E_n(z) \ \delta_{nn'} \tag{32}$$

where $E_n(z)$ is the zone centre energy for the nth band for the material at z. This will be a much better approximation than may be supposed at first sight because H_{nn}^{bulk} (z) will only be nonzero for basis functions U_n and $U_{n'}$ of the <u>same</u> symmetry and these approximate eigenstates are usually well separated in energy. With the approximation (32) used in (31) for n=s we obtain

$$- \frac{\hbar^2}{2m} \frac{d^2 F_s}{dz^2} - \frac{i\hbar}{m} \sum_r p_{sr}^z \frac{dF_r}{dz} + E_s(z) F_s = E F_s \qquad (33)$$

The F_r r≠s in (33) are approximately expressed in terms of F_s using the other envelope function equations. This procedure is just a real space version of the perturbation theory often carried out in k space (see eg Kane 1966). One obtains

$$- \frac{\hbar^2}{2} \frac{d}{dz} \left[\frac{1}{m_s(z)} \frac{dF_s}{dz} \right] + E_s(z) F_s = E F_s \qquad (34)$$

with
$$\frac{1}{m_s(z)} = \frac{1}{m} \left[1 + \frac{2}{m} \sum_{r \neq s} \frac{|p_{rs}^z|^2}{E_s(z) - E_r(z)} \right] \qquad (35)$$

which is the celebrated effective mass equation for a multilayer which includes the particle in a box model for quantum wells and the Kronig-Penney model for superlattices.

An interesting point about (34) is that it is satisfied throughout the multilayer structure including the abrupt interfaces. It follows that both F_s and $(1/m_s)$ (dF_s/dz) are continuous at interfaces. Since m_s will change abruptly at an interface this requires a discontinuity in dF_s/dz. Clearly this is at variance with our definition of the envelope functions which requires them to have continuous derivatives. The paradox can be removed by realising that the assumptions of small band offsets and similar zone centre eigenfunctions implies similar effective masses as can be seen from (35). So within the assumptions of the derivation, the inclusion or otherwise of the effective mass in the continuity condition for the derivative is equivalent. The ready interference from this conclusion is that, if (34) yields significantly different answers for E depending on whether or not the effective mass is included in the derivative boundary condition, then one is probably outside the regime of validity of the effective mass equation and certainly outside the assumptions on which this present derivation is based.

It is often assumed that the effective mass must be included in the derivative boundary condition in order to ensure current conservation. The inclusion is not necessary within the present formulation, however. The author has shown (Burt 1987, 1988b) that for envelope functions that have wavelengths of 4 lattice periods or longer the z component of the current J_z, is given by

$$J_z = \frac{1}{m} \iint dx dy \quad Re \left[\sum_n F_n^* \left[-i\hbar \frac{dF_n}{dz} \right] + \sum_{nm} p_{nm}^z F_n^* F_m \right] \qquad (36)$$

It is clear that F_n and dF_n/dz are continuous, J_z is also continuous as required.

In the effective mass regime one may approximately eliminate all the envelope functions save F_s using the same procedure as used to derive the effective mass equation and obtain

$$J_z \approx \frac{1}{m_s(z)} \iint dx dy \quad Re \left[F_s^* \left[-i\hbar \frac{dF_s}{dz} \right] \right] \qquad (37)$$

It should be stressed that this is an approximate expression. Hence the conclusion, that $(1/m_s(z))$ dF_s/dz should be continuous at boundaries to conserve J_z, is only approximate.

And, indeed, continuity of $(1/m_z(z)) \, dF_s/dz$ approximates the more fundamental requirement of continuity of dF_s/dz because our derivation of the effective mass equation essentially assumes small changes in the effective mass m_s.

It is worth noting that the inclusion of the effective mass in the derivative boundary condition leads to unphysical results in the ground state energy of a quantum well. Let E be the energy measured from the bulk band edge of the well material. Taking the origin of coordinates at the centre of the well, width L, the envelope function is proportional to $\cos kz$ where $E = \hbar^2 k^2/2m_w$, with m_w as the effective mass of the well material. In the barriers the envelope function is proportional to $e^{-q|z|}$ where $E + \hbar^2 q^2/2m_b = V$, the barrier height, and m_b is the effective mass of the barrier material. Matching logarithmic derivatives at the interfaces (ie <u>excluding</u> the effective mass from the boundary condition) gives the well known equation

$$k \tan kL/2 = q \qquad\qquad (38a)$$

Consider what happens when m_b is increased; $q \propto \sqrt{m_b}$ and so the RHS of (38a) increases for fixed E with increasing m_b. This causes the energy eigenvalue to increase because the LHS is an increasing function of energy while the RHS is a decreasing function of energy. This is physically very reasonable. The barrier 'hardens', allowing less penetration of the barrier, and the energy increases as one would expect from the uncertainty principle.

Now consider what happens if the effective mass is included in the derivative boundary condition. (38a) is replaced by

$$k \tan kL/2 = (m_w/m_b) \, q \qquad\qquad (38b)$$

Again consider what happens when m_b is increased. The RHS of (38b) is proportional to $1/\sqrt{m_b}$ and it <u>decreases</u> with increasing m_b. Hence as m_b is increased ie as the barrier 'hardens' the energy eigenvalue falls! This is a most unphysical result and contradicts the uncertainty principle.

The above observation suggests that if difference in the eigenvalues predicted by the effective mass equation (34) depend significantly on whether or not the effective mass is included in the derivative boundary condition, then one is outside the region of validity of effective mass theory or at least the region of validity of the assumptions on which (34) is based. In such situations it would prudent to evaluate the effect of the correction terms (Burt 1988a) using perturbation theory to see whether or not these are significant. If the correction terms are large, then one should return to the exact envelope function equations and attempt less severe approximations.

INTUITIVE JUSTIFICATION OF THE PARTICLE IN A BOX MODEL

While a formal justification of the effective mass approach to multilayer structures is very useful, an intuitive justification is also highly desirable and this is just what the approach to the envelope function method outlined in section 2 can provide. This intuitive justification, which has much in common with Altarelli's approach (1983a,b, 1986), will be outlined below for the case of a quantum well though the extension to more complex multilayer structures will be obvious.

The existence of an exact and unique envelope function expansion in which the envelope functions and their derivatives are continuous everywhere implies the existence of an exact set of envelope function equations. It is reasonable to suppose, provided the width L of the well is sufficiently large, that over most of the well region, the exact envelope function equations correspond to those for a bulk crystal made of the well material, save for some small region, of width w say, about each interface. Similarly, in the barriers we expect the exact equations to correspond to those of the bulk barrier material except in the interface regions. Hence in the well region the principal envelope function is a linear combination of planewaves, the wavevectors of which are determined by the bulk bandstructure of the well material. In the barrier regions the principal envelope function is a decaying exponential the decay length of which is determined by the bulk bandstructure for complex wavevector of the barrier material. One does not know for certain how the envelope function behaves in the interface region except that it and its derivative are continuous. If the bulk well and barrier solutions are slowly varying (such as for the ground state of a wide quantum well with small band offsets) over distances of order w, then it is reasonable to suppose that the envelope function and its derivative are sensibly constant over an interface region and one may require, to a good approximation, that the bulk solutions on either side of the interface region and their derivatives should be equal. This is approximately equivalent to matching the bulk solutions at the interface itself which is just the particle in a box model without the effective mass in the derivative boundary condition.

It should be pointed out that the above arguments apply equally well to cases where band extrema are degenerate; Each envelope function and its derivative are smoothly matched across each interface. All we require is that the envelope functions are slowly varying over distances of order w (a few lattice periods). This is expected to be the case for states in the region of band edges in structures with small band edge discontinuities.

CONCLUDING REMARKS

It is now clear that the envelope function approach can be formulated exactly for both latticematched and nonlatticematched microstructures, and can include spin orbit interaction, inhomogeneous strain, applied fields and nonlocal potentials. In the present paper we have concentrated on time independent problems. However, there would appear to be nothing to stop one recasting the time dependent Schroedinger equation in envelope function form and hence treating time dependent problems. There would also appear no need to restrict oneself to expansions involving periodic functions; the U_n could be chosen to all change with the same phase on translation by a latticevector. For instance, choosing U_n to have the same translational properties as the X point wavefunctions in bulk Si may well be a fruitful approach to Si/Ge structures.

The exact envelope function method allows one to make a systematic derivation of the particle in a box model. It is pointed out that the inclusion of the effective mass in the derivative boundary condition is not necessary to conserve the current and that it leads to unphysical results for the energy levels in quantum wells.

The exact formulation gives the envelope functions and their derivatives as finite and continuous everywhere from the outset. This has allowed us to derive the particle in a box type model intuitively in which slowly varying bulk envelope functions in adjoining layers are smoothly matched. This procedure should work for both lattice matched and nonlattice-matched structures.

ACKNOWLEDGEMENTS

First of all, I wish to thank Prof M Jaros and Dr R A Abram for inviting me to attend this workshop from which I have derived great benefit, and Dr R A Abram for many stimulating discussions over the past year. Thanks also to my colleagues Drs M A Gell, C Smith, D A H Mace and M Fisher for discussions and helpful comments. I would also particularly like to thank Dr N Apsley for inviting me to give a seminar on this topic at RSRE Malvern and to Drs D C Herbert and M S Skolnick for subsequent discussions, all of which helped to clarify my ideas. I thank, as well, Dr S Baroni for an interesting comment. This paper is published with the permission of the Director of Research, British Telecom.

REFERENCES

Altarelli M   1983a Physica <u>117B + 118B</u> 747
1983b Phys Rev <u>B28</u> 842
1986 in "Heterojunctions and Semiconductor Superlattices" Eds G Allan et al (Springer, Berlin) p12.

Burt M G   1987 Semicond Sci Technol <u>2</u> 460 and 701
1988a to appear in Semicond Sci Technol
1988b to be submitted for publication.

Herbert D C and Gell M A 1987 Phys Rev $\underline{B35}$ 9591.

Kane E O 1966 in "Semiconductors in Semimetals" $\underline{1}$ R K Willardson and A C Beer Eds (New York, Academic) p75.

Luttinger J M and Kohn W 1955 Phys Rev $\underline{97}$ 869.

Pikus G E and Bir G L 1959 Soviet Physics - Solid State $\underline{1}$ 1502.

Poetz W and Ferry D K 1986 Superlattices and Microstructures $\underline{2}$ 151.

A METHOD FOR CALCULATING ELECTRONIC

STRUCTURE OF SEMICONDUCTOR SUPERLATTICES: PERTURBATION

H.M. Polatoglou, G. Kanellis and G. Theodorou

Aristotle University of Thessaloniki
Physics Department, Thessaloniki 54006,
Greece

ABSTRACT

A similarity transformation of the superlattice Hamiltonian is presented. The transformed Hamiltonian includes two parts. One that describes the average crystal, and the other the interaction between average crystal states. This interaction is small and is treated as a perturbation. The method is applied to GaSb/InAs (100) superlattice, and its band structure is analyzed.

INTRODUCTION

In recent years considerable attention has been given to the various properties of artificially grown superlattices (SL) [1]. As a result, many theoretical approches for obtaining their band structure have been proposed [2-12]. In principle, there should be no problem in calculating the SL band structure by employing existing methods for pure compounds. However, the usually large unit cell of the SL structures imposes difficult computational problems. Besides the interpretation of the results is not obvious, even thought the bulk properties of the constituent pure compounds are more or less well known.

The purpose of this work is to present a method for the calculation of the band structure of lattice-matched superlattices. A similar method has been proposed by Kanellis [13] for the calculation of the phonon spectra of modulated structures. The advantage of this method is that the superlattice Hamiltonian is transformed into one that has two parts: One that describes the average crystal in terms of the primitive cell of the underlying simple structure and the other the interaction between average crystal states. The first part of the Hamiltonian is responçible for the folding effects while the second for the confinement effects.

DESCRIPTION OF THE METHOD

a. Choice of the base

We consider a superlattice built up from two binary compounds. The structure of these compounds is described in terms of identical primitive cells. The lattice produced by this primitive cell will be called the underlying lattice. The Brillouin zone associated with the underlying lattice is called the Original Brillouin Zone (OBZ). The SL is described in terms of a larger primitive cell. The latter will be called the supercell and the Brillouin Zone associated with it, the Superlattice Brillouin Zone (SBZ). Let us suppose that the volume of the supercell is equal to No times the volume of the primitive cell of the underlying lattice. Thus, each supercell contains No primitive cells. The position of an atom will be given by

$$\bar{R}_m + \bar{r}_n + \bar{\tau}_j$$

where $\bar{R}_m$ defines the position of the mth supercell, $\bar{r}_n$ the position of the nth primitive cell in the mth supercell and $\bar{\tau}_j$ the position of the j atom relative to the nth primitive cell.

In the reciprocal lattice, there are No SBZs contained in the first SBZ. For a wavevector k lying in the first SBZ, there exist No equivalent wavevectors lying in the first OBZ and these are given by

$$\bar{k}_\ell = \bar{k} + \bar{g}_\ell \ , \qquad\qquad (1)$$

where $\{\bar{g}_\ell\}$ are the No reciprocal lattice vectors of the SL, lying in the first OBZ.

The wavefunction of the superlattice satisfies the Bloch theorem with respect to translations corresponding to Bravais lattice vectors of the superlattice. If the original basis function are localized around the atoms, as in the case of atomic or Muffin-Tin orbitals, then the appropriate base for the crystal are the following Bloch sums

$$y^\alpha_{\bar{k},n,j} = \frac{1}{N_\bullet} \sum_m \exp[i\bar{k}.(\bar{R}_m + \bar{r}_n + \bar{\tau}_j)] \ \varphi^\alpha_{n,j}(\bar{r}-\bar{R}_m-\bar{r}_n-\bar{\tau}_j) \quad (2)$$

where $N_\bullet$ is the number of supercells in the SL, $\bar{k}$ a wave vector in the first SBZ and $\varphi^\alpha_{n,j}(\bar{r}-\bar{R}_m-\bar{r}_n-\bar{\tau}_j)$ a localized orbital centered at $\bar{R}_m + \bar{r}_n + \bar{\tau}_j$. We keep in mind, that for each $\bar{k}$ in the SBZ, there are No equivalent wavevectors of the SL lying in the OBZ and given by (1). Instead of using the function (2) as the basis set we have selected to use the following linear combination

$$\Psi^\alpha_{\bar{k},\bar{g}_\ell,j} = \frac{1}{N_o N_\bullet} \sum_{m,n} \exp[i(\bar{k}+\bar{g}_\ell).(\bar{R}_m+\bar{r}_n+\bar{\tau}_j)] \ \varphi^\alpha_{n,j}(\bar{r}-\bar{R}_m-\bar{r}_n-\bar{\tau}_j)$$

$$(3)$$

The above choice is reminiscent of the basis set appropriate for the case of the underlying lattice. That is, when all n

positions become equivalent, then $\Psi^\circ_{\bar{k},\bar{g}_{\perp},j}$ satisfies Bloch theorem with respect to translations corresponding to Bravais lattice vectors of the underlyed lattice. The new basis set facilitates the calculations and leads to a more clear interpretation of the results.

In the new representation, the Hamiltonian can be considered as consisting of diagonal blocks labeled $(g_{\perp},g_{\perp})$ and off-diagonal blocks $(\bar{g}_{\perp},\bar{g}_{\perp}')$. The elements of the diagonal blocks are given by

$$\langle \bar{k},\bar{g}_j,j,\alpha|H|\bar{k},\bar{g}_{\perp},j'\beta\rangle = \delta_{\alpha\beta}\,\delta_{jj'}\cdot\langle\varepsilon^\alpha_{nj}\rangle_n +$$

$$\sum_s \exp[-i(\bar{k}+\bar{g}_{\perp}).\bar{r}_{\bullet}]\langle t^{\alpha\beta}_{nj,\bullet}\rangle_n \qquad (4)$$

where the brackets $\langle\ \rangle_n$ denote averaged values with respect to n, over the supercell. The symbol ε^α_{nj} denotes the on-site matrix element of an α orbital at position (n,j) and the symbol $t^{\alpha\beta}_{nj,\bullet}$ denotes the Hamiltonian matrix element between an α orbital at site (n,j) and a β orbital at the s neighbor site. From equation (4) it becomes evident that each diagonal block of the Hamiltonian matrix contains the average interactions and describes subbands of the average crystal.

For the elements of the off-diagonal blocks $(\bar{g}_{\perp},\bar{g}_{\perp}\cdot)$ we get

$$\langle \bar{k},\bar{g}_{\perp},j,\alpha|H|\bar{k},\bar{g}_{\perp}\cdot,j',\beta\rangle =$$

$$= \delta_{\alpha\beta}\,\delta_{jj'}\cdot\ \frac{1}{N_o}\ \sum_n\exp[i(\bar{g}_{\perp}-\bar{g}_{\perp}\cdot).(\bar{r}_n+\bar{r}_j)]\delta\varepsilon^\alpha_{nj} +$$

$$+\ \frac{1}{N_o}\ \sum_s\sum_n\exp[-i(\bar{g}_{\perp}-\bar{g}_{\perp}\cdot).(\bar{r}_n+\bar{r}_j)]\exp[-i(\bar{k}+\bar{g}_{\perp}).\bar{r}_{\bullet}]\ \delta t^{\alpha\beta}_{nj\bullet} \quad (5)$$

where

$$\delta\varepsilon^\alpha_{nj} = \langle\varepsilon^\alpha_{nj}\rangle_n - \varepsilon^\alpha_{nj}$$

and

$$\delta t^{\alpha\beta}_{nj\bullet} = \langle t^{\alpha\beta}_{n,j,\bullet}\rangle_n - t^{\alpha\beta}_{n,j,\bullet}.$$

Relation (5) shows that the off-diagonal blocks depend only on the differences between the SL interactions and the average crystal values.

b. Perturbation

From Eqs. (4) and (5) one can notice that the matrix elements of the diagonal blocks are large, while that of off-diagonal blocks small. Therefore we can treat the problem as follows: First, we diagonalize the diagonal blocks, find the band structure of the average crystal and perform the folding in order to reduce the OBZ to the SBZ. Secondly, we take into account the off-diagonal blocks, which imply interaction between the states of the average crystal. We treat this interaction in the following way: To correct the energy of average state $|i\rangle$, we find all states $|j\rangle$ for which the ratio of matrix element <i|H|j> to the difference in energy between the two average states is, in absolute value, larger that 0.05. The set of these states define the truncated basis set. Then we calculate the corrected energies by solving the truncated secular equation.

ELECTRONIC ENERGY BANDS OF GaSb!InAs SL

As an application of the method we have studied the band structure of a GaSb!InAs (100) SL consisting of alternate layers, each 6 atomic layers thick. A nearest-neighbor approach is used within an sp^3s* basis [14] and a band-offset equal to -0.53 eV. To test the accuracy of the truncated basis set for the superlattice, we compare the eigenvalues at Γ point obtained by the truncated basis to those obtained by the full basis. The results are presented in table 1. It is apparent that the truncated basis gives very accurate rerults.

Table 1. Eigenvalues at Γ point calculated by the full and the truncated basis. Energies are in eV.

Full basis	Trancated basis
1.2514	1.2579
1.2220	1.2436
1.1862	1.1914
1.0211	1.0145
0.2551	0.2564
-0.1096	-0.1088
-0.1125	-0.1116
-0.2503	-0.2346
-0.4120	-0.4162
-0.4186	-0.4214
-0.6189	-0.6187
-0.6214	-0.6215
-1.0046	-1.0082
-1.0129	-1.0155
-1.0920	-1.0959
-1.0991	-1.1040
-1.3867	-1.3857

The dimension of the secular equation for the truncated basis is reduced, with respect to the full basis, by a factor of 7 for the upper valence bands and 2.5 for the conduction bands. Fig. 1 shows the subbands of the SL, as well as the subbands of the average crystal. From this figure it is clear that the eigenvalues for thin SL's are close to those of the average crystal. Consenquently, the average crystal can be considered as a good starting point for the description of the superlattice.

We also observe a splitting of the double degenerate upper valence band. This can be explained as follows: The symmetry group of GaSb!InAs SL is C^1_{2v}. All irreducible representations of this group are one-dimensional. Thus, from symmetry reasons, we expect removal of the degeneracy found

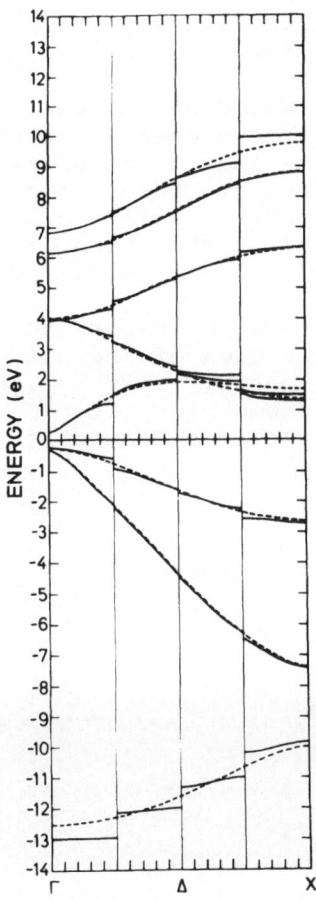

Fig. 1. Energy band structure of a 2x2 GaSb/InAs (100) SL
(solid lines), and band structure of the average
crystal (dashed lines).

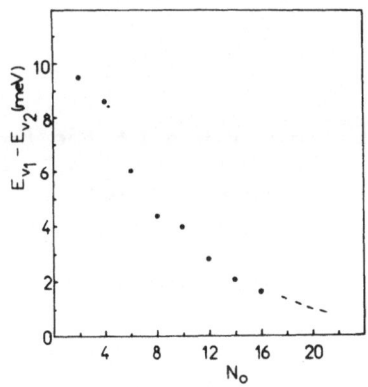

Fig. 2. Splitting of the upper valance band as a function of
the thickness of the superlattice.

in the bands of the average crystal. The splitting of the upper valence band as a function of the thickness of the superlattice is given in figure 2. This splitting diminishes for thicker superlattices.

This can be understood as follows: the atoms inside the layer satisfy locally the higher symmetry of the underlying lattice, while the interface atoms satisfy the lower symmetry of the SL. The reduction of the symmetry at the inteface atoms is responsible for the splitting. As a consequence the splitting will diminish as the ratio of interface to volume atoms decrease.

We have also calculated the gap $E_g = E_{c_1} - E_{v_1}$, where E_{c_1} and E_{v_1} are the energies of the bottom of conduction and top of valance bands respectively. The results for E_g as well as E_{c_1} and E_{v_1} are shown in figure 3. The calculations indicate a semiconductor-to-semimetal transition for No = 64.

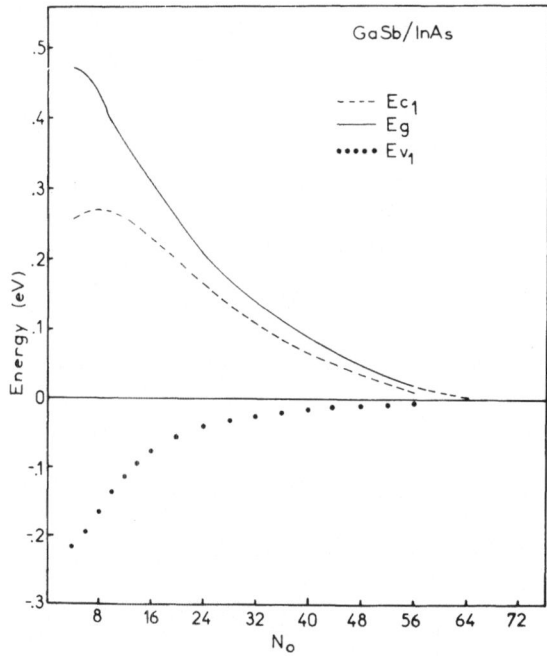

Fig. 3. Energy gap and positions of the bottom of conduction and top of valance bands, as a function of the thickness of the superlattice.

ACKNOWLEGMENTS

This work has been partially suported by the Greek ministry of Research and Technology. We are thankful to K. Tserbak for help on computing.

REFERENCES

1. L. Esaki in Proceedings of the 17th ICPS, edited by J.D. Chadi and W.A. Harrison (Springer-Verlag, New York, 1985).

2. G.A. Sai-Halasz, L. Esaki and W.A. Harrison, Phys. Rev. B. 18, 2812 (1978).

3. C. Tejedor, J.M. Calleja, F. Meseguer, E.E. Mendez, C.A. Chang and L. Esaki, Phys. Rev. B32, 5303 (1985).

4. J.N. Schulman and Y.C. Chang, Phys. Rev. 31, 2056 (1985).

5. K.B. Wong, M. Jaros, M.A. Cell and D. Ninno, J. Phys. C19, 53 (1986).

6. M.A. Cell, D. Nino, M. Jaros and D.C. Herbert, Phys. Rev. 34, 2416 (1986).

7. M.A. Cell, D. Ninno, M. Jaros, D.J. Wolford, T.F. Keuch and J.A. Bradley, Phys. Rev. 35, 1196 (1987).

8. D.M. Bylander and L. Kleinman, Phys. Rev. 34, 5280 (1986)

9. L. Brey and C. Tejedor, Phys. Rev. 35, 9112 (1987).

10. C. Mailhiot and D.L. Smith, Phys. Rev. B35, 1242 (1987).

11. D.Z.Y. Ting and Y. C. Chang, Phys. Rev. 36, 4359 (1987).

12. R.D. Graft, G.P. Parravicini and L. Resca, Solid State Commun. 57, 699 (1986).

13. G. Kanellis, Phys. Rev. B35, 746 (1987) and Solid State Commun. 58, 93 (1986).

14. P. Vogl, H.P. Hjalmarson and J.D. Dow, J. Phys. Chem. Solids, 44, 365 (1983).

THE EFFECTS OF ORDERING IN TERNARY SEMICONDUCTOR ALLOYS: ELECTRONIC AND STRUCTURAL PROPERTIES

Kathie E. Newman, Dan Teng, Jun Shen, and Bing–Lin Gu*

Department of Physics, University of Notre Dame, Notre Dame, Indiana 46556
*Department of Physics, Tsinghua University, Beijing, PRC

ABSTRACT

The relative strain energies of five types of ordered structures derived from a parent zinc-blende alloy $A_{1-x}B_xC$ have been investigated. The most stable $x = \frac{1}{2}$ and $x = \frac{1}{4}$ or $\frac{3}{4}$ structures are chalcopyrite and famatinite. Also investigated is the influence of order and strain on the bandstructure of the ordered compounds. Calculated tight-binding band gaps of ordered compounds of the $Al_{1-x}Ga_xAs$ family yield results not too different from those for the alloy. Band gaps for a same-cation family of compounds derived from $GaAs_{1-x}Sb_x$ exhibit a large bowing as a function of composition x similar to that reported experimentally for a metastable form of the alloy.

1. INTRODUCTION

There have been numerous recent reports of unusual ordered forms of ternary semiconductor alloys of the type $A_{1-x}B_xC$ and $AC_{1-x}D_x$.[1-5] The usual expected structure of these ternary alloys is zinc-blende (space group $F\bar{4}3m$), a disordered structure in which

ABC₂ Structures

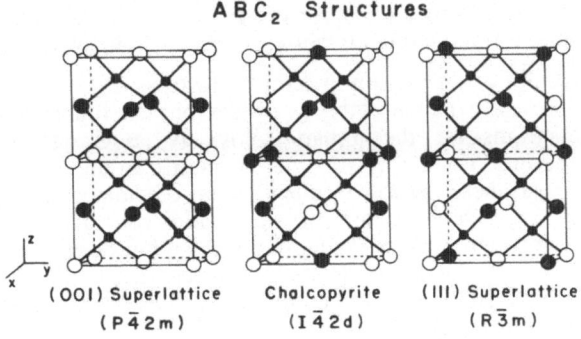

(001) Superlattice   Chalcopyrite   (III) Superlattice
($P\bar{4}2m$)   ($I\bar{4}2d$)   ($R\bar{3}m$)

FIGURE 1. Special-k point structures for ordered compounds ABC_2: $k = (001)$, 1×1 superlattice oriented along the z direction; $k = (0\frac{1}{2}1)$, chalcopyrite; and $k = (111)$, 1×1 superlattice oriented along the (111) direction. Also shown are the respective space-group designations of each crystal type. Note, each structure is shown undistorted from the parent zinc-blende form. Cations A and B are shown as large open and shaded circles, respectively, and the anions C are shown as smaller filled circles.

in the alloy $A_{1-x}B_xC$ atoms A and B randomly occupy one set of face-centered cubic (fcc) positions, and atoms C occupy the other set. The complement to this "disordered-cation" alloy is the "disordered-anion" alloy, $AC_{1-x}D_x$, in which the "anion" fcc positions are randomly occupied by atoms C and D, while the "cation" fcc positions are occupied just by atoms A. The newly reported ordered structures, shown in Figs. 1 and 2, all involve an ordering of either the cations (for the alloy $A_{1-x}B_xC$) or the anions ($AC_{1-x}D_x$) on the one fcc sublattice.

Now four different types of new ordered structures have been reported. For a composition $x \simeq \frac{1}{2}$, a (001) 1×1 superlattice phase (space group $P\bar{4}2m$) has been seen in the alloys $Al_{1-x}Ga_xAs$,[1] $GaAs_{1-x}Sb_x$,[2] and in $Ga_{1-x}In_xAs$.[3] Also found in $GaAs_{1-x}Sb_x$ for $x \simeq \frac{1}{2}$ is a chalcopyrite-like phase,[2] having the space group $I\bar{4}2d$. Very recently, a (111) 1×1 superlattice phase with space group $R\bar{3}m$ was found both in the ternary alloy $Ga_{1-x}In_xAs$ and in the quaternary alloy $Ga_{1-x}In_xAs_{1-y}P_y$.[4] And for $x \simeq \frac{1}{4}$, a famatinite structure, space group $I\bar{4}2m$, was found in the alloy $Ga_{1-x}In_xAs$.[5]

Since the ordering of these alloys occurs only on one fcc sublattice, we may initially understand the ordering of these materials in terms of the known ground states of fcc struc-

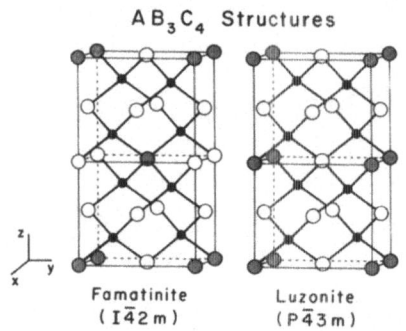

AB₃C₄ Structures

Famatinite
($I\bar{4}2m$)

Luzonite
($P\bar{4}3m$)

FIGURE 2. Special-k point structures for ordered compounds AB_3C_4: $k = (001)$, luzonite, and $k = (0\frac{1}{2}1)$, famatinite.

tures.[6] The disordered fcc structures have the space group Fm3m. If one ignores the "passive" fcc sublattice of the semiconductors (e.g., the C sites in $A_{1-x}B_xC$), then the $x \simeq \frac{1}{2}$ structures are exactly the three special-k point structures expected of fcc structures at $x = \frac{1}{2}$: (001), $(0\frac{1}{2}1)$, and (111) (space groups P4/mmmm, I4₁/amd, and R$\bar{3}$m). In fact, for fcc structures, it is known that a parameter $\alpha = J_1/J_2$, which is the ratio of a first-neighbor (negative) interaction to a second-neighbor (positive or negative) interaction, controls the type of structure found for the ground state, with ordering of the type (001) found for $\alpha < 0$, of the type $(0\frac{1}{2}1)$ found for $0 < \alpha < \frac{1}{2}$, and of the type (111) for $\alpha > \frac{1}{2}$.[7] The ground-state calculations for $x = \frac{1}{4}$ and $\frac{3}{4}$ also predict ordered (001) structures with space group Pm3m ($\alpha < 0$) and ordered $(0\frac{1}{2}1)$ structures with space group I4/mmm ($\alpha > 0$). In semiconductors, the latter structure corresponds with the famatinite form found in $Ga_{1-x}In_xAs$, while the former structure corresponds to a luzonite-type structure (space group $P\bar{4}2m$), and has not yet been seen in semiconductors.

Not known presently is the stability of the phases found. Mbaye et al.[8] have calculated phase diagrams for the ternary alloys $A_{1-x}B_xC$ and $AC_{1-x}D_x$ that show interesting narrow regions of stability of ordered phases around $x = \frac{1}{2}$ and $x = \frac{1}{4}$, $\frac{3}{4}$. (With the exception of these narrow regions, their calculated phase diagram has the expected miscibility gap characteristic of most alloys.) Their calculations do not distinguish the types of ordered phases found for $x \simeq \frac{1}{2}$, but do indicate the importance of strain energies in such calculations.[9] These strain energies both appear as a significant contribution to the heat of formation of the alloy as well as having a role in stabilizing the ordered $x \simeq \frac{1}{2}$ and $x \simeq \frac{1}{4}$, $\frac{3}{4}$ phases.

In this paper, we do not address the issue of stability of these phases, but instead attempt to characterize these new materials. It is apparent from the experimental reports that these new phases, existing either as equilibrium or metastable phases of the alloy, can be grown. As an ordered phases of the alloys, having a different structure (as indicated by the space group), we expect different properties, e.g., band gaps. These new properties should be derivable from what is known of the parent compounds, the zinc-blende materials AC and BC or AC and AD.

120

We present here selected results of such calculations for the five special-k point ordered structures of the alloys $A_{1-x}B_xC$ and $AC_{1-x}D_x$, the compounds ABC_2 and A_2CD (shown in Fig. 1) and the compounds A_3BC_4, AB_3C_4, A_4B_3D, and A_4BD_3 (shown in Fig. 2). We use two simple empirical theories of zinc-blende materials: A strain-energy formula due to Harrison,[10] and the Vogl tight-binding model.[11] We use the first theory to predict the effects of strain on bond lengths and other structural parameters. Then results of this calculation are then combined with the Vogl model to determine the effects of strain and the new crystal structure on the electronic properties of these new compounds. Specifically, we compare band gaps and the heavy-light hole splitting of the valence band edge of these new materials with those found in the disordered $A_{1-x}B_xC$ and $AC_{1-x}D_x$ alloys. We expect these calculations to be of aid to experimentalists searching for such new compounds.

2. THEORY

2.1 The Effects of Strain on Structure

To first approximation, the strain energy of a semiconductor is determined by the nature of the sp^3 bond.[10] In semiconductor alloys, e.g., $A_{1-x}B_xC$, we know from extended x-ray absorption fine structure (EXAFS) measurements that the alloy bond lengths d^{AC} and d^{BC} differ from those in the host compounds AC and BC.[12] We also expect angles to vary from those in the host compounds, e.g., the angles A-C-A and A-C-B are probably not those of a perfect tetrahedron (109.5°).

Harrison[10] has included these two effects in a simple phenomenological formula for the calculation of the strain energy of semiconductors:

$$E_{strain} = \frac{1}{2} C_0 \sum_{bonds} \frac{(d - d_0)^2}{d_0^2} + \frac{1}{2} C_1 \sum_{ijk} (\delta\theta_{ijk})^2 \ . \tag{1}$$

As is the case in the Keating model,[13] Eq. (1) schematically includes through the terms C_0 and C_1 the radial bond-stretching and angular bond-bending forces. (Here d_0 is the unstretched bond length, $\delta\theta_{ijk}$ is the deviation of the bond angle from tetrahedral angle, and C_0 and C_1 are fit to the elastic constants c_{11} and c_{12}.)

The strain energies of the ordered structures can be estimated from the Harrison formula (1) once the atomic positions in the ordered structures are determined. For example, consider the (001) superlattice shown in Fig. 1. Its Bravais lattice is simple tetragonal: The layering along the (001) direction allows the "cube" distance c along the z direction to be different in general from the distance a along the x or y directions, where c $\simeq$ a $\simeq$ a_0, with a_0 the "cube" distance in the average parent zinc-blende compound. In a same-anion compound ABC_2, the positions of the cations A and B are easily written in terms of a and c. Because of the symmetry of the structure, anions may distort only along $\pm z$, shortening one kind of bond length (e.g., A-C), and lengthening the other (B-C). Thus one more parameter, h, is required to specify the positions of the anions C. With all atomic positions known, we find all contributions to Eq. (1), and then minimize Eq. (1) with respect to the structural parameters a, c, and p. This determines the equilibrium values of the structural parameters as well as the size of the strain energy relative to the parent zinc-blende compounds.

Calculations for other ordered structures proceed similarly. The simplest calculation is for luzonite (2 parameters) and the most complex is for the (111) superlattice (5 parameters). In general, for structures with the same chemical formula, the complexity of the structure is not related to the final value of the strain energy.

2.2 Bandstructure Calculations

In our calculations of bandstructure, we employ a simple empirical tight-binding Hamiltonian that includes the 5 atomic orbitals s, p_x, p_y, p_z, and s^*. This Hamiltonian is reasonably accurate for the valence bands and the conduction-band edge of zinc-blende semiconductors. The parameters of this Hamiltonian for III-V and some II-VI zinc-blende compounds have been determined by fitting to experimental data by Vogl et al.[11] A particular virtue of the Vogl model is that the parameters exhibit proper chemical trends, e.g., Harrison's Law,[10] that the product of nearest-neighbor matrix elements V with the square of the bond length, Vd^2, should be nearly constant.

All calculations start from the standard postulates of tight-binding theory. Knowing the Bravais lattice and the positions of the atoms in the basis of an ordered structure, the bulk Hamiltonian matrix as a function of the wavevector k is easily found. We thus include the internal-strain effects of the previous section, e.g., the Slater-Koster direction cosines[14] must be modified from those of the parent zinc-blende compound. The required diagonal and off-diagonal matrix elements are found from those for the parent zinc-blende compounds. The diagonal matrix elements are modified by simple estimates of the valence-band offset between the two parent zinc-blende compounds.[15] The off-diagonal matrix elements are modified by Harrison's Law. For heavier compounds, such as $HgCdTe_2$, spin-orbit effects can be included by including three spin-orbit parameters λ and doubling the size of the Hamiltonian.[16]

3. RESULTS

3.1 Strain Results

We have computed the strain energies and effects of strain on five types of ternary III-V ordered compounds.[17] We describe here results for a typical semiconductor, $GaAs_{1-x}Sb_x$, compared to the parent zinc-blende (Z) compounds GaAs or GaSb at x = 0 or 1. Results similar to ours for the ordered structures famatinite, luzonite, chalcopyrite, and the (001) superlattice have reported elsewhere by others.[9, 18]

Summarizing the results for bond lengths shown in Fig. 3, in each ordered $GaAs_{1-x}Sb_x$ compound, bonds characteristic of either GaAs (squares) or GaSb (diamonds) are found. For example, in luzonite (L) structures with x = $^1/_4$, three Ga-As bonds are found for every Ga-Sb bond, while famatinite compounds (F) have two long and one short Ga-As bonds for each Ga-Sb bond. The famatinite bond lengths are on the average closer in value to those in the parent zinc-blende compounds, resulting in a lower value for the strain energy, as shown in Fig. 4.

Of the x = $^1/_2$ structures, both the (001) superlattice (S) and chalcopyrite (C) compounds have equal numbers of Ga-As and Ga-Sb bonds, two each per cation-centered tetrahedron (see Fig. 1). Whereas, in the (111) superlattice, due to the two types of

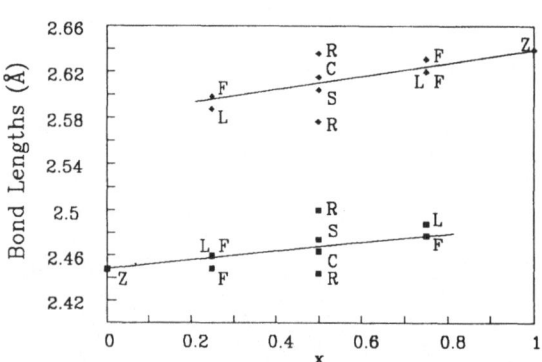

FIGURE 3. Calculated Ga-As bond lengths (squares) and Ga-Sb bond lengths (diamonds) in the $GaAs_{1-x}Sb_x$ family of compounds. For x = 0 or 1, the bond length is that of the zinc-blende (Z) compounds GaAs or GaSb. For x = $^1/_4$, the ordered compound luzonite (L) contains three Ga-As bonds for every Ga-Sb bond, while famatinite (F) has two long and one short Ga-As bonds for each Ga-Sb bond. A similar result holds for x = $^3/_4$. For x = $^1/_2$, (001)-oriented superlattices (S) and chalcopyrite compounds have equal numbers of Ga-As and Ga-Sb bonds, while the rhomobohedral (R) (111)-oriented superlattice contains three long and one short Ga-As bonds and three short and one long Ga-Sb bonds. The dashed line indicates Vegard's Law expected for the average bond length measured in the alloy by x-ray diffraction.

cation-centered environments (see Fig. 1), we find one *short* isolated Ga-As bond of approximately the same length in this structure as is found in zinc-blende GaAs and, on the other type of cation-centered tetrahedron, three *long* Ga-As bonds that have a length closer to the virtual-crystal length for $x = 1/2$ (shown in Fig. 3 by the dashed line). Thus, on the average, (111) superlattices are characterized by bond lengths that are long in comparison to the parent zinc-blende structures. This means that the rhombohedral (R) (111) superlattices have by far the *largest* strain energy, Fig. 4, of the $x = 1/2$ structures. Interestingly, a

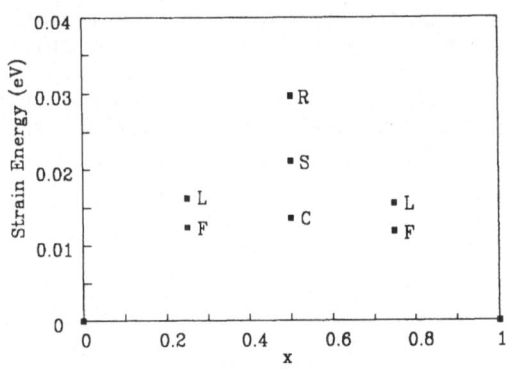

FIGURE 4. Calculated strain energies of the GaAs$_{1-x}$Sb$_x$ family of compounds.

structure having fewer strain parameters, chalcopyrite, is found to have always the lowest strain energy.

Calculations such as shown in Fig. 3 are also useful for the interpretation of EXAFS (extended x-ray absorption fine structure) data for bond lengths in random alloys. Assuming that the alloy is formed from a mixture of all possible orientations of tetrahedra GaAs$_n$Sb$_{4-n}$, with n = 0 to 4, portions of the alloy may show incipient order of the types shown in Figs. 1 and 2. The EXAFS data should then be viewed as a statistical average (weighted in proportional to the total energy of the state) of the types of configurations found in the alloy. We have obtained good agreement with the Mikkelsen and Boyce[12] results for the slopes of the bond lengths in the alloy by finding the slope of a line in our Ga$_{1-x}$In$_x$As plot that passes half-way between the chalcopyrite (C) and (001) superlattice (S) results.[17] Since we find that that average results for the strain calculations alone reproduce well the EXAFS data, this implies that charge-transfer effects are probably only a small correction for the III-V and II-VI families of compounds.[19]

3.2 Bandstructure Results

We have studied the effect of strain and of ordering on five possible types of ternary III-V ordered compounds. Again, here we describe results for the family of compounds derived from the alloy GaAs$_{1-x}$Sb$_x$.

As described in the previous section, the bandstructures of the ordered compounds shown in Figs. 1 and 2 can be determined from an empirical tight-binding theory. The Brillouin zones of these ordered compounds are shown in Fig. 5. In Figs. 6 and 7, we compare the bandstructures of the Ga$_2$AsSb and Ga$_4$As$_3$Sb ordered compounds. Notable is the obvious folding of the parent zinc-blende Γ to X band into Γ to Z in the (001) superlattice, Fig. 7(a). Bands in the considerably more complicated chalcopyrite structure, Fig. 7(b), can also be understood in terms of a remapping of points in the face-centered cubic Brillouin zone into the smaller (by a factor 4) body-centered tetragonal Brillouin zone.[20] Especially interesting is the splitting due to strain of the top of the valence band. In the similar structures, the (001) superlattice and chalcopyrite, the inequivalence of xy with z is manifest in Fig. 1. In both structures, we find that the ratio of the "cube" distance c along z is different than the distance a in the xy plane, and the ratio c/a is generally greater than one in the superlattices, and less than one, in chalcopyrite structures.[17] Thus the order of the two-fold and one-fold degenerate bands flips between these two structures. The (111) superlattice also exhibits a large strain splitting at the top of the valence band.

Of particular interest are the predicted band gaps of the ordered compounds in comparison with those for the alloy. We show in Figs. 8 and 9 results for an alloy family

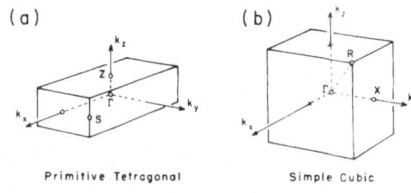

(a) Primitive Tetragonal

(b) Simple Cubic

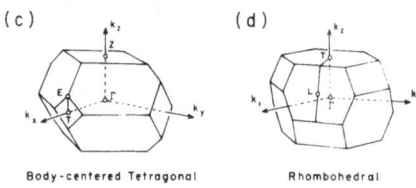

(c) Body-centered Tetragonal

(d) Rhombohedral

FIGURE 5. Brillouin zones for the ordered compounds of the types ABC_2 and AB_3C_4: (a) For (001) superlattice, primitive tetragonal; (b) For luzonite, simple cubic; (c) For chalcopyrite and famatinite, body-centered tetragonal; and (d) For (111) superlattice, rhombohedral.

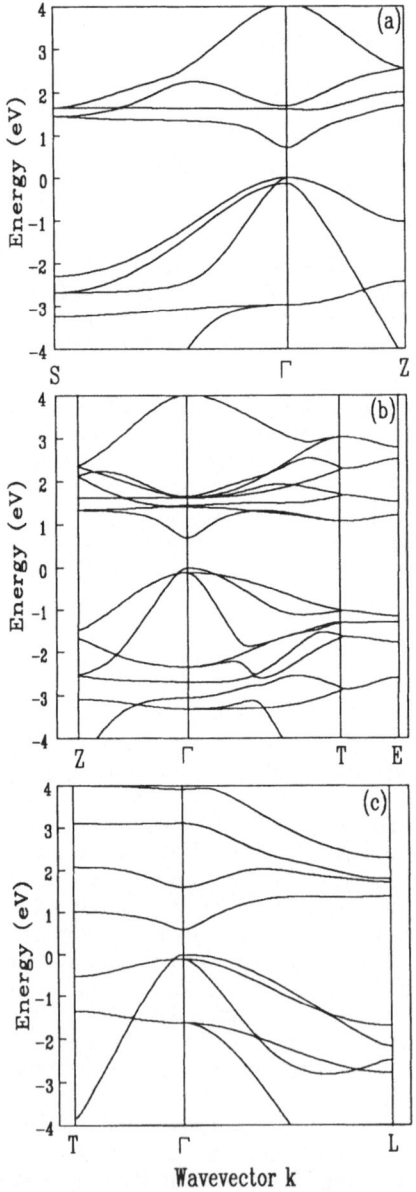

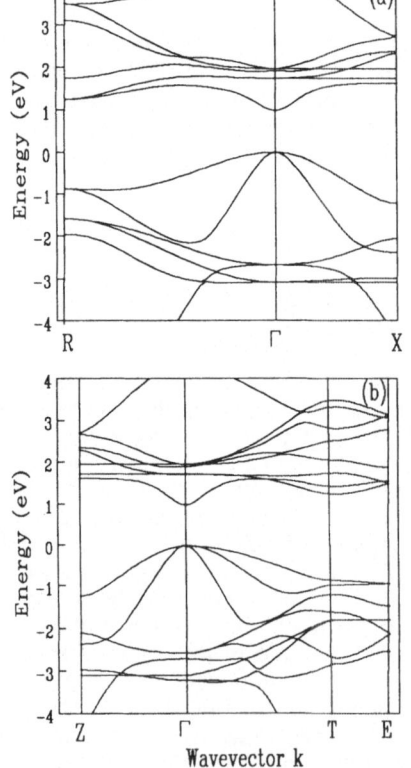

FIGURE 6. Band structures E(k) versus wavevector k for the ordered compounds Ga_4As_3Sb: (a) Luzonite and (b) Famatinite.

FIGURE 7. Band structures E(k) versus wavevector k for the ordered compounds Ga_2AsSb: (a) (001) superlattice; (b) Chalcopyrite; and (c) (111) superlattice.

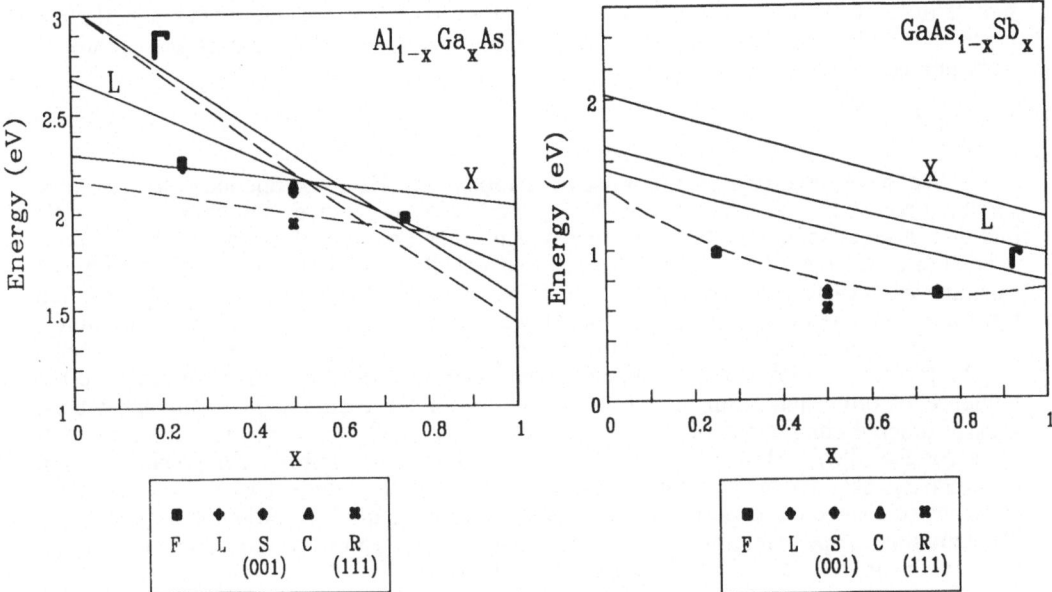

FIGURE 8. Band gap versus composition x for the $Al_{1-x}Ga_xAs$ family of compounds. Notation for the ordered compounds is that of Fig. 3. For comparison, the bowing of the alloy is also shown: Virtual-crystal approximation, solid lines, and experimental results for the direct gap, dashed line.

FIGURE 9. Band gap versus composition x for the $GaAs_{1-x}Sb_x$ family of compounds.

that shares the same anion, $Al_{1-x}Ga_xAs$, and for an alloy family that shares the same cation, $GaAs_{1-x}Sb_x$. In both figures, results determined from the virtual-crystal approximation (VCA) for the alloy (solid lines) are compared first with experimental results.[21,22] (Note that the theoretical results are fit to zinc-blende data taken at 4° K and the experimental data is for room temperature. Also note that the empirical fit of the tight-binding model is known to not be correct at the L point in the zinc-blende Brillouin zone.[11])

In the $Al_{1-x}Ga_xAs$ family of compounds, Fig. 8, the bowing of the alloy bandstructure is very slight. Results for the ordered compounds reflect this fact, being generally almost the same as those found for the alloy. That is, famatinite and luzonite points and (001) superlattice and chalcopyrite points almost overlap. The (111) result is particularly interesting in that it is found at an energy much lower than that for the other x = $1/2$ ordered compounds: The VCA alloy has a band gap of 2.19 eV, then chalcopyrite has a band gap of 2.14 eV, followed closely by the (001) superlattice, 2.10 eV,[23] and finally, for (111) superlattice, 1.94 eV.

Two effects cause a different set of results for the band gaps of the $GaAs_{1-x}Sb_x$ family of compounds, Fig. 9. First, in opposition to the same-anion compounds, the offset correction[15,24] for same-cation compounds causes a significant change in the calculated value of the band gap. Secondly, strain effects are large in this family of compounds. Again we find similar band gaps for similar structures, e.g., famatinite and chalcopyrite. But the positions of the ordered-structure band gaps are much lower than found in VCA, e.g., at x = $1/2$, we find a band gap of 1.14 eV for the VCA alloy, of 0.71 eV for the (001) superlattice, of 0.70 eV for chalcopyrite, and of 0.60 eV for the (111) superlattice. Interestingly, the experimental bowing of the direct gap in the alloy almost passes through these points.

4. SUMMARY

We have investigated the relative stability for III-V compounds of the three possible ordered $x = 1/2$ structures: the (001) and (111) superlattices and chalcopyrite. We find that in semiconductors the strain energy due to the bonding of anions to cations alone is enough to distinguish the three phases. Our calculations predict that the chalcopyrite structure should have a lower energy than either type of superlattice. Similarly, we find that for $x = 1/4$, the famatinite structure has lower strain energy than the luzonite structure.

Our strain calculations also allow us to predict structural parameters such as bond lengths for the new ordered III-V compounds. This in turn allows us to investigate the influence of order and strain on the bandstructure of these structures. Calculated band gaps of ordered compounds of the $Al_{1-x}Ga_xAs$ family yield results not too different from those for the alloy. However, band gaps for a same-cation family, compounds derived from $GaAs_{1-x}Sb_x$, are quite different than those for the alloy studied in VCA. Given the existence of a spread in experimental results for the band gap of this alloy for $x = 1/2$,[25-26] it is tempting to speculate, as others have,[27] that the band gap bowing of some metastable alloys is due to the incipient existence of alloy ordering.

ACKNOWLEDGMENTS — We are grateful to the Office of Naval Research, contracts no. N00014-85-J-0158 and N00014-84-K-0352 (J.S.), and also to AFOSR (AROSR-85-0331, J. S.) for their financial support. We acknowledge useful discussions with J. D. Dow, B. A. Bunker, and S. Y. Ren.

REFERENCES

1. T. S. Kuan, T. F. Kuech, W. I. Wang, and E. L. Wilkie, Phys. Rev. Lett. 54, 201 (1985).

2. H. R. Jen, M. J. Cherng, and G. B. Stringfellow, Appl. Phys. Lett. 48, 1603 (1986).

3. T. S. Kuan, W. I. Wang, and E. L. Wilkie, Appl. Phys. Lett. 51, 51 (1987).

4. M. A. Shahid, S. Mahajan, D. E. Laughlin, and H. M. Cos, Phys. Rev. Lett. 58, 2567 (1987).

5. H. Nakayama and H. Fujita, in *Gallium Arsenide and Related Compounds -- 1985*, edited by M. Fujimoto, IOP Conference Proceedings No. 79 (Institute of Physics, Bristol and London, 1986), p. 289.

6. For a review, see D. de Fontaine, in *Solid State Physics*, edited by H. Ehrenreich, F. Seitz, and D. Turnbull (Academic, New York, 1979), Vol. 34, p. 73.

7. M. J. Richards and J. W. Cahn, Acta Metall. 19, 1263 (1971); see also correction by S. M. Allen and J. W. Cahn, Scripta Metall. 7, 1261 (1973).

8. A. A. Mbaye, L. G. Ferreira, and A. Zunger, Phys. Rev. Lett. 58, 49 (1987). See also G. P. Srivastava, J. L. Martins, and A. Zunger, Phys. Rev. B31, 2561 (1985).

9. See A. A. Mbaye, D. M. Wood, and A. Zunger, Phys. Rev. B37, 3008 (1988) and references quoted therein.

10. W. A. Harrison, *Electronic Structure and the Properties of Solids* (W. H. Freeman, San Francisco, 1980).

11. P. Vogl, H. P. Hjalmarson, and J. D. Dow, J. Phys. Chem. Solids 44, 365 (1983).

12. J. C. Mikkelsen and J. B. Boyce, Phys. Rev. Lett. 49, 1412 (1982).

13. P. N. Keating, Phys. Rev. 145, 637 (1966); R. M. Martin, Phys. Rev. B6, 4546 (1972).

14. J. C. Slater and G. F. Koster, Phys. Rev. 94, 1498 (1954).

15. When experimental results are unavailable, we use theoretical estimates for the valence-band offset, as described in W. A. Harrison, Phys. Rev. B$\underline{24}$, 5835 (1981) and in E. A. Kraut, J. Vac. Sci. Technol. B$\underline{2}$, 488 (1984).

16. A. Kobayashi, O. F. Sankey, and J. D. Dow, Phys. Rev. B$\underline{25}$, 6367 (1982).

17. K. E. Newman, J. Shen, and D. Teng, unpublished.

18. A. Sher, M. van Schilfgaarde, A.-B. Chen, and W. Chen, Phys. Rev. B$\underline{36}$, 4279 (1987).

19. K. C. Hass and D. Vanderbilt, J. Vac. Sci. Technol. A$\underline{5}$, 3019 (1987).

20. U. Kaufman and J. Schneider, *Festkorperprobleme XIV*, edited by H. J. Queisser (Pergamon, 1974) p. 229.

21. M. B. Thomas, W. M. Coderre, J. C. Woolley, Phys. Status Solidi $\underline{2}$(a), K141 (1970).

22. A. Baldereschi, E. Hess, K. Maschke, H. Neumann, K. R. Schulte, and K. Unger, J. Phys. C$\underline{10}$, 4709 (1977).

23. Others have also estimated the (001) superlattice band gap. For example, D. M. Wood, S.-H. Wei, and A. Zunger, Phys. Rev. B$\underline{37}$, 1342 (1988) report a value of 2.14 eV, 0.10 eV below that for the alloy. The reported resonant Raman (0 0 1) superlattice band gap is 2.15 eV: M. Cardona, T. Suemoto, N. E. Christensen, T. Isu, and K. Ploog, Phys. Rev. B$\underline{36}$, 5906 (1987).

24. For $Al_{1-x}Ga_xAs$, we use the experimental offset 0.48 eV from D. J. Wolford, T. F. Kuech, J. A. Bradley, M. A. Grell, D. Ninno, and M. Jaros, J. Vac. Sci. Technol. B$\underline{4}$, 1043 (1986).

25. J. Klem, D. Huang, H. Morkoç, Y.-E. Ihm, and N. Otsuka, Appl. Phys. Lett. $\underline{50}$, 1364 (1987) and Y.-E. Ihm, N. Otsuka, J. Klem, and H. Morkoc, Appl. Phys. Lett. $\underline{51}$, 2013 (1987).

26. For a discussion of alloy band-gap variation in $Ga_{0.5}In_{0.5}P$, see A. Gomyo, T. Suzuki, K. Kobayashi, S. Kawata, I. Hino, and T. Ysasa, Appl. Phys. Lett. $\underline{50}$, 673 (1987).

27. For a discussion of alloy bowing in terms of chalcopyrite band gaps, see A. Zunger, International Journal of Quantum Chemistry: Quantum Chemistry Symposium $\underline{19}$, 629 (1986).

AB-INITIO MOLECULAR DYNAMICS STUDIES OF MICROCLUSTERS

Wanda Andreoni and Giorgio Pastore

IBM Research Division, Zurich Research Laboratory, 8803 Rüschlikon, Switzerland

Roberto Car and Michele Parrinello

International School for Advanced Studies, 34014 Trieste, Italy

Paolo Giannozzi

Institut de Physique Théorique, Université de Lausanne, 1015 Lausanne, Switzerland

The study of the structural and electronic properties of microclusters is a field of growing interest. Ab-initio molecular dynamics has provided a new and important tool for the theoretical approach to these questions. Here we present some very recent results on small semiconductor aggregates with special reference to calculations of equilibrium shapes and temperature effects. Results of simulations on alkali-metal microclusters are briefly mentioned.

I. INTRODUCTION

Synthesis as well as accurate characterization of size-selected clusters of a variety of elements are now possible owing to recent advances in technique.[1,2] However, the experimental investigation of both their structural and electronic properties is still in its infancy, and fundamental questions concerning equilibrium shapes and temperature effects remain open. On the theoretical side, on the other hand, it has recently been shown that both these questions can be explored on the same footing and from first principles.[3] This advance in the theoretical study of the physics of microclusters has been made possible by the use of a new method, developed by two of us, which unifies molecular dynamics and density-functional theory (DFT).[4]

In this paper, we shall give a brief summary of the salient results obtained so far, with special reference to calculations of equilibrium structures (Sect. II) and finite temperature properties (Sect. III). The former required application of the simulated annealing strategy, the latter consist of molecular dynamics simulations of the 'hot' system in equilibrium states. We have pursued studies on Si,[3] alkali-metals[5] and GaAs aggregates. Here, we shall concentrate more on semiconductor clusters and present some very recent results for Ge and GaAs aggregates.

Computational details have been reported in the papers mentioned.[3,5] Here, we recall that the electronic problem is treated within DFT in the local density approximation (LDA) for the exchange and correlation energies,[6] and with ab-initio pseudopotentials[7] to represent the core-valence interaction.

II. EQUILIBRIUM STRUCTURES

In this type of investigation, simulated annealing is used to explore the potential energy hypersurface of the clusters, and to search for the global minimum. Cooling rates are typically of the order of 10^{14} Ks^{-1}, which is very high compared to real time scales of formation. However, the interesting fact is that, in spite of this bottleneck of the computer experiments, the clusters calculated turn out to be ordered and to assume low-energy structures.[8] This may lead one to speculate that "real" clusters have the time to find their way to the ground state.

The first-principles, norm-conserving atomic pseudopotentials $V_{at}(r)$ we use have been obtained by a straightforward fitting procedure to the LDA valence energy levels and to the wavefunctions outside a "core radius" located between the outermost node and the outermost maximum. $V_{at}(r)$ is split into a local and a non-local part:

$$V_{at}(r) = V_{loc}(r) + V_{nl}(r) \tag{1a}$$

where

$$V_{loc}(r) = \frac{-Z_v}{r} \, \mathrm{erf}(r/R_c) + v_{l\,max}(r) \tag{1b}$$

and

$$V_{nl}(r) = \sum_{l=0}^{l\,max-1} P_l \left(v_l(r) - v_{l\,max}(r) \right) \tag{1c}$$

with

$$v_l(r) = \left(a_l + b_l r^2 \right) \, \exp(-(r/R_l)^2), \tag{1d}$$

where Z_v is the valence charge and the P_l's are projection operators. The analytical expression adopted is borrowed from Bachelet et al.,[9] and further simplified to include only one exponent for each v_l. The values of the parameters for Si, Ge, Ga and As are reported in Table 1.

In most calculations, we used only 's' non-locality, i.e. we assumed the $l = 1$ potential to represent the local part [$l\,max = 1$ in (1b)]. Here, we report on some calculations where, instead, the $l = 2$ pseudopotential was assumed to be the local component [$l\,max = 2$ in (1b)] and discuss the effect of 'p' non-locality. As expected, this is more relevant for Ge and GaAs and can easily be seen for the case of bulk solids. Table 2 contains the equilibrium structural parameters obtained with the potentials in Table 1 for the bulk solids of Si, Ge and GaAs, where total energies were calculated with 10 special k-points in the irreducible Brillouin zone[10] and an energy cutoff of 16 Ry.[11] The high accuracy is typical of the LDA-pseudopotential scheme for

Table 1. Pseudopotential Parameters [Eq. (1) — values are in a.u.]

	Si	Ge	Ga	As
R_c	1.09	1.10	1.12	1.077
R_0	0.81	0.88	0.91	0.866
a_0	10.152678	10.58149	10.339065	10.3094583
b_0	-5.239246	-6.013375	-5.713612	-6.1494476
R_1	0.92	0.97	1.00	0.938
a_1	2.817879	4.5355288	4.92587	4.5268280
b_1	-1.168488	-2.0156030	-2.225278	-2.1631682
R_2	1.00	1.09	1.14	1.00
a_2	-5.113543	-0.2536700	0.1034610	-0.1177821
b_2	1.327196	0.5228669	0.3745075	0.5116637

Table 2. Structural Parameters of Solids (experimental values are from Ref. 12c, Table 1, p. 3795)

	Si			Ge			GaAs		
	s-NL	sp-NL	exp	s-NL	sp-NL	exp	s-NL	sp-NL	exp
a_0 (a.u.)	10.37	10.24	10.26	10.47	10.64	10.68	10.35	10.53	10.66
B_0 (Kbar)	925	970	992	830	780	768	870	790	784
E (zb-rs, Ry)							0.075	0.079	—

these materials.[12] It is interesting to see that in GaAs, where interatomic distances are more sensitive to the neglect of 'p' non-locality, the energy difference between structures with different coordination is not altered in a significant way.

For the clusters, a supercell geometry with fcc periodicity was used in the calculations. Only the k = 0 point was taken as representative of the Brillouin zone. The edge of the fcc cell was chosen to be 40 a.u., and the energy cutoff in the plane-wave expansion of the wavefunctions was at 8 Ry. Effects of convergence have been discussed in Ref. 3 for Si_{10}.

Results for the smallest clusters can be compared with previous calculations using LDA[13] and also Hartree-Fock plus configuration interaction (CI) methods.[14] For example, Si_4 and Ge_4 turn out to be planar (rhombus), in agreement with previous predictions. Structural parameters, as well as the effect of 'p' non-locality, are reported in Table 3. For Ga_2As_2, we find that planar and three-dimensional isomers are very close in energy (see Table 3).

Our most interesting result is on Si_{10} (see Ref. 3), where we found a new and unforeseen structure, the tetracapped trigonal prism (TTP), as the one with lowest energy in our search for the global minimum of the potential energy hypersurface. Inclusion of the 'p' non-locality does not change the energy ordering of the significant local minima [TTP, tetracapped octahedron (TCO) and bicapped tetragonal antiprism (BTA)]. These results are reported in Table 4 and compared with those obtained for Ge_{10}. Interestingly, we find a different ordering for Ge with the BTA being almost degenerate with the TCO configuration (E(TCO) − E(BTA) < 0.01 eV/atom). We intend to investigate Ge_{10} further with more convergent calculations, with the aim of giving an interpretation of the fact that Si_{10} and Ge_{10} present different photoelectron spectra.[15]

Table 3. Tetramers [(a) from Ref. 13, (b) and (c) from Refs. 14a and b, respectively)

| | Si | | | | | Ge | | |
	s-NL	sp-NL	LDA (a)	CI (b)	CI (c)	s-NL	sp-NL	CI (c)
Bond length (a.u.)	4.42	4.40	4.35	4.35	4.72	4.40	4.49	5.02
Bond angle (degrees)	63	63	63	55	—	63	63	—

| | GaAs (2d) | | GaAs (3d) | |
	s-NL	sp-NL	s-NL	sp-NL
Bond length (a.u.)	4.25	4.42	4.25	4.41
Bond angle GaÂsGa (degrees)	61	63	84	86
E (eV/atom)	0	0	0.01	−0.02

Table 4. 10-Atom Clusters: Energy Ordering (E per atom in eV)

| | Si | | Ge | |
Structure	s-NL	sp-NL	s-NL	sp-NL
TTP	0	0	0	0
TCO	0.03	0.05	0.05	0.05
BTA	0.06	0.07	0.07	0.05

New results have also been obtained on the alkali-metals,[5] where in particular Na_{18} and Na_{20} turned out to assume structures with bulk-reminiscent features, e.g. the disposition of the atoms in layers and the presence of highly coordinated central atoms.

III. FINITE TEMPERATURE PROPERTIES

We have performed several simulations at finite temperatures. Calculations of the mean-square displacements of the atoms as a function of time indicate that, in the case of alkali-metals, the atoms diffuse in the cluster at temperatures as low as 200 K, while in both silicon and GaAs the atoms oscillate around definite positions up to high temperatures (of the order of the melting point of the solids).

Here, we report the results of a recent simulation of Ga_5As_5 at T = 2500 K. These calculations employ a cutoff of 7 Ry and an fcc cell of edge 54 a.u., and neglect 'p' non-locality. The results must be considered preliminary, although the qualitative information we obtain is certainly independent of the details of the computation. At this temperature, the cluster is in a diffusive regime. Fig. 1 shows the running coordination number for the three pairs of elements (Ga-As, Ga-Ga, As-As). The interesting feature to look at is the difference between the Ga-Ga and the As-As curves. This discrepancy is a sign of segregation and is also illustrated by the picture in Fig. 2 of the structure we obtain from quenching. Here the two average radii R (distances from the centre of the cluster) of the ionic distributions for the two species differ by ∼2 a.u., with the As ions clearly on the outside of the cluster. Fig. 3 shows instead the results of a computer experiment where, starting from the structure in Fig. 2 with the As and the Ga atoms exchanged, we allow the system to evolve freely in a microcanonical way. The quantity chosen as representative of the structural evolution of the cluster is again the average radius R. After 2000 steps of our integration process, the system reached an instantaneous temperature of 2000 K: This is the moment where the two radii cross, and the configuration with the As ions on the outside of the aggregate starts to form again.

The global information that we obtained from simulations of equilibrium states at different temperatures is that there is a specific tendency to segregate which increases with increasing temperature. We believe that this finding can be confronted

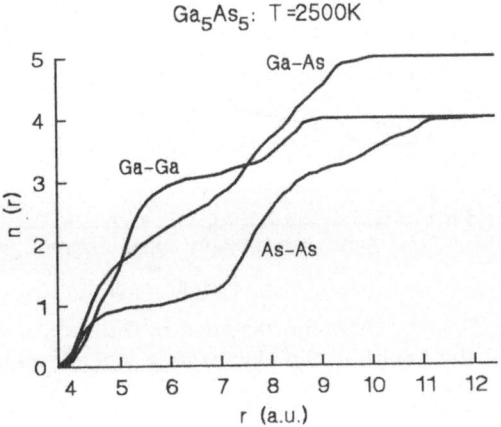

Fig. 1. Ga_5As_5: Average values of current coordination numbers $n_{ij}(r)$ [radial integral of two-body correlation functions $g_{ij}(r)$] at high temperature.

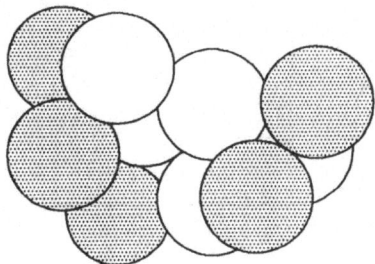

Fig. 2. Ga_5As_5: Structure corresponding to a local minimum obtained by quenching from high temperature.

with the experimental data on the Ga_NAs_N microclusters, indicating that they evaporate through loss of As.[16]

Segregation is also found in the binary mixtures of Na and K we have investigated. Here, this characteristic is present also in the low-energy structures. We believe it is due to the preference of the K ions to be localized in regions of low electron density. Again the tendency to segregate is reflected in experiments on neutral clusters, where K is found to evaporate first and the Na-rich clusters seem to be more stable.[17] The same tendency is found in LiNa mixtures, where again the aggregates rich in smaller-size ions are more abundant.[18]

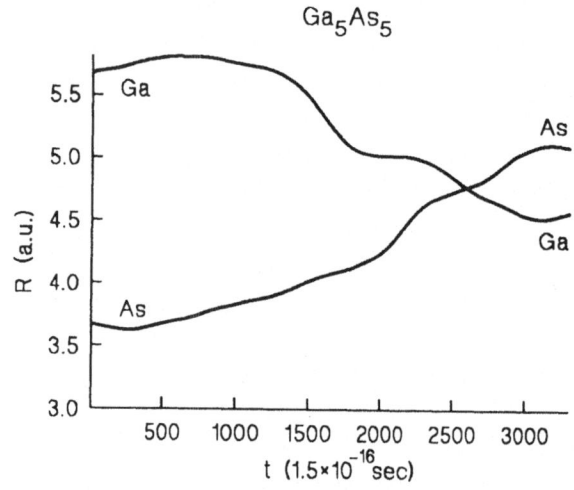

Fig. 3. Ga_5As_5: Average distance of Ga and As ions from the centre of the cluster as a function of time (see text).

CONCLUSIONS

Ab-initio molecular dynamics allows one to tackle new and interesting questions in the physics of microclusters. We have discussed two examples: calculation of equilibrium structures and high-temperature behaviour. Work is in progress to extend these studies to larger clusters and to investigate dissociation processes.

ACKNOWLEDGMENTS

It is a pleasure to thank Pietro Ballone for his invaluable contributions. One of us (PG) acknowledges support from the Swiss National Science Foundation.

REFERENCES

1. R. E. Smalley, Supersonic cluster beams: an alternative approach to surface science, in "Comparison of Ab Initio Quantum Chemistry with Experiment for Small Molecules: The State of the Art," R. J. Bartlett, ed., D. Reidel, Dordrecht, Holland (1985), p. 53, and references therein.

2. See also (a) "Microclusters," Springer Series in Material Sciences IV, S. Sugano, Y. Nishina, and S. Onishi, eds., Springer, Berlin (1987); (b) "Elemental and Molecular Clusters," Springer Series in Material Sciences VI, G. Benedek, T.P. Martin, and G. Pacchioni, eds., Springer, Berlin (1988).

3. P. Ballone, W. Andreoni, R. Car, and M. Parrinello, Equilibrium structures and finite temperature properties of silicon microclusters from ab-initio molecular dynamics calculations, *Phys. Rev. Lett.* 60:271 (1988).

4. R. Car and M. Parrinello, Unified approach for molecular dynamics and density-functional theory, *Phys. Rev. Lett.* 55:2471 (1985).

5. P. Ballone, W. Andreoni, R. Car, and M. Parrinello, Temperature and segregation effects in alkali metals microclusters (Preprint)

6. For the exchange-correlation local density functional, we have used the approximation given in J. P. Perdew and A. Zunger, Self-interaction correction to density-functional approximation for many-electron systems, *Phys. Rev. B* 23:5048 (1981).

7. D. R. Hamann, M. Schlüter, and C. Chiang, Norm-conserving pseudopotentials, *Phys. Rev. Lett.* 43:1494 (1979).

8. See also D. Hohl, R. O. Jones, R. Car, and M. Parrinello, The structure of selenium clusters − Se_3 to Se_8, *Chem. Phys. Lett.* 139:540 (1987).

9. G. B. Bachelet, D. R. Hamann, and M. Schlüter, Pseudopotentials that work: from H to Pu, *Phys. Rev. B* 26:4199 (1982).

10. D. J. Chadi and M. L. Cohen, Special points in the Brillouin zone, *Phys. Rev. B* 38:5747 (1973).

11. Both the lattice parameter a_0 and the bulk modulus B_0 were derived from fitting Murnaghan's empirical equation of state to the calculated curve E(V) at constant energy cutoff. This procedure is known to give reliable results, even when the number of plane waves in the expansion of the wavefunctions is not sufficient to give fully converged energy values.

12. (a) M. T. Yin and M. L. Cohen, Theory of static structural properties, crystal stability and phase transformations: Application to silicon and germanium, *Phys. Rev. B* 26:5668 (1982); (b) S. Froyen and M. L. Cohen, Structural properties of III-V zincblende semiconductors under pressure, *Phys. Rev. B*, 28:3258 (1983); (c) O. H. Nielsen and R. M. Martin, Stress in semiconductors: Ab-initio calculations on Si, Ge and GaAs, *Phys. Rev. B* 32:3792 (1985).

13. D. Tománek and M. A. Schlüter, Structure and bonding of small silicon clusters, *Phys. Rev. B* 36:1208 (1987).

14. (a) K. Raghavachari, Theoretical study of small silicon clusters: Equilibrium geometries and electronic structures of Si_n (n = 2-7, 10), *J. Chem. Phys.* 84:5672 (1986); (b) G. Pacchioni and J. Koutecký, Silicon and germanium clusters: A theoretical study of their electronic structures and properties, *J. Chem. Phys.* 84:3301 (1986).

15. O. Cheshnovsky, S. H. Yang, C. L. Pettiette, M. J. Craycraft, Y. Liu, and R. E. Smalley, Ultraviolet photoelectron spectroscopy of semiconductor clusters: silicon and germanium, *Chem. Phys. Lett.* 138:119 (1987).

16. S. C. O'Brien, Y. Liu, Q. Zhang, J. R. Heath, F. K. Tittel, R. F. Curl, and R. E. Smalley, Supersonic cluster beams of III-V semiconductors: Ga_xAs_y, *J. Chem. Phys.* 84:4074 (1986).

17. C. Bréchignac and P. H. Cahuzac, Evolution of photoionization spectra of metal clusters as a function of size, *Z. Phys. D* 3:121 (1987).

18. M. M. Kappes, M. Schär, and E. Schumacher, Are cluster abundances thermodynamic properties? Observation of lithium enrichment in Li_xNa_{n-x}, n ≤ 42, *J. Phys. Chem.* 91:658 (1987).

QUANTUM INTERFERENCE IN SEMICONDUCTOR DEVICES

M. Pepper

Cavendish Laboratory, Cambridge, England

ABSTRACT

A brief summary is presented on the observation of quantum interference in semiconductor devices and structures. It is shown how a magnetic field can be used to probe dimensionality and produce dimensionality transitions. The emphasis is on macroscopic effects and mesoscopic effects are not included.

INTRODUCTION

In discussing the transport properties of degenerate systems with $k_F l > 1$, where k_F and l are the Fermi wave vector and electron scattering mean free path respectively, it is normal to use the formula

$$\sigma = e^2 N(E_F) D \qquad (1)$$

$N(E_F)$ and D being the density of states at the Fermi energy and the electronic diffusivity respectively. Alternatively, we can write

$$\sigma = \frac{n e^2 \tau}{m} \qquad (2)$$

where n is the number of carriers, τ is the elastic scattering time and m is the effective mass. These formulae are based on the effective mass approximation and are relevant as the disorder increases to such an extent that a transition to Anderson localisation[1] occurs and transport is by hopping or excitation to extended states. Since the work of Abrahams et al[2], a number of papers have appeared showing that formulae (1) and (2) are subject to corrections arising from quantum interference[3,4,5].

The nature of the correction can be envisaged by considering the motion of an electron wave travelling between two points X and Y by motion along different paths where the distance between X and Y is greater than the elastic scattering length, ie. transport is diffusive. The amplitudes of the wavefunction at Y for waves travelling different routes are A_i, A_j ...., consequently the probability, w, of finding the electron at point B is given by

$$w = \left| \sum_i A_i \right|^2 = \sum_i |A_i|^2 + \sum_{i \neq j} A_i A_j^* \qquad (3)$$

The first term of equation 3 is the classical sum, ie particle behaviour, and the second term is the interference term. Under most circumstances, this second term sums to zero as the phases of the wavefunctions differ in a random manner. However, if two waves describe a self-intersecting loop, figure 1, but in different directions then on completion of the loop the two waves are again coherent and the second term of equation 3 has to be considered. We obtain

$$w = |A_i|^2 + |A_j|^2 + 2Re\ A_i\ A_j^* = 4\ |A|^2, \text{ where } A = A_i = A_j \qquad (4)$$

If, on the other hand, the waves were incoherent at the origin, then $w = 2\ |A|^2$; thus the probability of finding the electron at the origin has been increased by the interference and the probability of the electron diffusing elsewhere has been correspondingly diminshed.

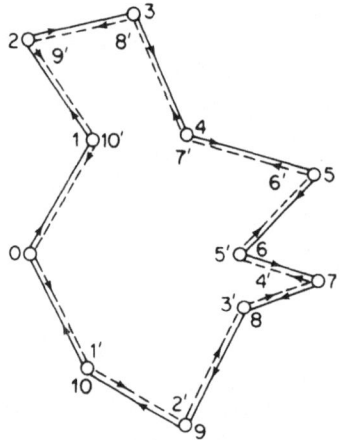

Figure 1. Diffusion loop illustrating the two different directions, 0,1 ..... 10, 0 and 0,1 .....10, 0 which give rise to coherent backscattering.

Evaluation of the conductivity correction due to the quantum interference has been considered by many authors[2-5]. Bergmann[6] has shown that a geometrical consideration of the problem produces the same results as diagrammatic techniques, and the magnitude of the interference correction depends on the number of states contributing to the process. If the electron enters the loop in the state $\underline{k}$, then interference is due to coherent backscattering into the state $-\underline{k}$. If the time of propagation of the electron before being scattered into $-\underline{k}$ is t then states within a radius $\underline{q}$ of $-\underline{k}$ contribute to the interference where $q = (Dt)^{-1/2}$. In 3D, the region of coherent backscattering in momentum space is a sphere of volume $(4/3)\pi\ (Dt)^{-3/2}$, in 2D a disc of area $\pi(Dt)^{-1}$ and in 1D a length $(Dt)^{-1/2}$. The resultant conductivity corrections come from integrating the evolution of the time dependence of the region of coherent backscattering. The interference is cut off by the phase relaxation time, in semiconductor systems at low temperatures this process is electron-electron scattering, electron-phonon scattering can be of importance in metals. In 3D, the conductivity correction to the Boltzmann value has the form

$$\sigma = \sigma_B - \frac{e^2}{2\pi^2\hbar}\left[\frac{1}{l} - \frac{1}{L_\phi}\right] \qquad (5)$$

where the phase relaxation length L_ϕ is given by $\sqrt{D\tau_{in}}$, l is the elastic scattering length.

As $L_{in} \to \infty$ as $T \to O$, this process is not significant at low temperatures. In 2D it is found that

$$\sigma = \frac{n e^2 \tau}{m} - \frac{e^2}{2\pi^2 \hbar} \ln (L_\emptyset/1) \tag{6}$$

In 1D, $\delta\sigma = \dfrac{n e^2 \tau}{m} - \dfrac{e^2 L_\emptyset}{\pi \hbar}$ \hfill (7)

The dimensionality of the system is determined by the value of L_{in}. Thus, if thickness and width are t and W (W > t), 3D behaviour is found for L_{in} < W,t, 2D when t < L_{in} < W and 1D when t, W < L_{in}. As corrections due to the electron-electron interaction can often be observed, the situation can arise in which the two sources of correction possess a different dimensionality. In general, as the temperature decreased, and L_{in}, increases so the dimensionality of the system becomes less.

It is to noted that in 2D and 1D all states are localised, and, in principle, at very low temperatures this weak localisation, (as quantum interference is often termed), passes into strong. However, as yet, this has not been found in experiment.

MAGNETO-RESISTANCE AND QUANTUM INTERFERENCE

Application of a magnetic field, B, results in the amplitudes of the wavefunctions traversing the loop acquiring an additional phase factor,

$$\exp\left[\frac{i \pi BA}{\emptyset_o}\right]$$

where $\emptyset_o$ is the flux quantum (h/e) and A is the area of the loop in the plane normal to the magnetic field. The phase difference, $\emptyset$, between waves traversing the loop in opposite directions becomes $2\pi\emptyset/\emptyset_0$. Reduction of the interference in this way gives rise to a negative magneto-resistance, accounting for the many observations in doped semiconductors which have been reported over a considerable time. The characteristic length scale is obtained from the condition that the phase difference is of order unity and is the magnetic length $L_c = (h/2eB)^{1/2}$.

The forms of the magneto-resistance are as follows:

3D

Kawabata[7] has found the following asymptotic limits

$$L_c \gg L_\emptyset, \delta\sigma(B) = \frac{e^2}{12\pi^2 \hbar} \left[\frac{eB}{\hbar}\right]^{1/2} \left[\frac{DeB\tau_\emptyset}{\hbar}\right]^{3/2} \tag{8}$$

$$L_c \ll L_\emptyset, \delta\sigma(B) = \frac{e^2}{2\pi^2 \hbar} \left[\frac{eB}{\hbar}\right]^{1/2} = 289.3 \sqrt{B} \ (\Omega m)^{-1/2} \ (Tesla)^{-1/2} \tag{9}$$

2D

Hikami et al[8] find that for a transverse field

$$\delta\sigma(B) = \frac{\alpha e^2}{2\pi^2 \hbar} f(4eBD \tau_\emptyset / \hbar) \tag{10}$$

α is a constant of order unity and

$$f(x) = L_n \, x + \psi \left[\frac{1}{2} + \frac{1}{x} \right]$$

where ψ is the Digamma function. $f(x)$ has asymptotic forms

$$x \ll 1, \ (L_c > L_\phi), \ f(x) = x^2/24$$

$$x \gg 1, \ (L_c < L_\phi), \ f(x) = Ln \, x$$

In the presence of a parallel field, Altshuler and Aronov[9] find that for $L_c \gg L_\phi$,

$$\delta\sigma(B) = \frac{e^2}{2\pi^2 \hbar} \ln \left[1 + t^2 \, L_\phi^2 \, B^2 \, e^2/3\hbar^2 \right] \qquad (11)$$

For $L_c \ll t$, the field effectively converts 2D behaviour to three and the strong field 3D formula should apply.

Dugaev and Khmelnitskii[10] have considered the effect of a paralled field on a 2D electron gas of thickness t under the condition $l \gg t$. They find that the Altshuler-Aronov equation holds (equation 11) but with t^3/l replacing t^2.

1D

As with 2D, it is necessary to distinguish between $l < W$ and $l > W$.

Transverse Field

For $l < W$ and $L_c > L_\phi$

$$\delta G \, (B) = \frac{e^2}{\pi\hbar} \left[\frac{1}{L_\phi^2} + \frac{W^2 4 e^2 B^2}{3\hbar^2} \right] \qquad (12)$$

For $L_c < W < L_\phi$, the system behaves as a 2D system with

$$\delta\sigma(B) \approx \frac{e^2}{2\pi^2\hbar} Ln \, B$$

For $l > W$ and $L_c > L_\phi$, it can be shown[11,12] that formula (12) holds, but with W^2 replaced by $\approx W^3/l$

Parallel Field

$$l < W, L_c, L_\phi$$

$$\delta G \, (B) = \frac{e^2}{2\pi^2\hbar} \frac{W^2 L_\phi^3}{L_c^4} \qquad (13)$$

For $l < W$, $L_c < W, t, < L_\phi$ - the field produces a dimensionality transition and the 3D law is obtained. In general, analysis of the low field magneto-resistance is a method of obtaining the phase relaxation rate, which in 3D and 2D is the inelastic scattering rate.

Spatial confinement of a system reduces the dephasing effects of a magnetic field. A field within a plane of a 2D system produces a significantly reduced negative magneto-resistance as the condition for the magnetic field to be the dominant influence is now $L_c \approx (L_\phi t)^{1/2}$ rather than $L_c \sim L_\phi$. At low temperatures, the area of the loop of coherent backscattering can be $\approx 10^{-3}$cms leading to a measurable magneto-resistance being found at fields ≈ 30 Gauss.

Theory suggests that the Hall effect is unaffected by the interference[4], however, the experimental situation is not clear[11].

Quite often, the negative magnetoresistance is obscured by the positive effect due to the enhancement of the electron-electron interaction[4]. This latter effect is principally a spin effect, whereas the suppression of the interference is orbital in origin. Consequently, the negative magneto-resistance is separated from the positive by the use of higher mobility samples[31].

3D SYSTEMS

Very often, 3D semiconductors show strong interaction effects which increase with decreasing temperature and the temperature dependence of the quantum interference is not easy to distinguish from the interaction correction. The negative magnetoresistance has been studied and phase relaxation rates extracted. In 3D, for $k_F l > 1$, the expected electron-electron scattering processes are Landau-Baber which involve scattering across the Fermi surface ($\tau^{-1} \propto T^2$) and a disorder induced process which is enhanced by the momentum indeterminacy. The total scattering rate is

$$\tau_\phi^{-1} = A\, T^2 + B\, T^{3/2} \text{ where A and B are constants}$$

However, near the metal-insulator transition, $k_F l \sim 1$, the predicted rate becomes

$$\tau^{-1} = CT$$

where C is a constant varying as $(k_F l)^2$. This law has been observed in both InP[13] and GaAs[30]. Kaveh et al[13] have shown that at very low temperatures, and near the transition, the shortest length scale is always determined by the interference.

The $T^{3/2}$ law has also been observed in 3D InGaAs as well as a temperature independent phase relaxation mechanism at low temperatures[14], the cause of which is not clear.

2D SYSTEMS

Early work on Silicon inversion layers established that quantum interference could be separated from interaction effects by the application of a magnetic field[31]. This led to a comprehensive investigation of the electron-electron scattering rate responsible for phase relaxation[15]. In 2D the disorder enhanced electron-electron scattering rate goes as T giving a total rate

$$\frac{1}{\tau_i} = A_1\, T + A_2\, T^2 \tag{14}$$

$$A_1 = \frac{2k}{\hbar k_F l}, \quad A_2 = \frac{a\, k^2}{\hbar E_F}$$

a is a constant of order unity. Experimentally a good fit was found to equation 14 with the parameter a taking the value 5, this is shown in figure 2. It has been suggested that additional terms are present in these electron scattering rates[16], in particular that

$$A_1 = \frac{k}{2\hbar k_F l}\, \ln (k_F l)$$

Experiments support the existence of this small correction term. The prediction of a small logarithmic term in A_2 cannot be excluded[17].

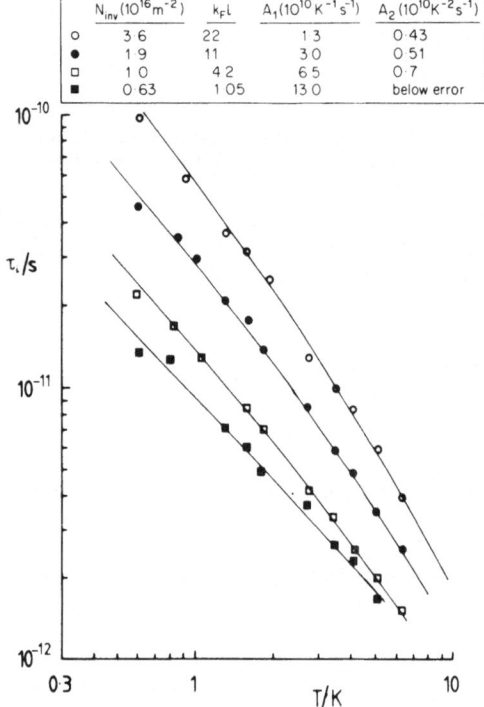

N_{inv} $(10^{16}$ m^{-2})	$k_F l$	A_1 $(10^{10}$ K^{-1} s^{-1})	A_2 $(10^{10}$ K^{-2} s^{-1})
○ 3·6	22	1·3	0·43
● 1·9	11	3·0	0·51
□ 1·0	4·2	6·5	0·7
■ 0·63	1·05	13·0	below error

Figure 2. The inelastic time, τ_i, versus temperature obtained from magnetoresistence measurements on inversion layers. Theoretical lines are drawn through the experimental points, the relevant parameters are shown; from reference 15

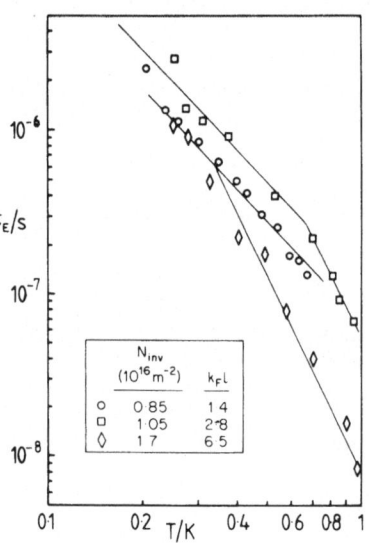

N_{inv} $(10^{16}$ m^{-2})	$k_F l$
○ 0·85	1·4
□ 1·05	2·8
◇ 1·7	6·5

Figure 3. The energy loss time, τ_E, is plotted as a function of temperature, the lines shown are $\tau_E \propto T^{-4}$ and $t_E \propto T^{-2}$; from reference 18.

The existence of a temperature dependent correction to the conductivity allows the electron gas to be used as its own thermometer. In this way the energy relaxation rate (arising from acoustic phonon emission) can be found, results suggest the emission rate varies as $AT^2 + BT^4$, where the T^4 term corresponds to the standard phonon emission process, but T^2 arises from an enhancement due to disorder. Reasonable agreement with theory has been found for this process[18], figure 3.

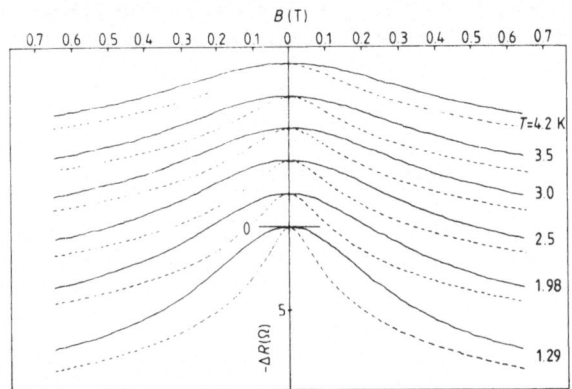

Figure 4. Parallel and perpendicular magnetoresistance (full and broken lines respectively) shown by a GaAs FET with thickness of conducting channel 440Å; from reference 20.

The GaAs Schottky gate FET has been very useful in investigating dimensionality changes and quantum interference. By controlling the gate voltage, the thickness of the n type conducting channel of doping $\sim 10^{17}$cm^{-3} can be altered with a consequent change in the dimensionality of the interference[19,20,21]. The thickness of the conducting channel can be calculated simply from the normal Schottky barrier expression. Figure 4 illustrates the parallel and transverse magneto-resistance as a function of temperature for a channel thickness t = 440Å, this value was sufficiently small to produce 2D interference. The low parallel field data is plotted in figure 5 to illustrate the B^2 dependence as proposed in equation (11). The data of figure 4 also allow extraction of the inelastic length from both directions of field. A comparison of the data is shown in figure 6 where it is clear that both orientations give virtually identical values of L_{in} as a function of temperature. L_{in} is proportional to T indicating that the disorder enhancement is dominant. Figure 7 shows the resistance as a function of B for a channel $\approx$ 1000Å thickness at 4.2K and 1.28K. At low fields, the B^2 behaviour is apparent, increasing B results in $B^{1/2}$ behaviour at 4.2K, but at 1.28K L_{in} exceeds the thickness and now lnB behaviour is found, characteristic of 2D.

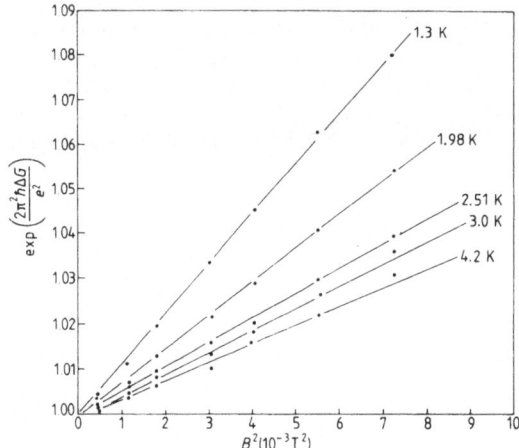

Figure 5. The low parallel field of figure 4 replotted to show the B^2 dependence; from reference 20.

The GaAs FET is a useful structure for investigating the magnetic field induced dimensionality transition. Figure 8 shows the resistance change in a transverse magnetic field as a function of temperature exhibited by a channel of 1300Å thickness, LnB behaviour is clear. The effects of a parallel field on this sample are shown in figure (9) where it is seen that for the same temperature range the magnetoresistance shows the 3D $B^{1/2}$ behaviour. The conductivity increase shown in this figure is 277.0 $(\Omega m)^{-1}$ per $Tesla^{1/2}$ which is within 5% of the theoretical value of 289.3 $(\Omega m)^{-1}$ per $Tesla^{1/2}$. Such agreement is within the error in the channel thickness and absolute value of conductivity and is one of the best examples of agreement with the predicted 3D behaviour.

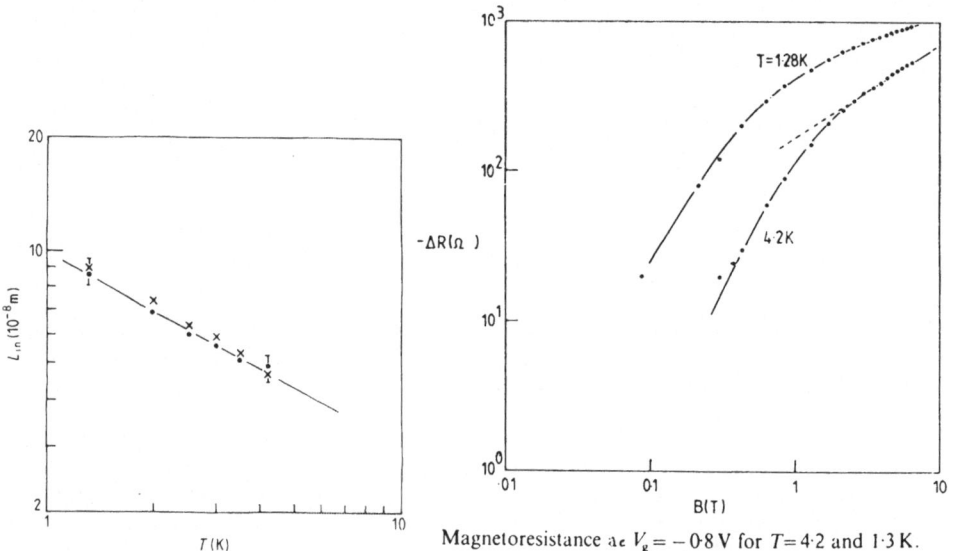

Magnetoresistance at $V_g = -0.8$ V for $T = 4.2$ and 1.3 K.

Figure 6. The inelastic length plotted as a function of temperature, obtained from the parallel X and transverse data of figure 4; from reference 20.

Figure 7. The magnetoresistance as a function of temperature for a channel thickness ≈ 1000Å, illustrating the B^2 behaviour at low fields; at high fields LnB behaviour is found at 1.28K and $B^{1/2}$ at 4.2K; from reference 19.

The use of this structure showed that there was an additional mechanism of parallel field magneto-resistance present between equation (11) and the 3D law, this is the low field data in figure 9. The conductivity correction was found to vary as $\ln T B^{1/2}$ ie the 2D temperature dependence, but a 3D magnetic field dependence, a case of hybrid dimensionality. It has been shown that the origin of this mechanism lies in the geometrical nature of the region of coherent backscattering[20,21].

The effect of spin-orbit coupling have been investigated, here a region of positive magnetoresistance is found at small values of field.

1D Behaviour

A number of experiments on 1D quantum interference and interaction effects have been reported using metals[4,29]. Semiconductor structures used have included narrow gate Si MOSFET's, MOSFET's in which the channel is electrostatically squeezed and split gate FET's fabricated from GaAs-AlGaAs heterojunctions[22]. Figure 10 illustrates the negative magneto-resistance seen on narrow electrostatically squeezed channel in a GaAs-AlGaAs heterojunction. The results are in agreement with the Altshuler-Aronov[4] expression, extraction of the electron-electron scattering rate shows that a 1D noise mechanism dominates in which electrons are scattered by electromagnetic fluctuations due to fluctuations in charge density[22,23]. This gives a $T^{-1/3}$ temperature dependence of L_ϕ, which has been observed, figure 11.

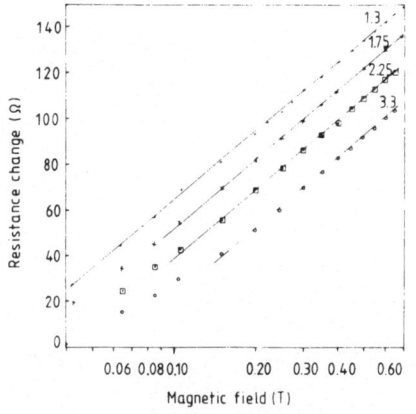

Figure 8. The resistance change of a 1300Å thickness channel in a perpendicular field plotted against LnB. The temperature in Kelvin is indicated; from reference 21.

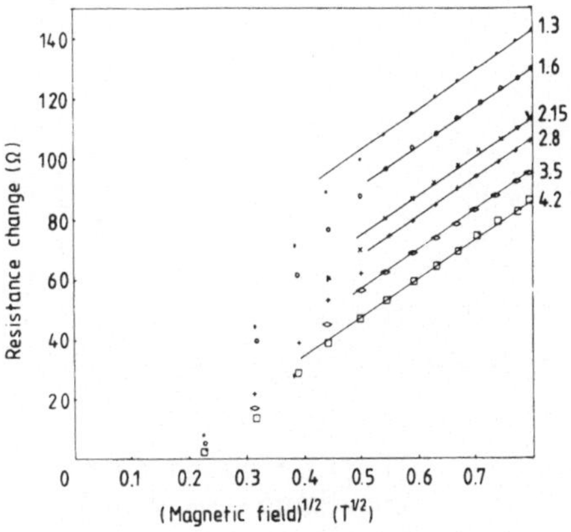

Figure 9. The resistance change of the same channel used for figure 8 plotted against $B^{1/2}$ in a parallel field. The temperature in Kelvin is indicated; from reference 21.

Electrostatic squeezing is of significance both for studies of quantum interference effects and transport. As the channel is squeezed, the one dimensional levels are forced through the Fermi energy and depopulation occurs. This is most marked in the presence of a magnetic field which increases the energy of the levels allowing observation of structure in the conductivity as the levels successively depopulate[24].

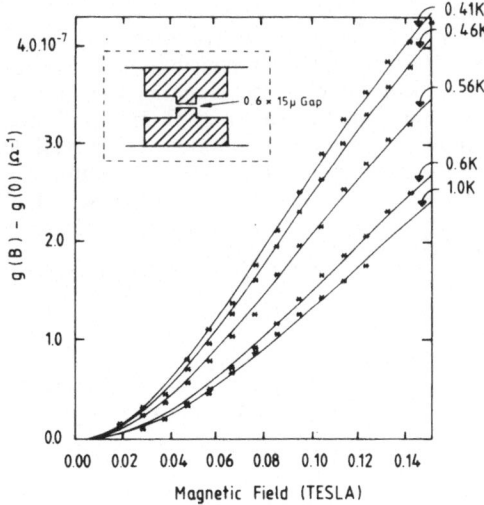

Figure 10. The measured value of conductance of a squeezed channel ($\approx$ 1000Å width) in a GaAs-AlGaAs heterojunction. The best fit to the theoretical 1D expression is indicated by the lines for each indicated temperature, an inset illustrates the split gate configuration; from reference 22.

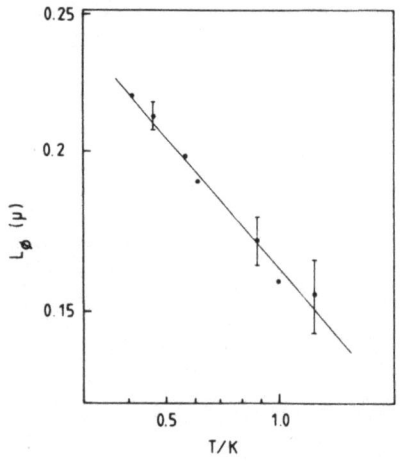

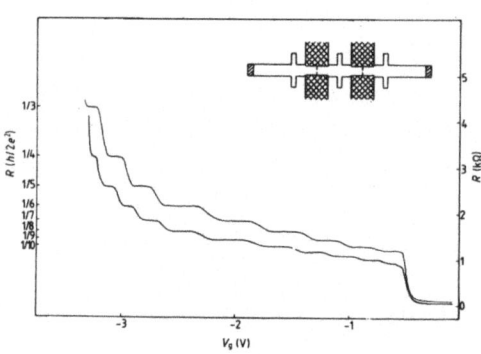

Figure 11. The phase relaxation time from figure 10 is plotted as a function of temperature.

Figure 12. The quantised resistance of a ballistic 1D channel is illustrated as a function of squeezing voltage, (increasing negative voltage corresponds to reducing the width which is always less than about 1000Å); from reference 25.

In the absence of disorder, transport through a 1D region is ballistic and it is possible to observe the quantised resistance of 1D subbands[25,26,27,28], $h/2e^2$. The total sample resistance due to i subbands becomes $h/2e^2i$ and as i is changed, the resistance varies rapidly between the quantised levels, figure 12. Further quantisation can be imposed by lifting the spin degeneracy with a parallel magnetic field, this avoids the depopulation effect produced by a transverse field[25].

CONCLUSION

A brief summary has been presented of quantum interference effects in semiconductor structures. It is shown how dimensionality effects can be observed and a magnetic field can be used to induce dimensionality transitions. Other aspects of interference not covered here include the Aharonov-Bohm oscillations in rings in which interference occurs between waves traversing a loop with a well defined area and the Universal Conductance Fluctuations. This latter effect occurs in small samples with few interference paths which do not average to zero. The resultant effect is fluctuations in conductivity with change of Fermi energy, magnetic field or width, this effect is discussed in the literature[28,29].

ACKNOWLEDGEMENTS

This brief review is based on experimental work by R A Davies, C McFadden, R Newbury, D J Newson, T J Thornton, M J Uren and D A Wharam. Theoretical work was in collaboration with K-F Berggren and M Kaveh. This work was supported by SERC.

REFERENCES

1. N F Mott and E A Davis, "Electronic Processes in Non-Crystalline Materials", 2nd Edition, Clarendon Press, Oxford, 1979.
2. E Abrahams, P W Anderson, D C Licciardello and T V Ramakrishnan, Phys. Rev. Lett. 42, 673, 1979.
3. M Kaveh and N F Mott, J. Phys. C 14, L177, 1981
4. B L Altshuler and A G Aronov in "Electron-electron interaction in disordered systems", edited by A L Efros and M Pollak, North-Holland, Amsterdam, 1985.
5. L P Gorkov, A I Larkin and D E Khmelnitskii, JEPT Lett. 30, 228, 1979.
6. G Bergmann, Phys. Rev. B 28, 2914, 1983.
7. A Kawabata, Solid State Comm. 38, 823, 1981, also J. Phys. Soc. Japan 53, 318, 1986
8. S Hikami, A I Larkin, Y Nagaoka, Progress Theor. Phys. 63, 707, 1980.
9. B L Altshuler and A G Aronov, Soviet Phys JEPT Letters, 33, 499, 1981.
10. L Dugaev and D E Khmelnitskii, Soviet Physics JETP, 59, 1038, 1984.
11. D J Newson, Ph.D Thesis Cambridge, 1986, also D J Newson, M Pepper, E Y Hall and G Hill, J. Phys. C 20, 4369, 1987
12. H van Houton, C W J Beenakker, B J van Wees and J E Mooij, Surface Science 196, 144, 1988.
13. D M Finlayson and G Mehaffey, J. Phys C 18, L953, 1985.
14. D J Newson, M Pepper, H Y Hall and J H March, J. Phys. C 18, L1041, 1985.
15. R A Davies and M Pepper, J. Phys C 16, L353, 1983.
16. E Abrahams in "Localisation and Metal-Insulator Transitions" ed H Fritzsche and D Adler, Plenum Press, 1985.
17. B L Altshuler, A G Aronov and D E Khmelnitskii, J. Phys. C 15, 7367, 1982.
18. M C Payne R A Davies, J C Inkson and M Pepper, J. Phys C 16, L291, 1983.
19. D J Newson, C McFadden and M Pepper, Phil. Mag. B 52, 437, 1985.
20. C McFadden, D J Newson, M Pepper and N J Mason, J. Phys. C 18, L383, 1985.
21. D J Newson and M Pepper, J. Phys. C 18, L1049, 1985.
22. T J Thornton, M Pepper, H Ahmed, D Andrews and G J Davies, Phys. Rev. Lett. 56, 1198, 1986.
23. B L Altshuler and A G Aronov, Solid State Comm. 46, 429, 1983.
24. K-F Berggren, T J Thornton, D J Newson and M Pepper, Phys. Rev. Lett. 57, 1769, 1986.
25. D Wharam, T J Thornton, M Pepper, H Ahmed, D Hasko, D C Peacock, D Ritchie, J Frost and G A C Jones, J. Phys. C 21, L209, 1988.

26. B J van Wees, H van Houten, C W J Benakker, J G Williamson, L P Kouwenhoven, D van der March and C T Foxon, Phys. Rev. Lett. 60, 848, 1986.
27. R Landauer, J. Phys. B 68, 217, 1987.
28. Y Imry in "Directions in Condensed Matter Physics" ed. G Grinstein and G Mazenko, World Scientific, 1986.
29. S Washburn and R A Webb, Adv. Phys. 35, 375, 1986.
30 M Kaveh, D J Newson, D Ben-Zimra and M Pepper, J. Phys. C 20, L19, 1987.
31. M J Uren, R A Davies and M Pepper, J. Phys C 13, L985, 1980, also M J Uren, R A Davies, M Kaveh and M Pepper, J. Phys C 14, 5737, 1981.

A REVIEW OF RECENT DEVELOPMENTS IN RESONANT TUNNELLING

L. Eaves, F. W. Sheard and G. A. Toombs

Department of Physics,

University of Nottingham, Nottingham NG7 2RD, U.K.

ABSTRACT

This chapter reviews some recent developments in our understanding of the physical properties of double barrier resonant tunnelling structures. Using the sequential theory of resonant tunnelling, the DC current-voltage characteristic of a double-barrier structure is calculated, taking into account the effect of space charge in the quantum well. A region of current bistability is found over a voltage range which is determined by the maximum space charge and the capacitance of the structure. Although there are good theoretical reasons to suggest that space charge build-up can cause intrinsic bistability, it is shown that the commonly observed bistability effect in the current-voltage characteristics of a typical resonant tunnelling device can be removed by connecting a suitable capacitance or resistance to the device. These measurements cast serious doubt on the recent observation and interpretation of a bistability in I(V) as an intrinsic space-charge effect. In the stabilised section of the I(V) curve, at voltages above the main resonant peak, the magnetoquantum oscillations observed with $\underline{B}||\underline{J}$ are used to investigate tunnelling assisted by LO phonon emission and by elastic scattering processes. Such processes have a deleterious effect on the peak/valley ratio which is commonly used as a figure of merit for resonant tunnelling devices. Resonant tunnelling devices with wide wells ($\sim$60 nm–1200 nm) exhibit a large number ($\geq$ 20) of regions of negative differential conductivity. The effect of a transverse magnetic field $\underline{J}\perp\underline{B}$ on the resonances in the I(V) characteristics of these wide well structures is investigated. At sufficiently high magnetic field, a transition is observed from tunnelling into magneto-electric box-quantised states to tunnelling into magnetically quantised cycloidal skipping states involving only the emitter barrier interface.

INTRODUCTION

The Resonant Tunnelling Structure has attracted considerable interest recently as a promising high speed device and as a system whose electrical properties are controlled by a fundamental quantum mechanical process. In this article, we report a series of investigations of the electrical properties of these devices, with emphasis on the application of high magnetic fields. The structures, based on n-(AlGa)As/GaAs were grown at Nottingham by molecular beam epitaxy using a Varian Gen II reactor. Devices exhibiting negative differential conductivity (NDC) with peak/valley ratio of up to 3.6/1 at room temperature and 25/1 at liquid helium temperatures have been fabricated. We describe three topics of current interest: (1) space charge build-up in the well at resonance and the question of whether or not these devices exhibit intrinsic bistability associated with this charge build-up; (2) the contribution of elastic and inelastic scattering processes to the tunnel current as revealed by analysis of magnetoquantum oscillations in the current-voltage characteristics for $\underline{J}||\underline{B}$; (3) the transition from tunnelling into electrically quantised states to tunnelling into magnetically quantised interface states in the presence of a large quantising magnetic field $\underline{J}\perp\underline{B}$.

CHARGE BUILD-UP AND INTRINSIC BISTABILITY; HAS INTRINSIC BISTABILITY BEEN OBSERVED?

Recently, Goldman et al[1], see also [2] reported the observation of bistability in the current-voltage characteristics I(V) of double barrier resonant tunnelling semiconductor heterostructures. This bistability occurs in the region of negative differential conductivity (NDC). It was interpreted as an intrinsic feature arising from the electrostatic effects due to the build-up of negative space-charge in the quantum well between the two barriers. Here we give a brief theoretical outline of how intrinsic bistability can arise.

There are two theoretical approaches to resonant tunnelling. In the first[3], the current is obtained from the global transmission

coefficient calculated for a coherent wave function throughout the
DBS. In the alternative sequential approach, proposed by Luryi[4],
the transmission is regarded as two successive transitions, from the
emitter into the bound state of the well and then from the well into
the collector contact. Weil and Vinter[5] have shown that both
approaches lead to the same results for the DC current in the absence
of charge build-up. Two of us (FWS and GAT) have recently used the
sequential model of resonant tunnelling since it provides a natural
framework for calculation of the stored charge[6].

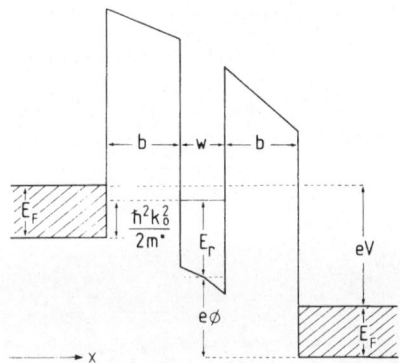

Fig. 1. Spatial variation of electron
potential energy through the
double-barrier structure,
showing bound state and Fermi
levels.

The model for a symmetric DBS under a bias voltage V is shown in
Fig. 1. The emitter and collector electrodes are heavily doped n^+
semiconductors with Fermi energy E_F. The quantum well (width w) is an
undoped layer of the same material and supports a quasibound state at
energy E_r from the bottom. For simplicity the screening charges in n^+
regions adjacent to the barriers (each of width b) and the stored
charge in the well are represented as sheets of infinitesimal
thickness. Electrostatically, the DBS is equivalent to two identical
capacitors C connected in series, the charge $-Q$ on the common central
plate corresponding to the charge per unit area in the well. The
potential drop across the right-hand capacitor is then

$$\varphi = \tfrac{1}{2}V + \tfrac{1}{2}(Q/C), \tag{1}$$

where $C = \varepsilon_r \varepsilon_o / (b + \tfrac{1}{2}w)$ and ε_r is the relative permittivity (assumed to be the same for barriers and well).

Taking the x axis perpendicular to the barrier interfaces, we label the electron states in the emitter by wave vectors $\underset{\sim}{k} = (k_x, \underset{\sim}{k}_{||})$ and in the collector by $\underset{\sim}{p} = (p_x, \underset{\smile}{p}_{||})$. In the well the longitudinal motion due to the bound state is the same for all states of transverse motion which are specified by two-dimensional wave vectors $\underset{\sim}{q} = (0, \underset{\sim}{q})$. For plane interfaces the transverse component of wave vector is conserved in tunnelling so that $\underset{\sim}{k}_{||} = \underset{\sim}{q} = \underset{\smile}{p}_{||}$. We describe the tunnelling in terms of transitions between the states $\underset{\sim}{q}$ of the quantum well and the states $\underset{\sim}{k}(\underset{\sim}{p})$ of the emitter (collector). Introducing the corresponding transition rates W_{kq} $(= W_{qk})$ and W_{pq} $(= W_{qp})$, the rate equations for the occupancies f_k and f_q of the emitter and well states are

$$\dot{f}_{\underset{\sim}{k}} = - \sum_{\underset{\sim}{q}} (f_{\underset{\sim}{k}} - f_{\underset{\sim}{q}}) W_{\underset{\sim}{k}\underset{\sim}{q}},$$

$$\dot{f}_{\underset{\sim}{q}} = - \sum_{\underset{\sim}{k}} (f_{\underset{\sim}{q}} - f_{\underset{\sim}{k}}) W_{\underset{\sim}{q}\underset{\sim}{k}} - \sum_{\underset{\sim}{p}} (f_{\underset{\sim}{q}} - f_{\underset{\sim}{p}}) W_{\underset{\sim}{q}\underset{\sim}{p}}.$$

The stored charge $Q = (e/A) \sum_{\underset{\sim}{q}} f_{\underset{\sim}{q}}$, where A is the interfacial area.

If the quasibound level is very narrow, conservation of energy and transverse momentum shows that an emitter electron can only tunnel into the well if its longitudinal kinetic energy $\hbar^2 k_x^2 / 2m^*$ is equal to

$$\hbar^2 k_o^2 / 2m^* = E_r + e\varphi - eV, \tag{2}$$

as shown in Fig. 1. We take $E_r > E_F$, so that at low temperatures when the Fermi surface is sharp, a finite bias is required before emitter states with $k_x^2 = k_o^2$ are occupied and resonant tunnelling into the well can occur. Tunnelling out of the well then takes place into empty states in the collector contact. Hence we may put $f_p = 0$. Furthermore, under DC conditions, $f_q = 0$. Solution of the rate equations then gives

$$Q = \frac{e}{A} \sum_{\underset{\sim}{k}, \underset{\sim}{q}} \frac{f_k W_{kq}}{\tau_e^{-1} + \tau_c^{-1}}, \tag{3}$$

152

and the current density is $J = Q/\tau_c$, where $\tau_e^{-1} = \Sigma_k W_{kq}$ and $\tau_c^{-1} = \Sigma_p W_{pq}$. By identifying τ_c^{-1} as the rate of decay of stored charge into unoccupied collector states, we may write

$$1/\tau_c = (v_r/2w)T_c,$$

where $v_r = (2E_r/m^*)^{\frac{1}{2}}$ and T_c is the transmission coefficient of the collector barrier. The expression (3) may be similarly evaluated using a zero-temperature Fermi-Dirac distribution for f_k. This involves counting the number of occupied emitter states which have the same value of $k_x^2 = k_o^2$ and gives (including a factor 2 for spin degeneracy)

$$Q = Q_m \left(\frac{k_F^2 - k_o^2}{k_F^2} \right) \Theta(k_o^2)\Theta(k_F^2 - k_o^2), \tag{4}$$

$$Q_m = \frac{ek_F^2}{2\pi} \frac{T_e}{T_e + T_c} \tag{5}$$

where T_e is the emitter-barrier transmission coefficient and $\Theta(s) = 1(s>0)$ or $0(s<0)$ is the unit step function.

But the tunnelling wave number k_o also depends on the stored charge Q through electrostatic feedback, and Eqs. (1) and (2) give

$$Q = \frac{2C}{e} \left(\frac{\hbar^2 k_o^2}{2m^*} - E_r + \frac{1}{2}\, eV \right). \tag{6}$$

The dependence of Q on bias voltage V is determined by the simultaneous solution of Eqs. (4) and (6) in which k_o^2, which gives the position of the bound level relative to the bottom of the emitter conduction band, appears as a parameter. We illustrate this for the simplified case in which T_e and T_c are treated as constants independent of bias voltage. In Fig. 2 we plot Q vs k_o^2 for Eq. (4), which gives a triangular-shaped curve with maximum charge Q_m, and for Eq. (6), which gives a straight line whose position is voltage dependent. The intersection points of the two plots give the required solution.

The onset of resonant tunnelling corresponds to $k_o^2 = k_F^2$, $Q = 0$ and occurs at a threshold voltage $V_{th} = 2(E_r - E_F)/e$. For a higher bias, such as V_1 in Fig. 2, there is one intersection point and the charge, given by explicit solution of Eqs. (4) and (6), is

$$Q = \frac{Q_m e}{2E_F} \frac{V - V_{th}}{1 + \alpha} , \qquad (7)$$

where $\alpha = Q_m e/2CE_F$ is a measure of the electrostatic feedback effect. When $V > 2E_r/e$, such as V_2 in Fig. 2, there is a range where three intersection points occur. The point with $0 < k_o^2 < k_F^2$ continues the above solution (7) while the point with $k_o^2 < 0$ gives $Q = 0$. The third solution is $k_o^2 = 0$ which indicates a pinning of the bound state at a level corresponding to the bottom of the emitter conduction band. This occurs for $2E_r < eV < 2(E_r + \alpha E_F)$. The pinning is sustained because a change in the applied voltage is electrostatically compensated by a change in the stored charge.

From these results we construct the explicit dependence of Q on V and hence, using $J = Q/\tau_c$, the DC current-voltage characteristic shown

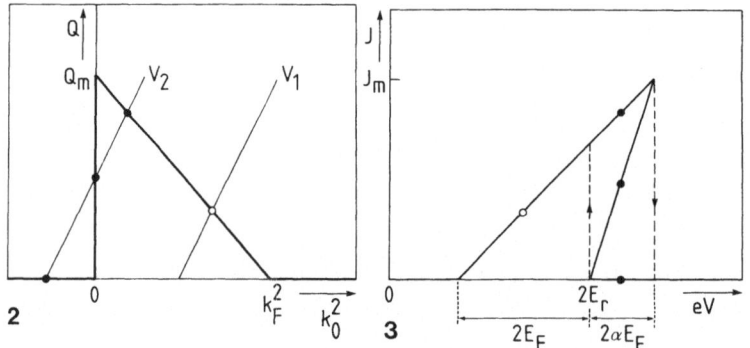

Fig. 2. (Left) Plots of Q vs k_o^2 for Eq. (4) (thick line) and Eq. (6) (thin lines) showing intersection points at two different voltages V_1 (open circle) and V_2 (solid circles).

Fig. 3. (Right) Current-voltage characteristics J vs eV, showing region of bistability. The open and solid circles refer to the intersection points of Fig. 2.

in Fig. 3. The dashed lines with arrows indicate the transitions between high-charge and zero-charge states which would occur on sweeping the voltage up and down through the region of resonant tunnelling. In the sequential model the time required for such transitions is clearly $\sim\tau_c$. Hence we find a region of intrinsic bistability in the voltage range for which pinning of the bound state occurs. However, this pinning is not directly observable. We note that the voltage width $2\alpha E_F$ of the bistable region is just Q_m/C.

The intrinsic bistability range is controlled by the feedback parameter $\alpha = Q_m e/2CE_F$. Using Eq. (5) for Q_m, we can express

$$\alpha = \frac{2b + w + 2\lambda_s}{a_0} \frac{T_e}{T_e + T_c} \tag{8}$$

where $a_0 = 4\pi\varepsilon_r\varepsilon_0\hbar^2/m^*e^2$ is the Bohr radius and we have used $C = \varepsilon_r\varepsilon_0/(b + \frac{1}{2}w + \lambda_s)$ to take approximate account of a finite screening length λ_s in the n^+ electrodes. For the GaAs/(AlGa)As DBS of Ref. 3 ($w = 5.6$ nm, $b = 8.5$ nm, $E_r \simeq 75$ meV, $E_F \simeq 20$ meV) we have estimated T_e and T_c from the WKB approximation and find $\alpha \simeq 0.8$. This gives an intrinsic bistability range ~32 mV in this structure. However, this range could be reduced by inhomogeneous broadening of the bound-state level due to structural imperfections.

Although evidence for space-charge buildup may be obtained from the magneto-oscillations in the current due to a magnetic field applied perpendicular to the barriers[7,8,9], observation of intrinsic bistability is a matter of controversy and has been challenged by Sollner[2] who claimed that the observed bistability in I(V) was a common characteristic of devices exhibiting NDC and was due to current oscillations in the device and in the external circuit. In this case, bistability arises because of the difference between the turn-on and turn-off points for oscillatory behaviour when sweeping the applied DC voltage up or down through the NDC region. The range of applied DC voltage over which the bistability occurs then depends on the external circuit parameters as well as on the intrinsic characteristics of the device itself.

Recently, we investigated the magnetic field dependence of the current-voltage characteristics[8,9] of resonant tunnelling double barrier structures based on n-(AlGa)As/GaAs. Our measurements confirmed the build-up of charge in the well at the resonant tunnelling voltages. In the region of NDC, the devices exhibited a current bistability similar to that originally reported by Goldman et al[1]. We also reported[9] that the form of the I(V) curves in the bistable region can be simulated quite well by numerical calculation of the current oscillations in an equivalent circuit consisting of a device exhibiting NDC in parallel with a capacitor and connected by resistive and inductive leads to a steady voltage source. This is consistent with Sollner's point of view. Sollner also pointed out that if the magnitude of the negative differential resistance $|dV/dI|$ is sufficiently large, the current oscillations should be suppressed by a capacitor placed in parallel with the device. In order to study this, we have made further measurements on double-barrier devices which have a higher impedance than those used by Goldman et al[1].

The measured device was a 100 μm-diameter mesa consisting of the following layers (structure A), in order of growth from the n^+GaAs substrate: (i) 1 μm of GaAs, $n = 2 \times 10^{17}$ cm^{-3} buffer layer; (ii) 8.8 nm of undoped (AlGa)As, [Al] = 0.33; (iii) 5.6 nm of undoped GaAs; (iv) 8.8 nm of undoped (AlGa)As, [Al] = 0.33; (v) 1 μm of 2×10^{17} cm^{-3} GaAs, top contact.

The I(V) characteristics shown in Fig. 4 were measured at 77 K with the mesa mounted on a standard transistor header. The upper curve of Fig. 4 was obtained with no external capacitance across the device and a region of hysteresis is clearly observable. An oscilloscope connected across the device indicated the presence of oscillations in the NDC region. The oscillations were not suppressed and the bistability persisted if a capacitor was connected across the leads emerging from the cryostat. When, however, a 0.28 μF chip capacitor was mounted directly on the transistor header so as to minimise the length of the inductive connecting leads, then the smooth I(V) curve shown in the lower curve of Fig. 4 was obtained. In this case, there was no bistability and no oscillations were observed throughout the NDC region.

An alternative method of suppressing the oscillations in the region of NDC is to connect, in parallel with the device, a small chip resistor r (25 Ω) of sufficiently low impedance that $r < |dV/dI|$ throughout the region of NDC. It is necessary to mount the resistor in

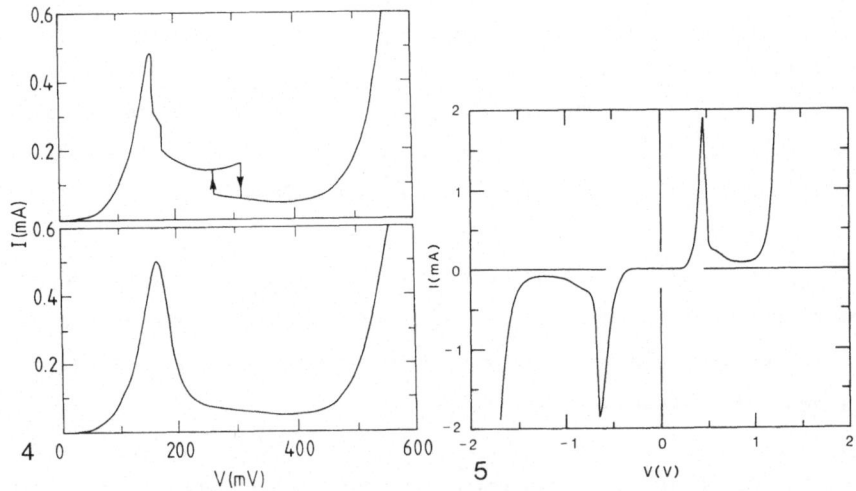

Fig. 4. (Left) The current-voltage characteristics at 77 K of a 100
         μm mesa fabricated from the double barrier structure A (5.6
         nm well) described in the text. The upper curve shows the
         d.c. current with no external capacitor across the mesa. In
         the voltage range corresponding to current bistability, the
         device is oscillating at a high frequency. In the lower
         curve, the oscillations and bistability are completely
         suppressed by placing a 0.28 μF capacitor directly across the
         mesa.

Fig. 5. (Right) The current-voltage characteristics at 77 K of a 5 μm
         diameter mesa fabricated from the double barrier structure B
         (5 nm well) described in the text. The oscillation and
         circuit bistability are suppressed by connecting a small
         resistor (25 Ω) in parallel with the device.

close proximity with the device with very short connecting leads. A
stable I(V) plot of the device and resistor in parallel can then be
obtained by applying a ramped voltage V. Even in the region of NDC,
the circuit does not break into oscillation. The stable I(V)
characteristics of the tunnelling device alone are calculated by
subtracting, from the total current, the current V/r flowing through
the parallel resistor. Typical results obtained at 77 K with this
simple procedure are shown in Figure 5 for a 5 μm diameter mesa device
(structure B) comprising the following layers, in order of growth from
the n$^+$GaAs substrate:

(i) 1.0 μm of n = 2 x 10^{18} cm^{-3} GaAs, (ii) 50 nm of n = 2 x 10^{16} cm^{-3} GaAs, (iii) 2.5 nm of undoped GaAs, (iv) 5.6 nm of undoped (AlGa)As, [Al] = 0.4, (v) 5 nm of undoped GaAs, (vi) 5.6 nm of undoped (AlGa)As, [Al] = 0.4, (vii) 2.5 nm of undoped GaAs, (viii) 50 nm of n = 2 x 10^{16} cm^{-3} GaAs and (ix) 0.5 μm of n = 2 x 10^{18} cm^{-3} GaAs.

The purpose of the undoped GaAs spacer layers on either side of the barriers was to prevent dopant diffusion into the barrier region and so to reduce ionised impurity scattering which, as discussed later, degrades the peak/valley ratio.

We conclude that our measurements support the objections raised by Sollner concerning the reported observation by Goldman et al of intrinsic bistability in double barrier resonant tunnelling devices. However, the magneto-oscillation studies[7,8,9] show clearly the existence of space-charge build-up in the quantum well. Thus intrinsic bistability driven by electrostatic feedback remains a theoretical possibility. Theoretical modelling[6] shows that the intrinsic bistability is removed by sufficient inhomogeneous broadening of the bound state level. A carefully designed structure would therefore be required to observe the intrinsic bistability effect.

ELASTIC AND INELASTIC SCATTERING AND THE VALLEY CURRENT

Either of the above methods of stabilising the devices against oscillations and circuit bistability allow us to investigate in detail the dynamics of tunnelling electrons in the voltage region beyond the main resonant peak in the current. As can be seen from Figures 4 and 5, there is a weak shoulder in the I(V) characteristics of both structures A and B in the voltage region beyond the main peak where the current is a minimum. Goldman et al[10] have attributed this feature to tunnelling of electrons, assisted by LO phonon emission, from the emitter contact into the quantum well. The emission of the LO phonon breaks the law governing conservation of k-vector component, $k_\perp$, perpendicular to the direction of the tunnel current. $k_\perp$ is generally assumed to be conserved for the majority of tunnelling electrons contributing to the main resonant peak of the I(V) curve. However inter-subband scattering due to LO-phonon emission has been identified in resonant tunnelling structures[11] from an analysis of

magnetoquantum oscillations. The LO phonon-related peak and the
contribution to the valley current due to elastic scattering can be
revealed more clearly by applying a quantising magnetic field B
parallel to the direction of the tunnel current. In the absence of a
magnetic field, $k_\perp$ is conserved if an electron does not scatter during
its transit of the barriers. In the presence of a magnetic field, the
energy due to motion in the plane of the barriers is quantised such
that the total electron energy in the accumulation layer of the
emitter and in the quantum well are given by

$$\text{Accumulation Layer } \varepsilon = \varepsilon_o + (n + \tfrac{1}{2})\hbar\omega_c$$
$$\text{Quantum Well } \varepsilon = \varepsilon_1 + (n' + \tfrac{1}{2})\hbar\omega_c$$

where $n(n')$ is the Landau level number, $\omega_c = eB/m^*$, and ε_o and ε_1 are
lowest bound state energies in the emitter and well respectively. The
conservation of k for B = 0 corresponds to the requirement that
$p = n' - n = 0$ in the presence of a quantising magnetic field, so that
in both cases resonant tunnelling occurs at an applied voltage for
which $\varepsilon_o = \varepsilon_1$.

Figure 6 shows the effect of a magnetic field $(\underline{B}\|\underline{J})$ on the I(V)
characteristics of a 100 μm diameter double barrier device with a well
width of 117 Å (Structure C). At zero magnetic field the oscillations
in the NDC region were suppressed by connecting a 25 Ω chip resistor
across the device. This structure comprises the following layers in
order of growth from the n^+ substrate: (i) 2 μm, 2×10^{18} cm^{-3}, n^+GaAs
buffer layer; (ii) 50 nm, 2×10^{16} cm^{-3} GaAs; (iii) 2.5 nm undoped
GaAs; (iv) 5.6 nm undoped (AlGa)As, [Al] = 0.4; (v) 11.7 nm undoped
GaAs; (vi) 5.6 nm undoped (AlGa)As, [Al] = 0.4; (vii) 2.5 nm undoped
GaAs; (viii) 50 nm, 2×10^{16} cm^{-3} GaAs; (ix) 0.5 μm, 2×10^{18} cm^{-3}
n^+GaAs top contact layer.

Figure 6 shows only the low voltage section of the I(V)
characteristics. Further peaks in the current are observed at 420 and
890 mV, corresponding to tunnelling into the second and third bound
states of the well. The following features of Figure 6 are
particularly noteworthy. Firstly, the magnetic field increases the
amplitude of the main LO phonon related peak. Secondly, a weak
secondary peak (E_1) emerges from the main resonant peak in the I(V)
curve as the magnetic field is increased. Similar subsidiary peaks

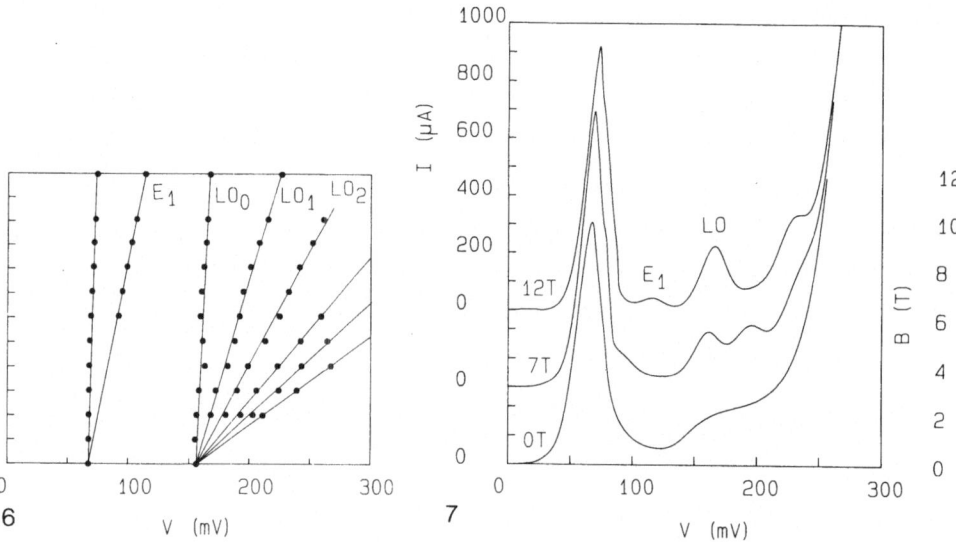

Fig. 6. (Left) The current-voltage characteristics at 4 K and at
various magnetic fields $\underline{B}||\underline{J}$ of a 100 μm diameter mesa
fabricated from structure C (11.7 nm well) described in the
text. The curves show only the region of the first
resonance. The oscillations and current bistability are
suppressed by a small resistor (25 Ω) in parallel with the
device. Positive V corresponds to the substrate biassed
positive.

Fig. 7. (Right) Fan chart showing the magnetic field dependence
($\underline{B}||\underline{J}$) of the peaks in I(V) shown in Figure 3. The elastic
and inelastic (LO phonon) scattering processes giving rise to
the oscillations are discussed in the text.

also evolve from the LO phonon feature with increasing B. Thirdly, the
magnetic field increases the peak/valley ratio. The fan chart in
Figure 7 shows the evolution of the magnetoquantum peaks in the I(V)
characteristics. Transitions for which $k_\perp$ (or n) is not conserved are
governed only by energy conservation. Therefore, for elastic and LO
phonon emission we have

$$\varepsilon_0 = \varepsilon_1 + \frac{p\hbar eB}{m^*} \, (+\hbar\omega_L) \ .$$

To a good approximation $\varepsilon_0 - \varepsilon_1$ increases linearly with total
applied voltage from its initial (negative) value at zero bias. Hence
the chart shows two clearly identifiable groups of lines. In the limit
$B \to 0$, peak E_1 extrapolates back to the main resonant peak at 70 mV.

160

The peaks LO$_p$ extrapolate back to the LO phonon satellite at 160 meV. The peak marked E$_1$ corresponds to non-resonant tunnelling involving an elastic (or quasi-elastic) scattering-induced transition from the n = 0 Landau level in the emitter contact to the n = 1 Landau level in the well. Such a transition could be induced by ionised impurities or interface roughness (elastic) or by acoustic phonon emission (quasi-elastic). The main resonant peak, which shifts only slightly with B, corresponds to tunnelling processes for which n is conserved. The peaks marked LO$_p$ correspond to transitions of electrons from the n = 0 Landau level in the emitter to the pth Landau level in the well, together with the emission of an LO phonon.

It is interesting to note that the effect of the quantising magnetic field is to suppress the (quasi-) elastic scattering induced transitions at certain voltages (below and above E$_1$) and enhance it at other voltages (at E$_1$). In zero magnetic field such processes, which change $k_\perp$, are energetically allowed for all voltages beyond the main resonant peak in the tunnel current. At these voltages the lowest energy bound state in the well is below the energy of the electrons in the emitter. However, a large magnetic field quantises the electronic motion in the plane of the barriers, thus giving rise to sharp peaks in the densities of states. At certain voltages beyond the main resonant peak, and in the presence of a quantising magnetic field, energy conservation inhibits (quasi-)elastic scattering into the well. This explains the enhancement of the peak/valley ratio with increasing B which is evident in Figure 6.

The difference in voltage between the main resonant peak and the LO phonon satellite and between the various peaks LO$_p$ in the presence of a magnetic field, are considerably larger than the LO phonon energy, $\hbar\omega_L/e$ (= 36 mV) and the cyclotron energy, $\hbar\omega_c$, respectively. This is fully consistent with the expected distribution of potential across the device: in the voltage range between ~100 and 300 mV only about 30% of the total applied voltage is dropped across the accumulation layer of the emitter, the emitter barrier and the half-well width. The remainder is dropped across the other half of the well, the collector barrier and the depletion layer of the collector contact.

We conclude this section by noting that the magnetoquantum peaks in the region of the valley current that were recently reported by Goldman et al[7] and tentatively attributed to elastic ionised impurity scattering transitions with non-conservation of $k_\perp$, are more likely to be due to LO-phonon emission (i.e. the LO_n series of magnetoquantum peaks reported here).

ELECTRICALLY AND MAGNETICALLY QUANTISED STATES IN WIDE WELLS

In this section we investigate resonant tunnelling in a double barrier structure whose well width, 60 nm, is considerably larger than those described in the previous sections. This structure (D) comprises the following layers in order of growth from the n^+GaAs substrate:

(i) 2 μm, 2 x 10^{18} cm^{-3} n^+GaAs buffer layer; (ii) 50 nm, 2 x 10^{16} cm^{-3} GaAs; (iii) 2.5 nm undoped GaAs; (iv) 5.6 nm undoped (AlGa)As, [Al] = 0.4; (v) 60 nm undoped GaAs; (vi) 5.6 nm undoped (AlGa)As, [Al] = 0.4; (vii) 2.5 nm undoped GaAs; (viii) 50 nm, 2 x 10^{16} cm^{-3} GaAs; (ix) 0.5 μm, 2 x 10^{18} cm^{-3} n^+GaAs top contact.

Figure 8 shows the I(V) characteristics of this structure obtained at 4 K. Note that the current is plotted logarithmically over 5 orders of magnitude. Sixteen distinct regions of negative differential conductivity are observed up to 1.0 V and several additional features are clearly observable at higher voltages. These resonances are associated with the large number of closely-spaced bound states that occur in a well of this width. The higher voltage peaks are probably associated with "over the barrier" resonances. A similar structure but with a well width of 120 nm shows up to 22 regions of NDC and 30 other resonances in I(V). The existence of well-defined "standing wave" resonances implies that some electrons traverse a distance of 0.24 μm (twice the well width) without appreciable scattering even when they are injected into the well with kinetic energies of several hundred meV.

Structures of this kind are ideal for studying the effect of a confining potential (quantum well) on the electron eigenstates in the crossed electric and magnetic field ($\underline{E} \perp \underline{B}$) configuration. The electric field in the well is perpendicular to the barriers so the magnetic field must be applied parallel to the barrier interfaces. The effect of a magnetic field on the current-voltage characteristics of this configuration ($\underline{B} \perp \underline{J}$) is shown in Figure 9. In order to enhance the

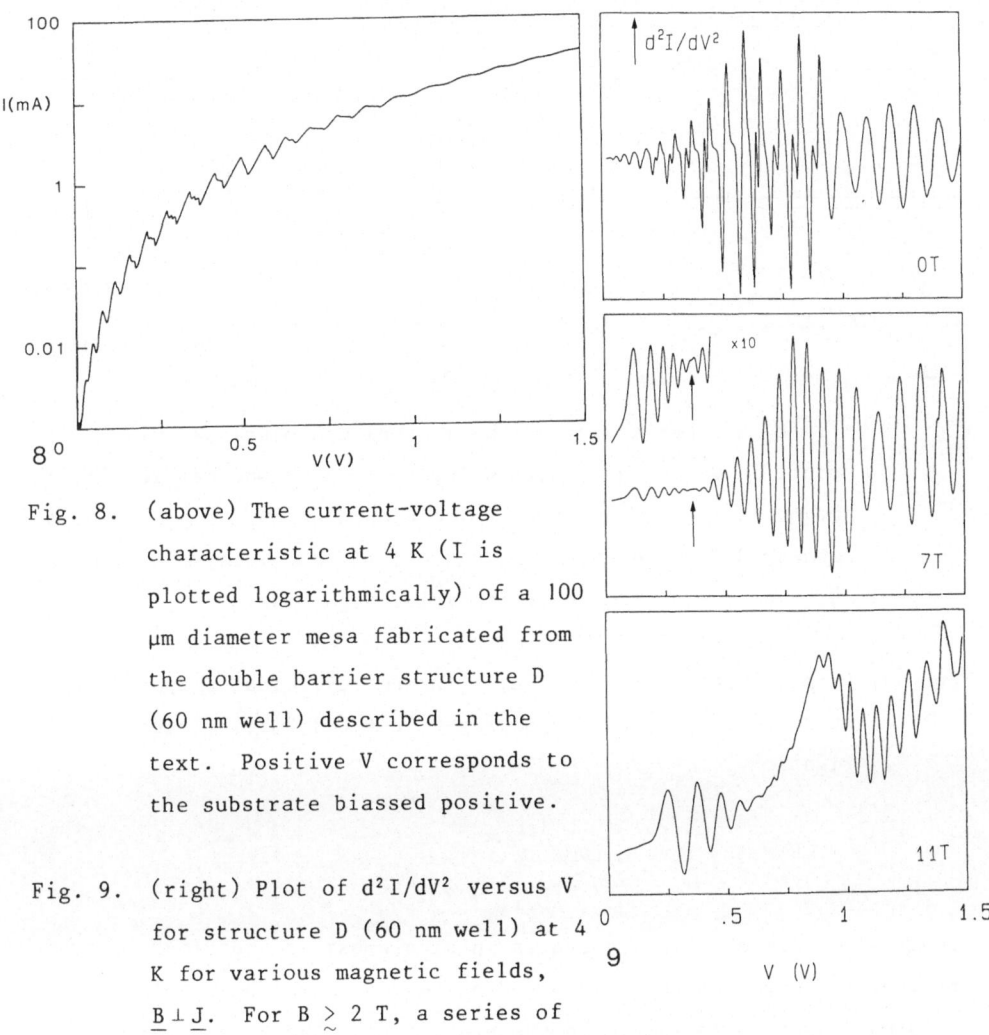

Fig. 8. (above) The current-voltage
characteristic at 4 K (I is
plotted logarithmically) of a 100
μm diameter mesa fabricated from
the double barrier structure D
(60 nm well) described in the
text. Positive V corresponds to
the substrate biassed positive.

Fig. 9. (right) Plot of d^2I/dV^2 versus V
for structure D (60 nm well) at 4
K for various magnetic fields,
$\underline{B} \perp \underline{J}$. For $B \gtrsim 2$ T, a series of
oscillations due to tunnelling into magnetically quantised
interface states can be observed at low voltages. For the
trace at 7 T, the critical voltage V_c for the disappearance
of the skipping orbits is indicated by an arrow.

resonant structure, the second derivative d^2I/dV^2 versus V is plotted
for various values of magnetic field. In high fields it can be seen
that the resonances form two groups, which are shown separated by an
arrow in the plot for B = 7 T in Figure 9.

At low magnetic field the electron orbits impinge on the barrier
interfaces on both sides of the well. But at sufficiently high
fields, skipping states develop in which the electron is repeatedly
reflected from only one interface. However, we note that the

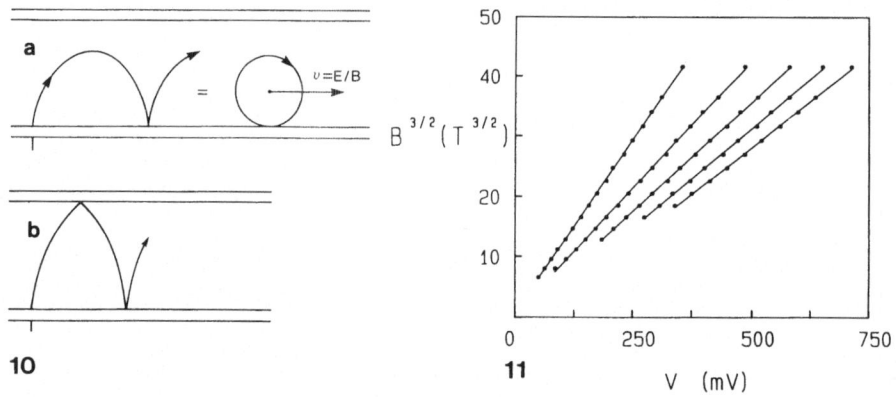

Fig. 10. (left) Schematic diagram showing (a) cycloidal skipping
        orbit in which the electron interacts with one barrier (high
        magnetic field, low bias) (b) orbit in which the electron is
        weakly deflected by the magnetic field and interacts with
        both barriers (low magnetic field, high bias).

Fig. 11. (right) Plot of $B^{3/2}$ versus V for the cycloidal skipping
        state resonances corresponding to n = 0,1,2,3,4 (left to
        right).

previously observed skipping states in single-barrier heterostructures[12] were in a heavily-doped n^+ region in which the electric
field was essentially zero. In the present case there is a large
electric field in the quantum well. The skipping motion is then
cycloidal with a uniform drift velocity v_d = E/B parallel to the
interface, superposed on the circular motion of angular velocity ω_c =
eB/m* in the plane perpendicular to $\underline{B}$ as shown schematically in Figure
10.

    We assume that the electrons tunnelling through the emitter
barrier emerge into the well with a kinetic energy which is small
compared with the kinetic energy gained during the subsequent
cycloidal motion. The circular component of the cycloidal motion then
has a radius R = v_d/ω_c and the orbit centre is located the same
distance R from the emitter-barrier interface. The width of such an
orbit along the direction of the electric field is 2R = $2v_d/\omega_c$ =
2m*E/eB². The critical case, when the cycloidal skipping orbit
adjacent to the emitter-barrier interface just touches the collector

164

barrier, is given by 2R = w, where w is the well width. We may put
E = V/d, where d = w + 2b + λ_L + λ_R and λ_L, λ_R are screening layer
widths in the emitter and collector electrodes. Thus tunnelling into
cycloidal skipping states only occurs for V < V_c, where the critical
voltage V_c = eB²wd/2m*. This B² dependence of V_c is observed
experimentally. For B = 7 T and taking λ_L = λ_R ≃ 10 nm (d = 91 nm,
m* = 0.067m_e) we find V_c ≃ 350 mV. This value is shown by an arrow in
Figure 9 and clearly lies at the division between the group of
resonances due to single-interface skipping states (V < V_c) and the
group due to states in which the electron motion is restricted by
reflection at the interfaces at both sides of the well (V > V_c).

We have analysed the positions of the resonances (minima in
d²I/dV² versus V) for the cycloidal skipping states over a wide range
of magnetic fields and voltages. When plotted as $B^{3/2}$ versus V, we
obtain the linear dependences shown in the fan chart of Figure 11.
These results may be understood in terms of the quantisation of the
cycloidal orbits discussed above. When quantised, the energy of
circular motion ½m*ω_c^2R² takes allowed values (n + φ)ℏω_c, where the
phase factor φ = 3/4 in the WKB approximation. Putting R = v_d/ω_c
gives ½m*v_d^2 = (n + φ)ℏω_c, which can be written

$$\frac{m^{*2}E^2}{2e\hbar B^3} = n + \varphi. \tag{9}$$

This shows that for a given applied voltage (fixed E) the resonances
are periodic in 1/B³. This is in contrast to the results obtained for
skipping states in n⁺ material in single-barrier structures[9] where
the resonances were periodic in 1/B. Eq. (9) is consistent with the
observed linear variation of $B^{3/2}$ with V given in the fan chart of
Figure 11. Also, using the WKB phase factor, the slopes of the
fan-chart lines calculated from Eq. (9) agree with experiment to
within ~15%. This is satisfactory bearing in mind the uncertainty in
the distance d over which the applied voltage is dropped.

Acknowledgement

This work is supported by SERC.
This paper reviews work done in collaboration with colleagues in the
Nottingham University MBE Research project and in the High Magnetic
Field Laboratory, Grenoble (CNRS).

References

1. V. J. Goldman, D. C. Tsui and J. E. Cunningham, <u>Phys. Rev. Lett.</u> 58:256 (1987); ibid. 59:623 (1987).

2. T. C. L. G. Sollner, <u>Phys. Rev. Lett.</u> 59:622 (1987).

3. B. Ricco and M. Ya Azbel, <u>Phys. Rev. B</u> 29:970 (1984)

4. S. Luryi, <u>Appl. Phys. Lett.</u> 47:90 (1985).

5. T. Weil and B. Vinter, <u>Appl. Phys. Lett.</u> 50:1281 (1987).

6. F. W. Sheard and G. A. Toombs, <u>Appl. Phys. Lett.</u> 52:1228 (1988).

7. V. J. Goldman, D. C. Tsui and J. E. Cunningham, <u>Phys. Rev.</u> <u>B</u>35:9387 (1987).

8. C. A. Payling, E. S. Alves, L. Eaves, T. J. Foster, M. Henini, O. H. Hughes, P. E. Simmonds, F. W. Sheard, G. A. Toombs and J. C. Portal, Proc. 7th Int. Conf. on the Electronic Properties of Two Dimensional Systems, Santa Fe, 1987. <u>Surf. Sci.</u> 196:404 (1988). See also C. A. Payling, E. Alves, L. Eaves, T. J. Foster, M. Henini, O. H. Hughes, P. E. Simmonds, J. C. Portal, G. Hill and M. A. Pate, Proc. 3rd Int. Conf. on Modulated Semiconductor Structures, Montpellier, France. <u>J. Physique</u> C5:289 (1987).

9. G. A. Toombs, E. S. Alves, L. Eaves, T. J. Foster, M. Henini, O. H. Hughes, M. L. Leadbeater, C. A. Payling, F. W. Sheard, P. A. Claxton, G. Hill, M. A. Pate and J. C. Portal, 14th Int. Symposium on Gallium Arsenide and Related Compounds, Crete, 1987. <u>Institute of Physics Conference Series</u> 91:581 (1988).

10. V. J. Goldman, D. C. Tsui and J. E. Cunningham, <u>Phys. Rev.</u> <u>B</u>36:7635 (1987). See also Proc. 3rd Int. Conf. on Modulated Semiconductor Structures, Montpellier, France. <u>J. Physique</u> C5:467 (1987).

11. L. Eaves, G. A. Toombs, F. W. Sheard, C. A. Payling, M. L. Leadbeater, E. S. Alves, T. J. Foster, P. E. Simmonds, M. Henini and O. H. Hughes, <u>Appl. Phys. Lett.</u> 52:212 (1988).

12. B. R. Snell, K. S. Chan, F. W. Sheard, L. Eaves, G. A. Toombs, D. K. Maude, J. C. Portal, S. J. Bass, P. Claxton, G. Hill and M. A. Pate, <u>Phys. Rev. Lett.</u> 59:2806 (1987).

OBSERVATION OF BALLISTIC HOLES[+]

M. Heiblum[*], K. Seo[*], H. P. Meier[**] and T. W. Hickmott[*]

[*] IBM Research Division, T. J. Watson Research Center,
Yorktown Heights, New York 10598.
[**] IBM Research Division, Zurich Research Laboratory - 8803
Rüschlikon, Switzerland

ABSTRACT

We report the first direct observation of ballistic hole transport in semiconductors, via energy spectroscopy experiments. Light holes are preselected and injected via tunnelling into 31 nm thick p^+ GaAs layers. About 10% of the injected holes have been found to traverse ballistically maintaining distributions $\cong$ 35 meV wide, with a mean free path of about 14 nm. Resonances in the injection currents, resulting from quantum interference effects of the ballistic holes, are used to support the light nature of the ballistic holes.

Ballistic transport of carriers in solids was previously inferred via a variety of indirect techniques[1-3]. Recent energy spectroscopy experiments in GaAs have demonstrated directly the existence of ballistic electrons[4], however holes were never directly observed to transport ballistically. In GaAs there are two valence bands[5]: a light hole band containing about 5% of the total hole population, with a curvature effective mass m_{lh} = 0.082m_e, where m_e is the free electron mass, and a heavy hole band with m_{hh} = 0.51m_e. Since the mean free path (mfp) is approximately proportional to the inverse of the mass, ballistic transport of heavy holes is unlikely in practical structures. In order to look for ballistic hole transport we have used a tunnel barrier, which has large transmission for light holes, injecting $\cong$ 30 meV wide energy distributions of light holes into heavily doped p^+ GaAs layers, 31 nm

---

[+] This paper was published previously in Physical Review Letters (Phys. Rev. Lett. 60, 828 (1988)) and is reproduced by kind permission of the American Physical Society.

thick. With spectroscopy performed after traversal, we have measured similar distribution widths and peak energies, with about 10% of the holes being ballistic. The ballistic transport and the light nature of the holes are supported by observing resonances in the injection currents due to quantum interference effects of the ballistic light holes in the thin GaAs layers[6].

Experiments were done with a novel three terminal structure (hot hole transistor, HHT) grown by MBE on a p^+ (100) GaAs substrate, described in Fig. 1. A tunnel injector composed of a p^+ GaAs layer (called emitter), an intrinsic $Al_xGa_{1-x}As$ barrier layer with x=0.5 (12 nm thick), and p^+ GaAs layer (31 nm thick, called base), was used to select and inject light holes. When biased with V_{EB} it injected a quasi-monoenergetic distribution of light hot holes ($\cong$ 30 meV wide), favored by the tunnelling process (by a factor of $\cong 10^7$), with most holes emerging into the base layer with excess energy near eV_{EB}. The 31 nm base layer was terminated with a spectrometer made of a relatively thick, intrinsic $Al_yGa_{1-y}As$ layer with y=0.31 (47 nm thick, called collector barrier), followed by a thick p^+ GaAs layer (called collector). The AlAs mole fraction in the collector barrier was linearly graded down to y=0.17 over 6 nm on the base side to minimize quantum mechanical reflections. The GaAs layers were doped with acceptors (Be) to a level of 1.6×10^{18} cm^{-3}. The structure was selectively etched to expose the base layer and alloyed ohmic contacts were made to the emitter, base, and collector layers.

Arriving hole distributions were analyzed by the thick AlGaAs spectrometer barrier. Upon the application of a potential difference, V_{CB}, the potential height of the collector barrier, Φ_c, changes, affecting the collected current density $J_c = e \int_{\Phi_c}^{\infty} n(E_\perp) v_\perp(E_\perp) dE_\perp$, where e is the electronic charge, $n(E_\perp)$ is the number of holes per unit normal energy, an energy associated with the normal component of the velocity, $v_\perp(E_\perp)$. The normal energy distribution can be found from $ev_\perp(E_\perp)n(E_\perp) = dJ_c/d\Phi_c$, or $v_\perp(E_\perp)n(E_\perp) \propto e^{-2}\eta^{-1}dJ_c/dV_{CB}$, where $\eta = e^{-1}d\Phi_c/dV_{CB}$ is a proportionality factor. If our graded barrier were ideal, for $V_{CB}>0$ $\eta \cong 1$ (with peak potential at collector side), and for $V_{CB} < 0$ $\eta \cong 6/47 \cong 0.13$ (with peak 6 nm away from the base side). However, some barrier parameters such as the density of any unintentional charges[7], or the extent of acceptor (Be) segregation from the collector[8], are important and difficult to determine accurately. Thus, a study of the spectrometer barrier, leading to the actual Φ_c and η, has to be done.

This study is done, with our HHT structure, by finding a low temperature threshold injection voltage, V_{EB}^*, for some V_{CB}, for which an onset in the collector current occurs. Then the Fermi level in the emitter is at the same height as the peak potential of the collector barrier, and some ballistic holes can graze and pass the top of the barrier[9]. More accurately, the collector barrier height is $\Phi_c = eV_{EB}^* + \zeta_B + \delta$, where ζ_B is the Fermi energy in the base (Fig. 1) and $\delta \cong 10\text{-}20$ meV is a correction due to holes tunnelling somewhat below the barrier

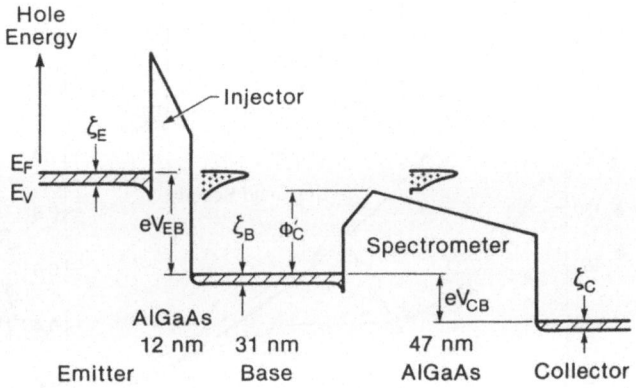

Fig. 1   The valence band in the HHT with hole energy plotted upwards. The tunnel injector and spectrometer barriers have AlAs mole fractions of 0.5 and 0.31, respectively. The collector is shown biased negatively. The emitter, base and collector are p^+ GaAs doped to a level of 1.6×10^{18} cm^{-3} .

top. The results of such experiments, done at high current sensitivities, are seen in Fig. 2(a). From here, neglecting δ, the collector barrier height above the Fermi level in the base $\Phi_c' = eV_{EB}^*$, is plotted in Fig. 2(b). With no bias applied $\Phi_c' = 170$ meV, $\zeta_B \cong 9$ meV for our doping, and thus $\Phi_c = 179$ meV $+ \delta$, a value higher by at least 20 meV than the published band discontinuity values[10]. This difference is most probably related to unintentional positive charges in the barrier, which we address below.

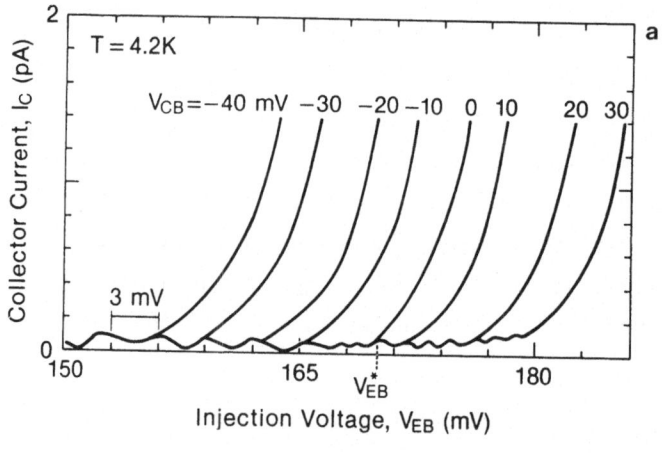

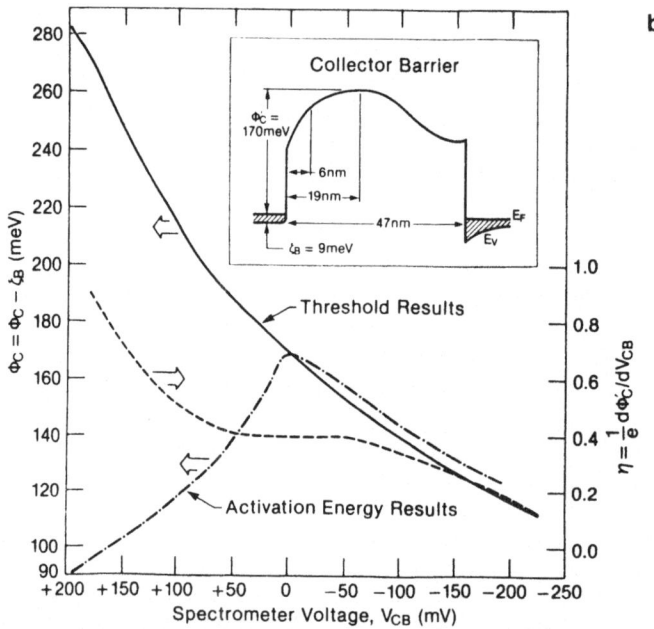

Fig. 2   (a) Collector current onsets for different V_{CB}, and an
example of determining the threshold eV_{EB}^* for $V_{CB}= 0$. Note
the very small currents rising from the noise line. (b) Φ_c
and the derivative η as a function of V_{CB}. Also plotted are
the barrier heights from separate measurements of activation
energy for thermionic emission (for $V_{CB} > 0$ (<0) the barrier
is that for holes in the collector (base)). The inset
describes a probable collector barrier shape.

Figure 2(b) gives also the calculated η, which together with Φ'_c are sufficient for the analysis of our spectroscopy experiments; however, studying Φ_c and η in some detail gives us a physical insight of the barrier shape, as related to actual barrier parameters. In the range -50 mV $< V_{CB} < 50$ mV, $\eta=0.4$, suggesting a barrier peak at some 0.4×47 nm$=19$ nm away from the base side. This, and the increase in the barrier height from the value given in Ref. 10 (seen above), can result from unintentional positive charges in the barrier; charges which we independently measured[11]. One can also see that η approaches unity only at large positive V_{CB}, an effect that can be attributed to barrier lowering on the collector side, most probably due to Be segregation into the barier during the growth[8].

This was verified by measuring activation energies for thermionic emission, with results given in Fig. 2(b)[10]. The method proved credible as seen from the good agreement between the activation energies results for $V_{CB} \leq 0$, and Φ'_c determined from the threshold measurements. For large positive V_{CB} we find a rather low barrier height on the collector side $\Phi_c - \zeta_c \cong 90$ meV.

From these results a more realistic shape of the collector barrier is shown in the inset in Fig. 2(b). This barrier has a potential peak that moves from near the base side to the collector side as V_{CB} increases.

Spectroscopy was done by measuring, at 4.2 K, the collector current I_c versus the collector voltage V_{CB}, at different injection voltages V_{EB} (Fig. 3(a)). The current rises steeply up to a knee where the slope clearly changes. The hole energy distributions, shown in Fig. 3(b), are derived by dividing dI_c/dV_{CB} curves by the η determined above, and converting the horizontal scale to excess normal energy above ζ_B (using Fig. 2(b)).

A few interesting features can be noticed. For injection energies in the range 190-210 meV, a clear ballistic behaviour is observed. The distributions are extremely sharp with widths at half maximum of $\cong 35$ meV, and peaks tracking exactly the injection energy, eV_{EB} (10 meV apart). At the lower energy end decaying tails of distributions, peaking somewhere closer to the base Fermi level, are seen, masking the corresponding lower energy ballistic peaks. These tails are likely to originate from holes excited up from the equilibrium hole population in

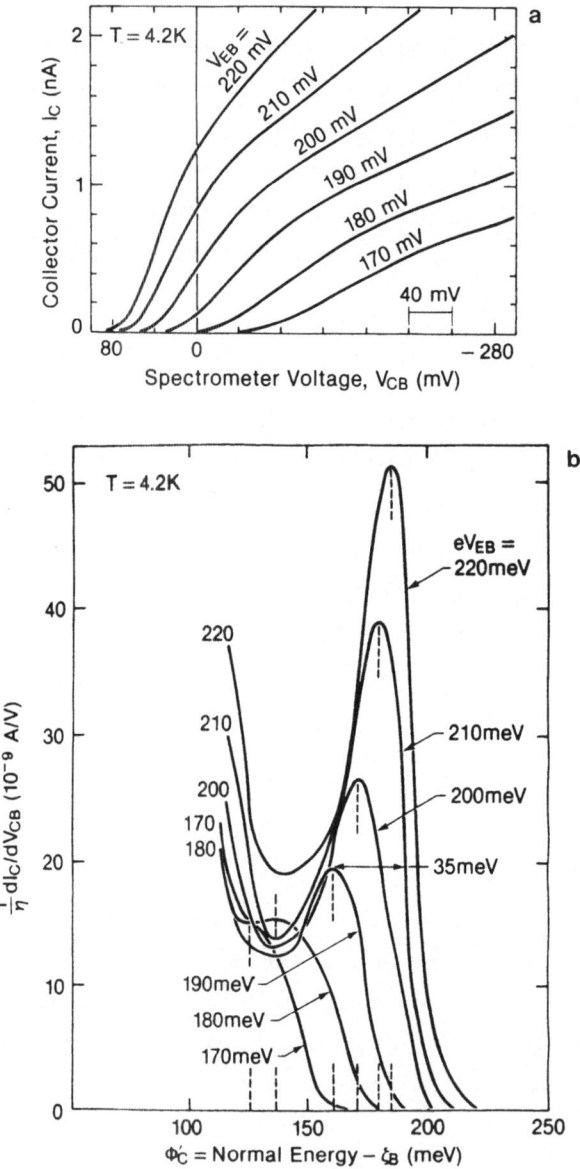

Fig. 3 (a) Collector current I_c versus spectrometer voltage V_{CB}, for different injection voltages V_{EB}. (b) Hole energy distributions deduced from the above $I_c - V_{CB}$ curves and the η given in Fig. 2(b). Ballistic peaks are seen for V_{EB} = 190,200 and 210 mV. For lower V_{EB}, upper energy tails of non-ballistic hole distributions are dominant.

the base[12]. For $eV_{EB} > 220$ meV peaks do not shift in energy any more, as the spectrometer potential peak moves rapidly toward the collector side. Then holes, most probably, can not traverse ballistically the full width of the AlGaAs barrier against a relatively strong retarding electric field, resulting in a drop of the collector current and in an artificial peak in dI_C/dV_{CB}.

The observed hole distributions are narrower by a factor of two compared to those of ballistic electrons measured by a similar technique[4]. This is due to the narrower supply function in the emitter being determined by the heavy holes: ζ_E + band bending (calculated classically) $\cong$ 27 meV for $V_{EB} = 200$ mV. The small displacement in the peak positions below the Fermi level in the emitter is fully accounted for by noting that the injected distribution $n(E_\perp) \sim (E_F - E_\perp) D(E_\perp)$ at 0 K, where $D(E_\perp)$ is the one-dimensional tunnelling probability, peaks

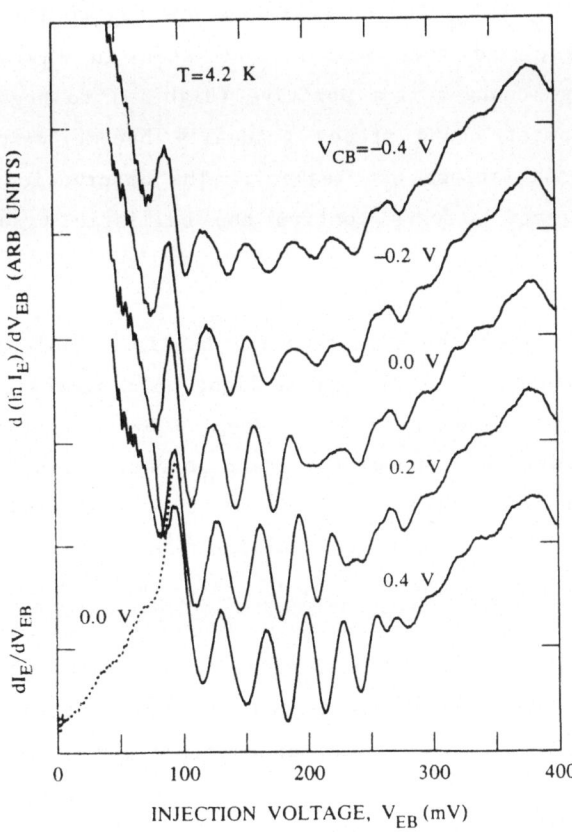

Fig. 4   Solid curves are the derivative of the logarithm of the injected current (high frequency oscillations at the low V_{EB} are due to noise). As the collector voltage changes from positive to negative, fewer bound states are visible. The dotted line is dI_E/dV_{EB} for $V_{CB} = 0$, showing the lower energy resonances.

some 15 meV below E_F[4,13], and that the actual Φ_c' is higher by δ than the one used in Fig. 3(b). The above results are consistent with ballistic transport of the narrow distributions. Integrating the area under the distributions we find that 8-11% of the injected hole current is ballistic, with a calculated mfp$\cong$14 nm (using $0.1 \cong \exp - (31/\text{mfp})$).

Even though it is highly likely that the observed ballistic holes are light, the spectroscopy measurements described above are not sufficient to prove it. Confirmation is provided by analyzing strong resonances resulting from hole interference effects, observed in the injection tunnelling currents, at certain energies that are particularly sensitive to the effective mass of the holes[6]. These resonances are due to a faster increase in the injection current whenever the Fermi level in the emitter crosses the bottom of a quasi 2D hole band formed in the base. Figure 4 shows the derivatives $d(\ln I_E)/dV_{EB} - V_{EB}$, for different V_{CB}. As long as $eV_{EB} \leq \Phi_c'$, strong oscillations in the derivative are observed because the participating 2D states are strongly bound. However, when $eV_{EB} > \Phi_c'$, the oscillations are due to the less confined (virtual) states, and thus weaker. As seen in Fig. 4, when the collector voltage changes from positive (high Φ_c) to negative (low Φ_c) values, the topmost bound states gradually become virtual, and the corresponding oscillations get weaker[6]. The observation of the peaks due to interference effects confirm the ballistic transport of the holes.

In order to find the corresponding ballistic hole mass, we have estimated the positions of the bound light hole states in a symmetric square well, 31 nm wide and 200 meV deep. Obviously this is a gross simplification since it ignores the exact potential distribution in the base and the nonparabolicity effects of the hole band; however, it distinctly shows the nature of the ballistic holes. For $m_{lh} = 0.082 \, m_e$ we find bound levels at 3, 15, 35, 62, 97, 138, and 182 meV[14]. From Fig. 4, for $V_{CB} = 0$, the observed bound peak positions including ζ_B are at 41, 72, 102, 133, and 168 meV, agreeing well with the calculations (the first two calculated levels are not seen since they are just at or below the Fermi level). Note that at energies higher than 100 meV the well widens and non-parabolicity increases the light hole effective mass[5], causing the last two observed peaks to be at somewhat lower energies. For comparison, heavy holes with $m_{hh} = 0.51 m_e$, would have sixteen bound states in the base, excluding them from the observed transport.

In summary, our results show, for the first time, unambiguous spectroscopic observation of ballistic light hole transport in GaAs. Ballistic hole distributions, 35 meV wide, with a mean free path of about 14 nm, have been measured. Lower energy distributions, that might be due to excited holes from the Fermi bath, are observed too. Quantum interference effects in the thin transport regions provide added evidence for the ballistic transport.

ACKNOWLEDGEMENTS

We thank L. F. Alexander and C. Lanza for their technical help. We also thank E. E. Mendez and A. C. Warren for their help in some of the calculations, and J. Batey, M. V. Fischetti, A. B. Fowler, S. P. Keller, C. J. Kircher, and F. Stern for their comments on the manuscript. The work was partly supported by DARPA and administered by ONR, contract #N00014-87-C-0709.

REFERENCES

1. R. J. von Gutfeld and A. H. Nethercot Jr., Phys. Rev. Lett. 18, 855 (1987).

2. L. F. Eastman, R. Stall, D. Woodard, N. Dandekar, C. E. C. Wood, M. Shur, and K. Board, Electron. Lett. 16, 525 (1980).

3. R. Trzcinski, E. Gmelin and H. J. Queisser, Phys. Rev. B 35, 6373 (1987).

4. M. Heiblum, M. I. Nathan, D. C. Thomas and C. M. Knoedler, Phys. Rev. Lett. 55, 2200 (1985); M. Heiblum, I. M. Anderson and C. M. Knoedler, Appl. Phys. Lett. 49, 207 (1986).

5. J. C. Blakemore, J. Appl. Phys. 53, R123 (1982).

6. M. Heiblum, M. V. Fischetti, W. P. Dumke, D. J. Frank, I. M. Anderson, C. M. Knoedler and L. Osterling, Phys. Rev. Lett 58, 816 (1987).

7. T. W. Hickmott, P. M. Solomon, R. Fischer and H. Morkoc, J. Appl. Phys. 57, 2844 (1985).

8. D. L. Miller and P. M. Asbeck, J. Appl. Phys. 57, 1816 (1985).

9. We may assume that some ballistic holes, few as they may be, always exist in the structure. We later prove their existence and find their number from spectroscopy measurements.

10. J. Batey and S. L. Wright, J. Appl. Phys. 59, 200 (1986). The valence band discontinuity ΔE_v = 5.5x, where x is the AlAs mole percent. For our AlGaAs collector barrier, x=31% and ΔE_v = 171 meV.

11. Capacitance measurements done on separate p^+-intrinsic AlGaAs-p structures show a voltage shift in the flat band condition consistent with positive charges in the barrier in the low 10^{16} cm^{-3} range.

12. Similar low energy electron distributions had been reported by A. F. J. Levi, J. R. Hayes, P. M. Platzman and W. Wiegmann, Phys. Rev. Lett. 55, 2071 (1985).

13. M. Heiblum, Solid St. Electron. 24, 343 (1981).

14. For 31 nm base thickness, light and heavy band mixing at k=0 is insignificant. See for example S. Brand and D. T. Hughes, Semicond. Sci. Technol. 2, 607 (1987).

QUANTUM TRANSPORT THEORY OF RESONANT TUNNELING DEVICES

William R. Frensley

Central Research Laboratories
Texas Instruments Incorporated
Dallas, Texas 75265

INTRODUCTION

The ability to fabricate semiconductor heterostructures on the scale of a few atomic layers has led to the development of devices which exploit the quantum-mechanical wave properties of electrons in their operation. The quantum device which has recieved the most attention recently is the quantum-well resonant-tunneling diode (RTD).[1,2] This device shows a negative-resistance characteristic which is quantum-mechanical in origin, and is potentially a very fast device. Most of the theoretical work on this device has employed the formal theory of scattering, focusing on the behavior of pure quantum states which are asymptotically plane waves. While this approach should adequately describe the device under stationary conditions, it is poorly equipped to treat any sort of time-varying behavior. The reason for this is that the behavior of the RTD, and indeed any electronic device, is manifestly time-irreversible, and a proper notion of irreversibility cannot be introduced into pure-state quantum mechanics. A pure quantum state cannot evolve time-irreversibly. Models which attempt to introduce such behavior inevitably violate some fundamental physical law, usually the continuity equation. However, *transitions* between quantum states may proceed irreversibly if the system of interest interacts with an external system having a continuum of states. Such processes may be consistently described in terms of statistically mixed states, which are represented most simply by the single-particle density matrix.[3] A description of a many-particle system in terms of such a single-particle distribution is generally termed a kinetic theory.[4] The present paper describes such a theory of electron devices which incorporates quantum coherence effects (including tunneling).

QUANTUM KINETIC TRANSPORT THEORY

A satisfactory transport theory must adequately treat two fundamental aspects of electron devices. First, an electron device is necessarily an open system; it is useless unless connected to an electrical circuit and able to exchange electrons with that circuit. If one wishes to study the behavior of the device apart from that of the circuit, it is convenient to represent the effects of the external circuit by ideal electron reservoirs attached to the terminals of the device. Secondly, a device is also a time-irreversible system, as evidenced by the set of nonequilibrium steady states which constitute the $I(V)$ characteristic. An elegant and consistent kinetic model of a device can be obtained which incorporates *both* openness and irreversibility into the model via boundary conditions applied to the Wigner distribution function.[6] The Wigner function is simply a mathematical transform of the density matrix $\rho(x,x')$ so that it is defined in the phase-space (x,k_x):

$$f(x,k) = \int_{-\infty}^{\infty} dy\, e^{-iky}\, \rho\left(x+\tfrac{1}{2}y, x-\tfrac{1}{2}y\right) . \qquad (1)$$

The Wigner function is often invoked to derive the correspondence between quantum and classical statistical mechanics. In the present case it provides the means by which an essentially classical model of the coupling of an open system to external reservoirs can be introduced into a quantum calculation. To describe purely ballistic transport of electrons (that is, neglecting collisions within the device) the Liouville equation for the time evolution of the Wigner function can then be written

$$\frac{\partial f}{\partial t} = -\frac{\hbar k}{m}\frac{\partial f}{\partial x} - \frac{1}{\hbar}\int_{-\infty}^{\infty}\frac{dk'}{2\pi}\, U(x, k-k')f(x, k') \equiv \frac{L}{i\hbar}f \qquad (2)$$

where L is the Liouville super-operator. The form of the Liouville equation is quite similar to that in the classical case, with the exception that the effect of the potential is now non-local. This is how quantum interference effects enter the present model. The kernel of the potential operator is given by

$$U(x,k) = 2\int_{0}^{\infty} dy\, \sin(ky)\,[V(x+\tfrac{1}{2}y) - V(x-\tfrac{1}{2}y)] . \qquad (3)$$

The open system boundary conditions can be obtained in a physically appealing way by assuming that the reservoirs to which the device is connected have properties analogous to those of a black body: the distribution of electrons emitted into the device from the reservoir is characterized by the thermal equilibrium distribution function of the reservoir, and all electrons impinging upon a reservoir from the device are absorbed by the reservoir without reflection. To implement this picture, we must be able to distinguish the sense of the velocity of an electron at the position of the boundary. Thus the Wigner function is the natural representation for an open system, because it involves both the position and the velocity. Let the interface between the device and the left-hand reservoir occur at $x=0$, and the interface between the device and the right-hand reservoir occur at $x=l$. Then we may write the open-system boundary conditions as

$$f(0, k)\big|_{k>0} = F(k, \mu_l, T_l)$$
$$f(l, k)\big|_{k<0} = F(k, \mu_r, T_r) \qquad (4)$$

where F is the Fermi distribution function (integrated over the transverse momenta), $\mu_{l,r}$ are the Fermi levels, and $T_{l,r}$ are the temperatures of the respective contacts. Note that these boundary conditions are in themselves time-irreversible, because under time-reversal they would map into a specification of the distribution of outgoing particles. While the Liouville equation (2) contains only the ballistic transport of electrons within the system, the irreversibility due to the coupling to the contacts is both necessary and sufficient to obtain a meaningful description of a device. Of course it will eventually be desirable to include those irreversible processes attributable to random scattering events, and a first approximation to such processes is described below.

The boundary conditions (4) are inhomogeneous. It is readily shown that the Liouville operator (2), subject to boundary conditions of the form (4) is non-singular, and its eigenvalues are confined to the lower half of the complex plane, corresponding to stable solutions.[7] Because L is non-singular, any choice of the boundary distribution F leads to a well-posed problelm. To obtain quantitative results, the Wigner function is evaluated within a discrete (finite-difference and finite-sum) approximation.[5]

STEADY-STATE BEHAVIOR

The steady-state Wigner function is obtained by numerically solving the Liouville equation for the condition $\partial f/\partial t = 0$. The $I(V)$ characteristic for the RTD was obtained by calculating the steady-state Wigner function for each of a large set of bias voltages, and the current density was evaluated by averaging over the Wigner function. The results of such a calculation are shown in Fig. 1, along with the $I(V)$ curve obtained from a more conventional scattering-theory calculation for comparison. The agreement between these calculations is quite good in the vicinity of the peak tunneling current, and is somewhat poorer in the vicinity of the valley.

This agreement between the transport and scattering theories is considerably better than what was reported earlier.[5,7] The scattering calculations shown in the earlier work were in error, because the wrong velocity was used to evaluate the current density contribution from each state. The incorrect formula, which has been widely quoted,[8] involves the velocity of the electron on the incoming side of the barrier, so that this velocity cancels the "density of states" factor. The correct formula, as pointed out by Coon and Liu,[9] involves the velocity of the electron on the *outgoing* side of the barrier, which is not the same as the incoming velocity when there is a nonzero bias voltage. Correcting this error brings the scattering calculation into much better agreement with the predictions of the present quantum transport theory. Recent work by Mains and Haddad[10] indicates that modifications to the method of evaluating the potential operator (3) can have the effect of reducing the magnitude of the valley current. Such modifications should improve the agreement between the transport and scattering theories in this region of the $I(V)$ curve. These modifications have not yet been incorporated into the present calculations.

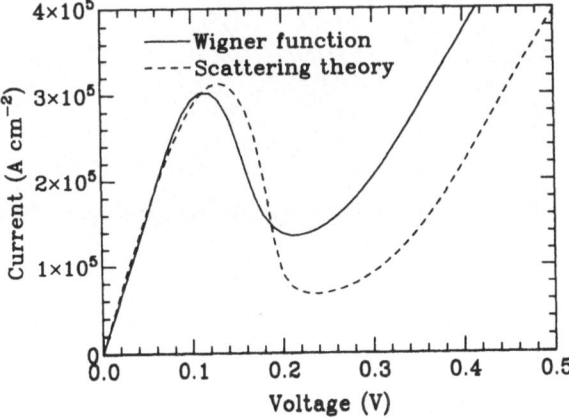

Fig. 1.   Current density *vs.* voltage for a resonant-tunneling diode consisting of 2.8 nm layers of $Al_{0.3}Ga_{0.7}As$ bounding a 4.5 nm GaAs well, at a temperature of 300 K. The current derived from a calculation of the Wigner function (solid line) is compared to that derived from a more conventional scattering calculation (dashed line).

The device structure assumed in the present calculations consists of a 4.5 nm wide quantum well of GaAs bounded by identical 2.8 nm wide barrier layers of $Al_{0.3}Ga_{0.7}As$. The conduction-band discontinuity was taken to be 0.60 of the total bandgap discontinuity. 17.5 nm of the GaAs electrode layer was included in the simulation domain on each side of the device. Because Hartree self-consistency was not incorporated into the present calculations, the applied bias voltages were assumed to be dropped uniformly across the well and barriers. The electron density assumed in the boundary reservoirs was 2×10^{18} cm^{-3}. All calculations were performed at a temperature of 300 K.

TRANSIENT RESPONSE

The time-dependent response of the RTD to changes in the applied voltage is readily evaluated by integrating the Liouville equation (2), using the numerical procedures described in Ref. 5. The results of such a calculation are shown in Fig. 2. Since the negative-resistance characteristic is the interesting feature of this device, the transient response calculation was performed for a switching event across this region of the $I(V)$ curve. Figure 2 shows the current density in the device as a function of position and time for an event in which the initial bias of 0.11 V (corresponding to the peak in the current) was suddenly switched to 0.22 V (corresponding to the bottom of the valley) at $t = 0$. More specifically, the steady-state Wigner function for a bias of 0.11 V was used as an initial value, and the time evolution under the Liouville operator for 0.22 V bias was evaluated. The response of the current is complex, as might be expected, but shows

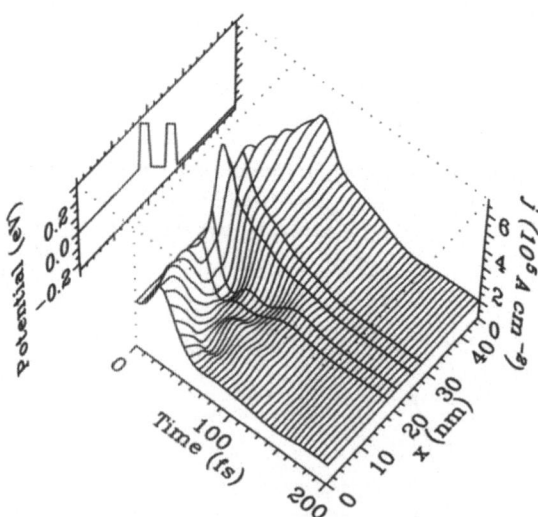

Fig. 2.   Transient response of the resonant-tunneling diode of Fig. 1. Current density is plotted as a function of time and position within the device. The potential profile illustrates the device structure. At $t = 0$, the voltage was suddenly switched from 0.11V (corresponding to the peak current) to 0.22V (corresponding to the valley current). After an initial peak, the current density approaches the lower steady-state value in 100-200 fs.

some features that are readily interpreted. The current density initially increases throughout the structure, so that the device displays a positive resistance over a short time. The destructive interference which underlies the negative resistance takes some tens of femtoseconds to manifest itself. The current has settled quite near to its steady-state value after 200 fs. Of course the response of real devices will be limited by the time required to charge the device capacitance through the parasitic series resistance of the contacts. Such effects were deliberately omitted from the present model in order to observe the intrinsic response of the tunneling process itself.

SMALL-SIGNAL RESPONSE

In order to obtain the small-signal ac response,[11] we assume that a small ac signal of amplitude v is superimposed upon the dc bias V. For a fixed V, the conduction current density j through the intrinsic device can be expanded in a power series in v, and to second order it is given by:

$$j(t) = j_0(V) + \tfrac{1}{2}(yve^{i\omega t} + cc) + \tfrac{1}{2}a_{rect}v^2 + \tfrac{1}{4}(a_{2\omega}v^2e^{2i\omega t} + cc) + \dots, \tag{5}$$

where cc denotes the complex conjugate. Here ω is the angular frequency, j_0 is the dc current density, and y is the linear admittance (which equals dj_0/dV at $\omega = 0$). The nonlinear coefficients a_{rect} and $a_{2\omega}$ describe rectification and second-harmonic generation, respectively, and both are equal to d^2j_0/dV^2 at $\omega = 0$.

To obtain the small-signal ac response, we apply a simple form of perturbation theory to equation (2). The Liouville operator can be written as:

$$L = L_0 + \tfrac{1}{2}\lambda(L_\omega e^{i\omega t} + cc). \tag{6}$$

The dc part L_0 includes the kinetic energy term and the dc potential. The ac part L_ω includes only the effect of the time-varying potential and thus is proportional to v. λ is a perturbation parameter introduced solely to keep track of the order of the perturbation, which will ultimately be set equal to unity. The Wigner function f can be expanded in a perturbation series, which to second order is given by

$$f = f_0 + \tfrac{1}{2}\lambda(f_\omega e^{i\omega t} + cc) + \lambda^2 f_{rect} + \tfrac{1}{2}\lambda^2(f_{2\omega}e^{2i\omega t} + cc) + \dots. \tag{7}$$

Here f_0 is the dc part of the Wigner function, f_ω contains the linear ac response, and again f_{rect} and $f_{2\omega}$ describe rectification and second-harmonic generation, respectively. The perturbation equations are obtained by inserting (6) and (7) into (2) and collecting terms of equal frequency and equal order in λ. The resulting equations are:

$$L_0 f_0 = 0. \tag{8a}$$

$$f_\omega = -\frac{1}{L_0 + \hbar\omega}L_\omega f_0. \tag{8b}$$

$$f_{rect} = \frac{1}{2L_0}Re\left(L_\omega^*\frac{1}{L_0 + \hbar\omega}L_\omega f_0\right). \tag{8c}$$

$$f_{2\omega} = \tfrac{1}{2}\frac{1}{L_0 + 2\hbar\omega}L_\omega\frac{1}{L_0 + \hbar\omega}L_\omega f_0. \tag{8d}$$

These equations resemble those of the conventional perturbation series, but differ in detail primarily because quantum-mechanical convention for the time-dependence (e^{-iLt}) of f has been mixed with the electrical engineering convention ($e^{i\omega t}$) for the time-dependence of the applied signal. In particular, that is the reason the "+" sign appears in the denominators. The resolvent expressions in (8b-d) are readily evaluated within the discretization approximation by ordinary matrix operations.

The contribution of a component f_i of the Wigner distribution to the terminal current density is obtained by averaging the current operator over the momentum, and over the active region of the device in accordance with the Ramo-Shockley theorem:[12,13]

$$J[f_i] = \frac{1}{x_r - x_l} \int_{x_l}^{x_r} dx \int_{-\infty}^{\infty} \frac{dk}{2\pi} \frac{\hbar k}{m^*} f_i(x,k). \qquad (9)$$

The coefficients in (4) are thus given by:

$$j_0 = J[f_0]. \qquad (10a)$$

$$y = J[f_\omega]/v. \qquad (10b)$$

$$a_{rect} = \tfrac{1}{2} J[f_{rect}]/v^2. \qquad (10c)$$

$$a_{2\omega} = \tfrac{1}{2} J[f_{2\omega}]/v^2. \qquad (10d)$$

The small-signal response was evaluated for an assumed structure which was the same as that described above, except that the doping in the contact layers was taken to be 2×10^{17} cm^{-3}. This structure is similar to the sample number 2 of Sollner et al.[14] The linear admittance was evaluated from (8b) and (10b), for a bias voltage near the center of the negative resistance region. The resulting admittance as a function of frequency is shown in Fig. 3. The real conductance is negative at lower frequencies, as expected. The negative conductance "rolls off" in the THz region and goes positive at about 6 THz. The imaginary part of the electronic admittance is negative and proportional to ω at lower frequencies, and thus resembles an inductance. This is due to the phase shift resulting from the electrons' inertia.[15] The rather complex behavior of the

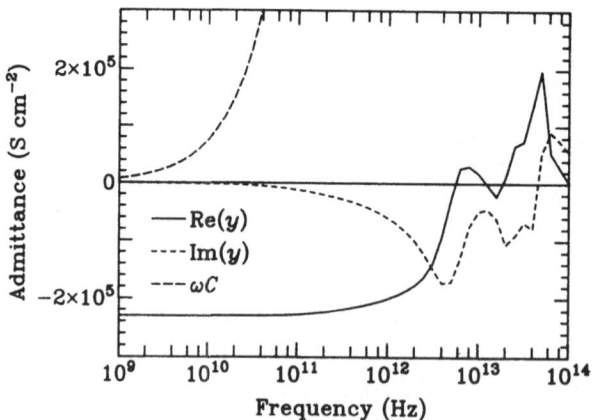

Fig. 3. Electron admittance as a function of frequency. The electron conductance is Re(y) and the electron susceptance, due to inertial effects, is Im(y). The negative conductance at lower frequencies is apparent. The susceptance due to the parasitic capacitance ωC is shown to provide a measure of the effect of the parasitic elements.

electronic admittance above 10 THz reflects other resonant processes in the system. In this frequency range the current response is quite nonlocal.[16]

An estimate of the susceptance of the parasitic capacitance of the RTD is also plotted in Fig. 3, for comaprison. The effects of this capacitance will become dominant when the magnitude of its admittance exceeds that of the tunneling current, which occurs, for the present model, somewhat below 100 GHz. This is the practical limit for the observation of a linear negative conductance.

Some nonlinear effects are observable to much higher frequencies than the linear effects, however. To examine the behavior of such processes, the nonlinear coefficients were evaluated from (8c,d) and (10c,d) for that dc voltage at the resonant peak of the $j(V)$ curve. The modulus of a_{rect} and of $a_{2\omega}$ are plotted in Fig. 4 as functions of frequency. The interesting point is that the calculations predict an enhancement in the coefficient for rectification between 1 and 8 THz. This agrees with the observations of Sollner $et~al.$[2] of rectification at 2.5 THz in their experi-mental devices. The quantity a_{rect} is the same as that which is denoted I'' in Ref. 2.

EFFECTS OF PHONON SCATTERING

The effects of the electron-phonon interaction may be easily incorporated into the present transport theory by adding an appropriate collision super-operator C to the Liouville equation:

$$\frac{\partial f}{\partial t} = \frac{L}{i\hbar}f + Cf \tag{11}$$

The existing numerical machinery can handle such a term so long as C can be treated as local in space and time. The obvious first step toward obtaining such an operator is to employ the classical Boltzmann equation form:

$$[Cf](x,k,t) = \int dk'\,[W_{k\,k'}\,f(x,k',t) - W_{k'\,k}\,f(x,k,t)], \tag{12}$$

where $W_{k\,k'}$ is the transition rate from k' to k, etc. The work of Levinson[17] and

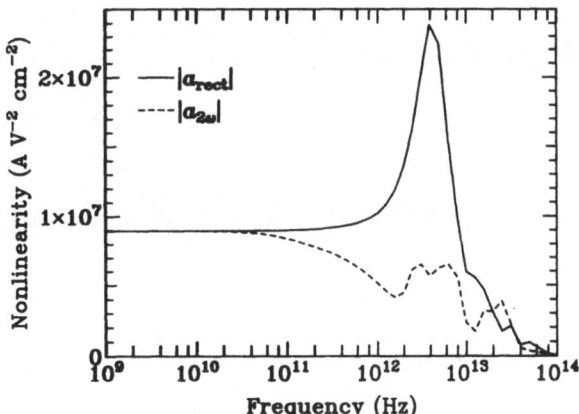

Fig. 4. The nonlinear response coefficients as functions of frequency. Rectification is described by a_{rect} and second-harmonic generation is described by $a_{2\omega}$. The persistence of the rectification effect to terahertz frequencies is in agreement with the experimental results of Ref. 2.

that of Lin and Chu[18] suggests that this is an appropriate approximation in the semi-classical case (that is, when the Fermi golden rule may be used).

Because the RTD is in reality a three-dimensional system, the integral in (12) must be three-dimensional. However, the model is one dimensional, so the operator in (12) must be projected onto the one-dimensional subspace by integrating over the transverse wavevectors $\mathbf{k}_\perp$ and $\mathbf{k'}_\perp$. Invoking once more the assumption that the distributions with respect to the transverse momenta are Boltzmann, the expression for the transition rates projected into one dimension is:

$$W_{k\,k'} = \frac{2\pi}{\hbar} \frac{V}{(2\pi)^3} \int d^2\mathbf{k}_\perp \int d^2\mathbf{k'}_\perp |\langle \mathbf{k}|H'|\mathbf{k'}\rangle|^2 \delta(E_{\mathbf{k}} - E_{\mathbf{k'}} \mp \hbar\omega) \frac{2\pi\beta\hbar^2}{m} exp\left[-\frac{\beta\hbar^2\mathbf{k'}_\perp^2}{2m}\right] . \quad (13)$$

Here H' is the Hamiltonian which describes the particular electron-phonon interaction. The numerical collision operators obtained from (12) and (13) were checked for consistency with the requirements of detailed balance by applying the operator to an equilibrium distribution function and verifying that the result was zero.

In the present calculations the deformation potential interaction was included for scattering with acoustic phonons and the Fröhlich interaction was included for scattering with longitudinal optical (LO) phonons.[19] The effects of these phonon scattering mechanisms on the $I(V)$ curve of the RTD are shown in Fig. 5. Acoustic phonon scattering has a nearly negligible effect on the $I(V)$ curve. The effect of LO phonon scattering is rather more pronounced, primarily in the reduction of the peak current. When both acoustic and LO phonons are included in the calculation, the resulting $I(V)$ curve is indistinguishable from that obtained with LO scattering only.

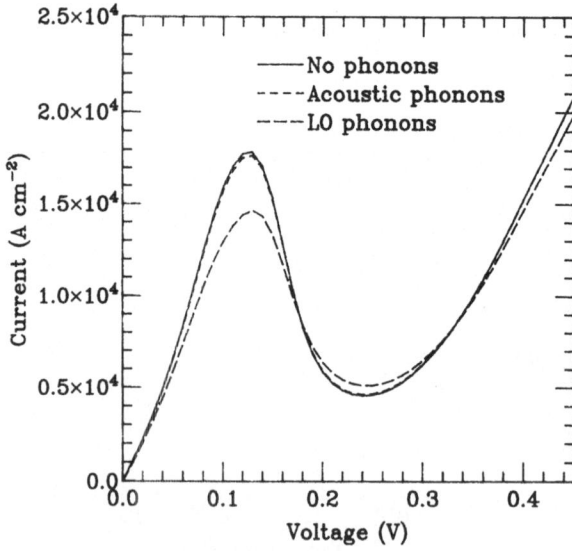

Fig. 5.   Effect of semi-classical phonon-scattering operators on the $I(V)$ characteristic of a resonant-tunneling diode.

CONCLUSIONS

The present quantum kinetic transport theory has proven to be exceptionally effective in modeling the behavior of resonant-tunneling devices. This success may be attributed to several key elements: First, it incorporates a simple and explicit notion of irreversibility through the coupling of the device to its contacts. This permits a simple treatment of irreversible phenomena such as the transient response, which are beyond the scope of theories which do not explicitly include irreversibility. The second element is that it corresponds as closely as possible to a classical model, departing only as required to include quantum interference effects [which enter via the nonlocal potential (3)]. This permits a direct computation of classically measurable quantities, such as the small-signal admittance. Finally, in contrast to the more sophisticated techniques of many-body theory, it suppresses enough of the complexities of the system so as to remain computationally tractable.

ACKNOWLEDGEMENT

This work was supported by the U.S. Office of Naval Research.

REFERENCES

1. L. L. Chang, L. Esaki and R. Tsu, Appl. Phys. Lett. **24**, 593 (1974).
2. T. C. L. G. Sollner, W. D. Goodhue, P. E. Tannenwald, C. D. Parker and D. D. Peck, Appl. Phys. Lett. **43**, 588 (1983).
3. U. Fano, Rev. Mod. Phys. **29**, 74 (1957).
4. Kubo, R., M. Toda, and N. Hashitsume, *Statistical Physics II. Nonequilibrium Statistical Mechanics.* (Springer-Verlag, Berlin, 1985).
5. W.R. Frensley, Phys. Rev. **B36**, 1570 (1987).
6. E. Wigner, Phys. Rev. **40**, 749 (1932).
7. W.R. Frensley, Phys. Rev. Lett. **57**, 2853 (1986).
8. R. Tsu and L. Esaki, Appl. Phys. Lett. **22**, 562 (1973).
9. D.D. Coon and H.C. Liu, Appl. Phys. Lett. **47**, 172 (1985).
10. R.K. Mains and G.I. Haddad, "Numerical Considerations in the Wigner Function Modeling of Resonant-Tunneling Diodes," to be published.
11. W.R. Frensley, Appl. Phys. Lett. **51**, 448 (1987).
12. S. Ramo, *Proc. IRE* **27**, 584 (1939).
13. W. Shockley, *J. Appl. Phys.* **9**, 635 (1938).
14. T.C.L.G. Sollner, E.R. Brown, W.D. Goodhue, and H.Q. Le, *Appl.Phys. Lett.* **50**, 332 (1987).
15. K.S. Champlin, D.B. Armstrong, and P.D. Gunderson, *Proc. IEEE* **52**, 677 (1964).
16. W.R. Frensley, to be published in Superlattices and Microstructures.
17. I.B. Levinson, "Translational invariance in uniform fields and the equation of motion for the density matrix in the Wigner representation," Soviet Physics JETP **30**, 362-7 (1970).
18. J. Lin and L.C. Chiu, "Quantum theory of electron transport in the Wigner formalism," J. Appl. Phys. **57**, 1373-6 (1985).
19. E.M. Conwell, *High Field Transport in Semiconductors*, (Academic Press, New York, 1967) ch. 5.

HOT ELECTRON EFFECTS IN MICROSTRUCTURES

P. Lugli

Dipartimento di Ingegneria Meccanica, II Università di Roma

Via O. Raimondo, 00173 Roma, Italy

ABSTRACT

A series of new devices generically called "hot electron transistors" is based on the idea of improving the device performance by injecting fast electrons into thin base regions. We present here a theoretical study, based on a Monte Carlo simulation, of the characteristic energy and momentum losses of hot electrons injected into a doped region, as found for example in the planar doped barrier (PDB) and in the tunneling hot electron transfer amplifier (THETA) devices. The interaction between the injected electrons and the background of cold carriers is shown to be a very effective channel of dissipation. The full self-consistent simulation of the THETA device is also presented.

INTRODUCTION

In recent years, ultrafast phenomena in semiconductors have become of great importance. Devices are now built with typical lengths of their active regions well below the micron. On the other side, ultrashort laser pulses can be obtained with resolution of few hundreds femtoseconds. Many interesting effects involving hot electrons have surfaced in these situations. The possibility of achieving ballistic transport, with electrons moving at very high velocities, virtually without scattering, has opened new horizons in the area of very fast semiconductor devices. One of the most important trends in semiconductor device technology consists in the reduction of the switching time. As a result of this achievement it is possible to obtain high computing speed, improving computer performance independently of architecture. The most common technology used in the fastest switching logic circuits is the MESFET technology (Metal Semiconductor Field Effect Transistor) using, as the semiconductor material, gallium arsenide (GaAs). Within this class of devices it is possible to achieve switching times of the order of $50ps$. Further improvements have been obtained with the HEMT (High Electron Mobility Transistor) which is capable of a switching time of the

---

Supported by Consiglio Nazionale delle Ricerche (CNR) under Progetto Finalizzato "Materiali e Dispositivi per l'Elettronica a Stato Solido" (MADESS)

order of $10ps$. Very high speeds are obtained by minimizing the energy and momentum losses that electrons undergo as they cross the active area of the device. In this context, a very promising possibility resides in the concept of Hot Ballistic Electrons. In a standard device, such as a MESFET or HEMT, electrons are injected into the channel with a thermal energy distribution and a small initial velocity. In order to reduce the transit time through the channel or base, it is adventageous to increase their initial velocity; this is accomplished in a class of devices called Hot Electron Transistors, such as the PDB [1] and the THETA [2,3] transistors. There, electrons are injected into the base with energies some hundreds of meV greater than the thermal energy. If the hot electrons could mantain their high speed through the active region of the device, then very good performance (fast switching , high cut-off frequencies) should be expected. On the contrary, it has been suggested, see for example [4], that the interaction of the injected electrons with the base electrons might be responsible for the far-from-ideal characteristics of planar-doped barrier (PDB) transistors. As well as in metals, it has been pointed out [5,6] that plasma phenomena can be of great importance also in polar semiconductors. Electrons injected into highly doped regions can be scattered by the collective excitations of the electron gas, as well as through normal binary collisions with the other electrons.

We will start by presenting in Section 2 a detailed theoretical study of carrier- carrier scattering, separating short range (screened electron-electron interaction) and long range (electron-plasmon interaction) contributions. Section 3 will then present a Monte Carlo study of the dynamics of high energy electrons injected into a doped GaAs region. This situation is typical of submicron devices such as the PDB and THETA transistors. The importance of intercarrier scattering (including the long range electron-plasmon interaction) is also discussed. The results of a direct Monte Carlo simulation of the THETA device is reported in Section 4.

CARRIER-CARRIER SCATTERING

We present in this section a general treatment of electron-electron scattering, that will be the basis of the Monte Carlo study of hot electron injection.

The interaction among carriers in semiconductors can be analysed starting from the electronic Hamiltonian

$$H_e = \sum_i \frac{p_i^2}{2m} + \frac{e^2}{2\kappa} \sum_{i,j i \neq j} \frac{1}{|\mathbf{r}_i - \mathbf{r}_j|} \tag{1}$$

The first term on the right-hand-side represents the free-electron energy, and the second one the bare Coulomb interaction among electrons. By following the procedure outlined by Bohm and Pines [7] it is possible to rewrite H_e in the random phase approximation (RPA) as the sum of four different contributions:

$$H_e = H_e^{(k)} + H_e^{(Sc)} + H_e^{(pl)} + H_e^{(e-pl)} \tag{2}$$

which can be explicitly written (in the occupation number representation) as [8]:

$$H_e^{(k)} = \sum_{\mathbf{k}} \frac{\hbar k^2}{2m} c_{\mathbf{k}}^+ c_{\mathbf{k}}, \tag{3}$$

$$H_e^{(Sc)} = \sum_{\mathbf{k} > \mathbf{q}_o} \frac{2\pi e^2}{V \kappa k^2} \sum_{\lambda \mu} c_{\mathbf{k}_\lambda + \mathbf{k}}^+ c_{\mathbf{k}_\mu - \mathbf{k}}^+ c_{\mathbf{k}_\mu} c_{\mathbf{k}_\lambda}, \tag{4}$$

$$H_e^{(pl)} = \sum_k \hbar\omega_p \left(a_\mathbf{k}^+ a_\mathbf{k} + \frac{1}{2}\right), \tag{5}$$

$$H_e^{(e-pl)} = \sum_{\mathbf{k}<\mathbf{q}_c} \left(\frac{\pi e^2 \hbar^2}{2V\kappa\omega_p m^2 k^2}\right)^{1/2} \sum_{\mathbf{k}'} (2\mathbf{k}-\mathbf{k}'+k^2)(a_\mathbf{k} c_{\mathbf{k}'+\mathbf{k}}^+ c_\mathbf{k}^+ + a_{-\mathbf{k}} c_{\mathbf{k}'-\mathbf{k}}^+ c_{\mathbf{k}'}) \tag{6}$$

The form of Eq. (2) is obtained after several transformations of the original Hamiltonian, which involve a Fourier transform of the interparticle term and the introduction of additional canonical conjugate variables. Details can be found in reference [8].

The four terms in Eq.(2) have a precise physical meaning. $H_e^{(k)}$ (Eq. 3) represents the free electron kinetic term, with $c_\mathbf{k}^+$ and $c_\mathbf{k}$ respectively creation and annihilation electron operators. The second term $H_e^{(Sc)}$ (Eq. 4) describes a screened Coulomb interaction in which the momentum $\mathbf{k}$ is transferred from the μth to the λth electron (annihilation of the two electrons with momentum $\mathbf{k}_\mu$ and $\mathbf{k}_\lambda$, and creation of two electrons with momentum $\mathbf{k}_\lambda + \mathbf{k}$ and $\mathbf{k}_\mu - \mathbf{k}$). A characteristic inverse screening length is defined by the cut-off wavevector q_c which is introduced arbitrarily to separate long range and short range components into the Fourier expansion of the Coulombic term of Eq. (1). The physical significance of q_c will be discussed below. The term $H_e^{(pl)}$ represents the energy of a gas of free bosons, which are here identified with the plasmons, i.e., with the quantized oscillations of the electron gas. In Eq. (5) $a_\mathbf{k}^+$ and $a_\mathbf{k}$ are respectively the creation and annihilation operators of a plasmon of wavevector $\mathbf{k}$, and $\hbar\omega_p$ is the energy of the plasmon. The last term, given by Eq. (6), describes the electron-plasmon interaction, through a process in which a momentum $\mathbf{k}$ is transferred to an electron with absorption of a plasmon $\mathbf{k}$ or emission of a plasmon $-\mathbf{k}$. Processes involving more than one plasmon are neglected in the RPA.

According to the previous analysis, two main contributions to the carrier -carrier scattering can be identified:
- the individual carrier-carrier interaction via a screened Coulomb potential of the form

$$V(r) = \frac{e^2}{\kappa r} e^{-\beta_c r} \tag{7}$$

which accounts for two-body short-range interaction;
- the electron-plasmon interaction, which accounts for the collective long-range behaviour of the electron gas.

In semiconductors, the plasmon energy at a reasonable electron density can be of the same order of magnitude as the characteristic phonon energies. A determination of the value of the cut-off wavevector q_c can be obtained from an independent analysis of the complex wavevector and frequency-dependent dielectric function [9]. It is found that it is reasonable for most cases to assume that q_c is equal to the inverse Debye screening lenght β for non-degenerate conditions.

Screened electron − electron interaction

Using the screened potential of Eq. (7) with $q_c = \beta$, the transition probability of two electrons from wave-vectors $\mathbf{k}$ and $\mathbf{k}_o$ to $\mathbf{k}'$ and $\mathbf{k}_0'$ is obtained (using the Fermi Golden rule) as:

$$S_{\mathbf{k}_0,\mathbf{k}\to\mathbf{k}_0',\mathbf{k}'} = \frac{2\pi}{\hbar}|M|^2 f_{\mathbf{k}_0} f_\mathbf{k}(1-f_{\mathbf{k}_0'})(1-f_{\mathbf{k}'})\delta(\epsilon_{\mathbf{k}_0'}+\epsilon_{\mathbf{k}'}-\epsilon_{\mathbf{k}_0}-\epsilon_\mathbf{k}), \tag{8}$$

189

where the f's and ϵ's are the occupation probabilities and the energy at each wave-vector, respectively. A parabolic energy-momentum dispersion is used. The transition matrix element is given by

$$M = \langle \mathbf{k}_0, \mathbf{k}|V(r)|\mathbf{k}_0', \mathbf{k}'\rangle = \frac{4\pi e^2}{V\kappa} \frac{\delta_{\mathbf{k}_0+\mathbf{k}, \mathbf{k}_0'+\mathbf{k}'}}{|\mathbf{k}'-\mathbf{k}_0|^2 + \beta^2}, \tag{9}$$

where V represents the volume of the crystal and the electron wave functions are approximated by simple plane waves. The matrix element describes a two- body collision, where the total momentum is conserved, as indicated by the Kroenecker δ-function. The total scattering rate $\Gamma_{ee}(\mathbf{k}_0)$ due to the electron-electron interaction is obtained from (8) by summing over $\mathbf{k}, \mathbf{k}'$ and $\mathbf{k}_0'$. For non-degenerate situations $(f_\mathbf{k} = f_{\mathbf{k}'} = 0)$ we have [9]:

$$\Gamma_{e-e}(\mathbf{k}_0) = \frac{me^4}{\hbar^3 V \kappa^2} \sum_\mathbf{k} f_\mathbf{k} \frac{|\mathbf{k}-\mathbf{k}_0|}{\beta^2(|\mathbf{k}-\mathbf{k}_0|^2 + \beta^2)}. \tag{10}$$

The choice of the wave vector $\mathbf{k}$ determines the value of the wave-vector difference $\mathbf{g} = \mathbf{k} - \mathbf{k}_0$ of the two electrons before the collision. After the collision the relative wave vectors will be changed to $\mathbf{g}' = \mathbf{k}' - \mathbf{k}_0'$. Since the scattering conserves the magnitude of $\mathbf{g}$, the new vector $\mathbf{g}'$ can be determined from the angular distribution probability $P(\theta)d\theta$ with θ equal to the angle between $\mathbf{g}$ and $\mathbf{g}'$. The angular probability $P(\theta)d\theta$, determined from the differential scattering probability $S(\mathbf{g}, \mathbf{g}')$ is given by

$$P(\theta)d\theta = \frac{sin\theta d\theta}{g^2 sin^2(\theta/2) + \beta^2}, \tag{11}$$

with $g = |\mathbf{g}|$.

Electron $-$ Plasmon Interaction

As it was shown earlier, the splitting of the Hamiltonian of an electron gas gives a term describing the interaction of the electrons with the quasi- particles associated with the collective behaviour of the electron gas, i.e. plasmons. The interaction Hamiltonian for the electron-plasmon scattering is given by

$$H_{e-pl} = \sum_{q<q_c} \left(\frac{\pi e^2 \hbar^3}{2V\kappa\omega_p m^2 q^2}\right)^{1/2} \sum_\mathbf{k}(2\mathbf{k}\cdot\mathbf{q} + q^2)$$
$$\cdot (a_\mathbf{q}c_{\mathbf{k}+\mathbf{q}}^+ c_\mathbf{k} + a_{-\mathbf{q}}c_{\mathbf{k}-\mathbf{q}}^+ c_\mathbf{k}). \tag{12}$$

The squared matrix element can be easily calculated for plane waves and the following expression is obtained [8]

$$|M|^2 = \frac{\pi e^2 \hbar^3}{2V\kappa\omega_p q^2 m^2}(2\mathbf{k}\cdot\mathbf{q} + q^2)^2[N_q\delta(\epsilon_{k+q} - \epsilon_k - \hbar\omega_p)$$
$$+ (N_q + 1)\delta(\epsilon_{k-q} - \epsilon_k + \hbar\omega_p)] \tag{13}$$

Here, N_q is the Bose-Einstein equilibrium distribution population for the plasmon population and $\mathbf{q}$ is the wave vector of the plasmon involved in the interaction. Equation (13) describes a process in which an electron with wave vector $\mathbf{k}$ absorbs or emits a plasmon $\mathbf{q}$, with an energy exchange equal to $\hbar\omega_p(q)$. If the small-q limit is taken, the energy involved in the transition is just $\hbar\omega_p$, where ω_p is the plasma frequency. The upper limit q_c in (13) defines the maximum wave vector for which plasmons exists as an independent excitation of the electron gas. Its value can be determined from the analysis of the complex dielectric function. The final closed form for Γ_{e-pl} obtained from Fermi golden rule, is fairly complex, since different regions are defined by the requirements of energy and momentum conservation [9]:

$$\Gamma_{e-pl}(k) = \frac{e^2 \omega_p m^*}{\pi \kappa \hbar k}[N_q ln(\frac{q_a^+}{q_a^-}) + (N_q + 1)ln(\frac{q_e^+}{q_e^-})u(E_k - \hbar \omega_p)], \qquad (14)$$

where $u(\epsilon_k - \hbar \omega_p)$, the step function, is zero for $\epsilon_k < \hbar \omega_p$ and is unity for $\epsilon_k > \hbar \omega_p$. The terms in the argument of the logarithmic functions are given by:

$$q_a^- = k[(1 + \frac{\hbar \omega_p}{\epsilon_k}^{1/2}) - 1],$$

$$q_e^- = k[1 - (1 - \frac{\hbar \omega_p}{\epsilon_k})^{1/2}],$$

$$q_a^+ = min\{q_c, k[(1 + \frac{\hbar \omega_p}{\epsilon_k})^{1/2} - 1]\},$$

$$q_e^+ = min\{q_c, k[(1 - \frac{\hbar \omega_p}{\epsilon_k})^{1/2} + 1]\},$$

where $min\{\ \}$ indicates the minimum between the two quantities enclosed in the brackets and the suffix a and e indicate respectively absorption and emission. If the wave vector k_p, associated with the plasma frequency, is introduced (with $\hbar^2 k_p^2/2m = \hbar \omega_p$), Eq.(14) can be rewritten in the following form: for the emission process with $k > k_p$,

$$k_p > q_c; \Gamma_{e-pl}(k) = \begin{cases} C(N_q + 1)ln\frac{q_c/k}{1-(1-\hbar \omega_k/\epsilon_k)^{1/2}}, & k > \frac{q_c^2+k_p^2}{2q_c} \\ 0, & k < k < \frac{q_c^2+k_p^2}{2q_c} \end{cases} \qquad (15a)$$

$$k_p < q_c; \Gamma_{e-pl}(k) = \begin{cases} C(N_q + 1)ln\frac{1+(1-\hbar \omega_p/\epsilon_k)^{1/2}}{1-(1-\hbar \omega_p/\epsilon_k)^{1/2}}, & k_p < k < \frac{q_c^2+k_p^2}{2q_c} \\ C(N_q + 1)ln\frac{q_c/k}{1-(1-\hbar \omega_p/\epsilon_k)^{1/2}}, & k > \frac{q_c^2+k_p^2}{2q_c} \end{cases}$$

For the absorption,

$$k_p > q_c; \Gamma_{e-pl}(k) = \begin{cases} CN_q ln\frac{q_c/k}{-1+(1+\hbar \omega_p/\epsilon_k)^{1/2}}, & k > \frac{-q_c^2+k_p^2}{2q_c} \\ 0, & k < \frac{-q_c^2+k_p^2}{2q_c} \end{cases}$$

$$k_p < q_c; \Gamma_{e-pl}(k) = \begin{cases} CN_q ln\frac{1+(1+\hbar \omega_p/\epsilon_k)^{1/2}}{-1+(1+\hbar \omega_p/\epsilon_k)^{1/2}}, & k < \frac{q_c^2-k_p^2}{2q_c} \\ CN_q ln\frac{q_c/k}{-1+(1+\hbar \omega_p/\epsilon_k)^{1/2}}, & k < \frac{q_c^2-k_p^2}{2q_c} \end{cases} \qquad (15b)$$

Here,

$$C = \frac{me^2 \omega_p}{\kappa \hbar^2 k}.$$

These expressions are similar, but more complete, than the ones derived by Kumenov and Perel [10].

In the limit of sufficiently small q_c, the condition $k_p > q_c$ is easily satisfied. Taking q_c for the non-degenerate case as $\alpha \beta$, with $\alpha < 1$,the condition $(k > (q_c^2 + k_p^2)/2q_c)$ for the emission process reduces to

$$\epsilon_k > \frac{2K_B T}{\alpha^2}.$$

For $\alpha = 0.5$ we recover the result found by Davidov [11]for plasmon losses in ionized plasmas: only electrons with energy higher than twice the thermal energy can emit plasmons.

If the electron energy ϵ_k is much higher than the plasma energy $\hbar\omega_p$, the term $1 - [1 - (\hbar\omega_p/\epsilon_k)]^{1/2} = \frac{1}{2}\hbar\omega_p$, and the scattering rate for plasmon emission is

$$\Gamma_{e-pl}^{(em)} = \frac{m^* e^2 \omega_p}{\kappa \hbar k}(N_q + 1)ln(\frac{\hbar k}{m^*}\frac{q_c}{\omega_p}). \tag{16}$$

From the expression for the squared matrix element $|M|^2$, and the fact that $2\mathbf{k}\cdot\mathbf{p}+q^2 = \frac{2m}{\hbar}\omega_p$, it follows that the electron-plasmon differential scattering probability $S(\mathbf{k},\mathbf{k}_0)$ is of the same form as the one obtained for the polar optical phonons (at least as far as the q dependence is concerned). The angular dependence of the scattering is also of the same type, and it is given for parabolic bands by

$$P(\beta)d\beta = \frac{\sin\beta d\beta}{\epsilon + \epsilon' - 2\sqrt{\epsilon\epsilon'}\cos\theta}, \tag{17}$$

where ϵ and ϵ' are the electron energies before and after the scattering.

Plasmon − phonon coupling

In polar materials, such as GaAs, the interaction between the electrons and lattice vibrations introduces novel features in the properties of the material. When electrons are present in the conduction band, the screening effects modify both the optical mode frequency and the electron-phonon interaction. For non-degenerate conditions, Ehrenreich [12] found that the optical-phonon frequency was raised for wavelengths longer than the thermal wavelength of the electrons. If carriers are degenerate, the frequencies of the charge density fluctuations are comparable to the optical frequencies. The longitudinal phonon is no longer a normal mode, since the longitudinal electric field, which accompanies such phonons, must couple strongly to the charge-density fluctuations of the electron gas. The coupling of the phonons to the plasmons leads to important changes in the nature of the collective modes. Varga [5] showed that, in the RPA, the polarizabilities of electrons and ions are additive. Thus, the total dielectric function for the coupled system is given by

$$\epsilon_T(q,\omega) = \epsilon_e(q,\omega) + \epsilon_L(q,\omega) - 1, \tag{18}$$

where ϵ_e is the electronic dielectric function. $\epsilon_L(q,\omega)$ is the lattice contribution given by

$$\epsilon_L(q,\omega) = \kappa_\infty \frac{\omega_{lo}^2 - \omega_{to}^2}{\omega^2 - \omega_{to}^2} \tag{19}$$

where ω_{lo} and ω_{to} are the frequencies of the longitudinal and transverse optical phonon respectively. The dispersion of the optical phonon modes has been neglected because, at the concentration of interest here, the Fermi wave vector is a few percent of the Brillouin-zone width. The transverse + longitudinal optical frequencies are connected by the Lyddane-Sachs-Teller relation

$$\omega_{lo}^2 = (\kappa_0/\kappa_\infty)\omega_{to}^2$$

κ_0 being the static dielectric constant, and equal to 10.3 in GaAs. In the long-wavelength limit, Eq.(18) can be written as

$$\epsilon_T(q,\omega) = \kappa_\infty + \kappa_\infty \frac{\omega_{lo}^2 - \omega_{to}^2}{\omega^2 - \omega_{to}^2} - \kappa_\infty (\frac{\omega_p}{\omega})^2, \tag{20}$$

where the $q \to 0$ limit of the electronic dielectric function

$$\epsilon_e(q,\omega) = \kappa_\infty \left(1 - \frac{\omega_p^2}{\omega^2}\right) \tag{21}$$

has been used. The zeros of Eq. (19) give the dispersion curve for the coupled modes. At $q = 0$ the coupled modes are described by two branches, denoted ω_+ and ω_-, given by

$$\omega_\pm^2(q=0) = \frac{1}{2}[\omega_p^2 + \omega_{lo}^2] \pm [(\omega_p^2 + \omega_{lo}^2)^2 - 4\omega_p^2\omega_{to}^2]^{1/2}. \tag{22}$$

Such equation compares very well with the data of Mooradian and Wright [13]. For a small concentration, the lowest root starts out as a plasmon of frequency $(ne^2/\kappa_0 m)^{1/2}$, then at higher concentration it becomes more phonon-like. In the limit of large density, it is the longitudinal optical phonon mode of the crystal. The frequency is ω_{to} and not ω_{lo}, as it would be expected, because the charge carriers have shielded out the ionic field, which, in a pure crystal, raise ω_{lo} compared to ω_{to}. Conversely, the root ω_+ starts out at the LO phonon and turns into a plasmon (now at frequency ω_p) as n increases. The coupled modes are a mixture of collective electronic and ionic motion.

The analysis of the total dielectric function shows very interesting properties for polar semiconductors with of high electron densities. In particular, it has been shown [5,13] that the scattering strength of the coupled plasmon-phonon modes is proportional to $Im(1/\epsilon_T)$. In the limit of very low density the unscreened Froehlich scattering lifetime is recovered from the analysis of $Im(1/\epsilon_T)$. For very high carrier concentrations the highly screened LO phonon limit [14] is obtained

ENERGY LOSSES IN DOPED REGIONS

We present the results of an ensemble Monte Carlo (EMC) study of the effect of the complete e-e (including the contribution of the long range part of the interaction, i.e. plasma oscillations) interaction in GaAs, with conditions appropriate to the base of the PDB under high injection [15,16]. A two valley (Γ and L) model is used, that includes polar optical, non-equivalent (Γ-L) and equivalent (L-L) intervalley, acoustic phonons, as well as ionized impurities. Electrons with an initial energy of a few meV's are injected into a low-field region with a density of cold ambient electrons of $10^{17} cm^{-3}$. Limiting the study to the concentrations considered here allows us to neglect the coupling between the plasmon and phonon modes. If the electrons are injected with energy below the threshold for intervalley scattering (0.3 eV), then energy and momentum losses will be mainly controlled by the interaction of the fast electrons with polar optical phonons, and with the ambient electrons and plasmons.

The scattering rates for electron-electron and electron-plasmon processes presented previously are tabulated at the beginning of the simulation. Carrier-carrier scatterings are then treated as any other mechanisms in the Monte Carlo algorithm. The initial conditions for the simulation correspond to the injection of electrons over a barrier, that is the EMC electrons are started as a monoenergietic beam at energy E_0 and momentum k_0 in the $< 100 >$ direction. A uniform electric field of 400 V/cm along the same direction is assumed for the base region which characterized by an impurity concentration of $10^{17} cm^{-3}$. It is also assumed that the ambient electrons are in thermal equilibrium with the lattice at 77 K, and are not perturbed by the incoming fast carriers.

Fig. 1 shows the average drift velocity as a function of the distance travelled by the injected packet of electrons, for two different values of the initial energy E_0. The dashed curve refers to the case where the total e-e interaction is ignored, while the case with the complete interaction included is depicted by the solid curves. The absence of ballistic motion under the simulated conditions is evident; the polar optical phonon scattering is efficient enough to reduce the average drift velocity by 30 percent over distances of the order of 200 nm for high injection energies, and 100 nm for low injection energy. At the same distances, it is found that e-e and e-pl scattering further reduces the average electron velocity to about one half of its initial value. It should be pointed out that, by applying a field of 10 kV/cm, high velocities can be sustained over much longer distances, in accordance with the result of Tang and Hess [17]. In such a case, the energy pumped into the electron system by the electric field causes a significant intervalley transfer, which again drastically reduces the drift velocity of the injected electrons for distances larger then 200 nm. The average energy loss as a function of distance is shown in Fig. 2. The lower slope of these curves, with respect to Fig. 1, is related to the gain of perpendicular momentum by the electrons, with also shows up as a fast increase of the electron effective temperature.

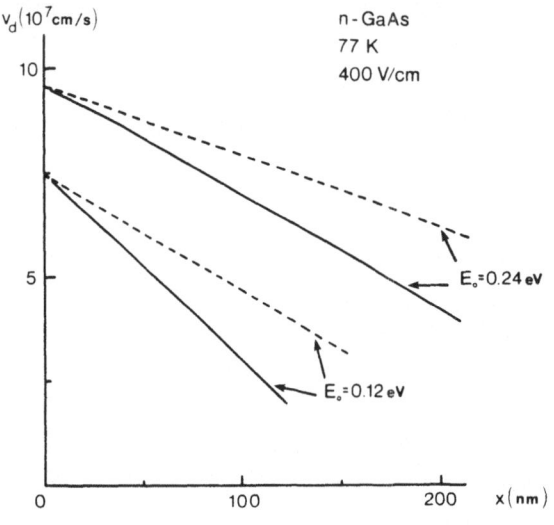

Fig. 1. Average drift velocity versus distance with (solid curve) and without (dashed curve) e-e interactions, for two different injection energies.

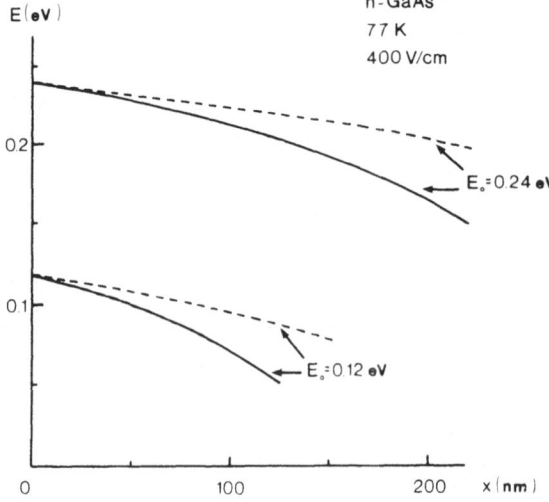

Fig. 2. Average energy versus distance with (solid curve) and without (continous curve) e-e interaction, for two different injection energies.

E-e and e-pl scattering have a quite dramatic effect on the distribution function of the injected electrons. Fig. 3 displays the distribution function for energy (curve a) and parallel momentum (curve b), evaluated when the center of the injected packet is 200 nm from the point of injection, for the highest injection level ($E_0 = 0.24$ eV). The distinct peaks in the energy distribution without e-e scattering (dotted curve) are related to the discrete energy losses by the emission of polar optical phonons. The contribution of e-e and e-pl interactions is evident from the dashed curve where, as a result of the much faster randomization of the hot electrons, the distribution functions are completely smeared out. A comparison between Figs. 1 and 2 indicates that, even if not terribly effective in dissipating energy, Coulombic processes are extremely effective in setting up a smoother distribution, with a long tail at lower energies and momenta. Temperature is also an important factor that affects the energy and momentum losses of fast electrons in highly-doped regions. A simulation performed at room temperature has indicated a faster randomization with respect to the nitrogen temperature case, but no significant changes in the energy dissipation rate.

It is fair to extrapolate those results by suggesting that the performances of submicrometer devices is going to be affected by the strong losses of momentum perpendicular to the contact (collector) barrier, for densities above $10^{17} cm^{-3}$.

THE THETA DEVICE

A complete device simulation, based on the model just described, has been performed for the THETA device [2-3,18-19]. The conduction band edge for a cross section of the device is shown in Fig. 4a. The active part of the device (Fig. 4b) consist of a GaAs-AlGaAs-GaAs quantum well, with a very thin (300 A) and highly doped ($10^{18} cm^{-3}$) GaAs base. Electrons are injected into the base by tunneling through the potential barrier between emitter and base. Because of the homogeneity of the electric field in the direction normal to the plane of Fig. 4b, it is possible to restrict the simulation to a two dimensional real space.

The main aspect of this work is that, in addition to the electron degeneracy effects, the quantum structure of the potential barrier at the base collector interface has been taken into account; this is an important point in order to compare the model predictions with experimental data.

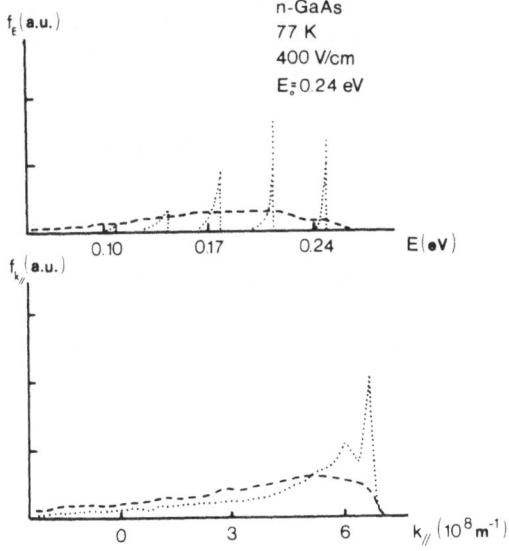

Fig. 3. Distribution function for energy (curve a) and parallel momentum (curve b) at an average distance of 200 nm for the high injection case. The dashed curve includes e-e interaction, the dotted one considers only phonon and impurity scatterings.

The GaAs model is the same as described earlier. The Monte Carlo algorithm is coupled here to a Poisson solver, so that the electrons move in a potential distribution which is solved self-consistently. In the THETA device, the quantum transport properties are strongly affected by the shape of the conduction band edge which is shown, for the particular device that has been studied, in Fig. 4b. This structure and the high doping density, introduced in order to obtain reasonable emitter currents with a bias smaller than the separation between Γ and L valleys, impose several quantum corrections necessary for an accurate description of the device dynamics. The quantum effects are included in scattering rates (through Pauli exclusion principle), in distribution functions and in transmission and reflection coefficients.

Some other important feature are listed below:

a) The doping in the base region produces a "sea" of cold degenerate electrons (in the simulated device about 10^5 electrons.) These electrons are confined in the base by the collector potential barrier. As pointed out earlier, the electrons are one component of the "plasma" which provides an important energy or longitudinal momentum loss mechanism. Electron-electron and electron-plasmon scattering are considered as expained in the previous section. Due to the high doping in the base region, the Thomas-Fermi screening length is used. To avoid unphysical long range correlations in the electron-electron scattering the partners are chosen, if there are any, in the range of two Thomas-Fermi lengths.

To account for the Pauli exclusion principle which affects to a remarkable degree the dynamics of the highly degenerate particles considered, we used the technique in which k-space is divided in to elementary cells of arbitrary volume, the maximum occupation number (i.e. the maximum number of particles allowed by the Pauli exclusion principle) is obtained by counting the number of states contained in each cell (see Refs. 21 and 22 for details).

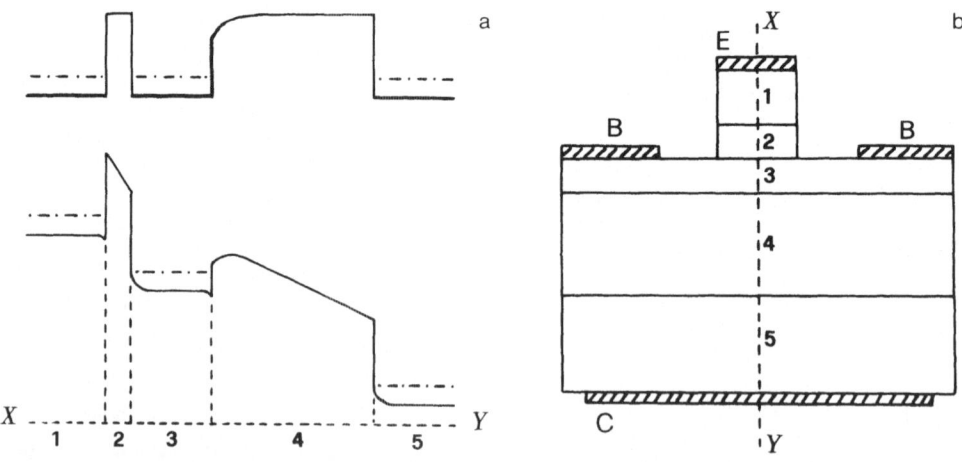

Fig. 4. a.) Conduction band edge along a cross section of the simulated device. Case a is for zero bias, case b for positive base-collector bias. (1) is the emitter region, (2) the emitter-base barrier, (3) the base region, (4) the base-collector graded barrier, (5) the collector region. b.) Schematic representation of the simulated device.

In order to consider particle inhomogeneities in the base, four different occupation numbers are computed in terms of the dimension of the region and assigned to a particular k-space matrix.

b) The second relevant point is the presence of the emitter and collector barriers. The spectrum of injected electrons is obtained integrating numerically the Schroedinger equation for incident plane wave functions. In this way one obtains the transmission coefficient as a function of total electron energy $T_e(E)$ and normal electron energy $T_e(E_n)$, where E_n is the energy associated with momentum component k_n normal to the collector barrier. Two approximations have been used: no alloy scatterings have been considered and the influence of the collector barrier has not been taken into account (resonant tunneling). Fig. 5a shows the distribution of injected particles $N(E_n)$ at 4.2 K which, in terms of current density distribution, is given by:

$$N(E_n) = \int_0^T \int_S J(E_n)dsdt \tag{23}$$

and $J(E_n)$ is current distribution per unit surface:

$$J(E_n) = \frac{m_e}{2\pi\hbar^3}T(E_n)\int_0^\infty f_E(E)[1 - f_B(E)]dE_t \tag{24}$$

where f_E and f_B are respectively the density of states in the emitter region and in the base region, and E_t is the energy associated with the momentum components parallel to the collector barrier.

c) Finally it is necessary to analyse is the effect of the collector barrier. As already mentioned, resonant tunneling has not been included in the model, thus it was possible to decouple the emitter and collector barriers and compute numerically the transmission and reflection coefficients $T_c(E_n)$ and $R_c(E_n)$. It is very likely that resonant tunneling has a considerable influence on HEIBD dynamics, and more work is needed.

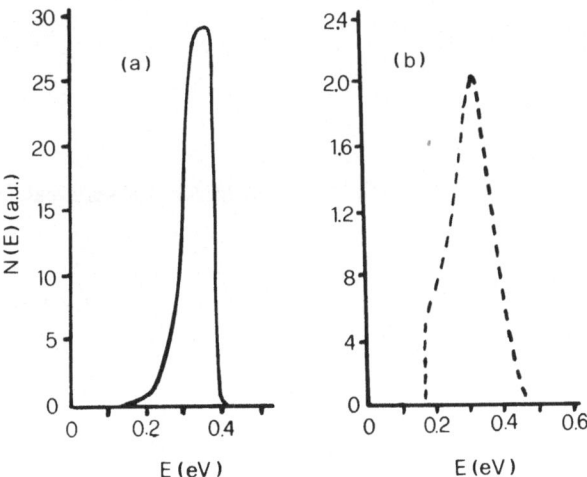

Fig. 5. a.) Energy spectrum of the injected particles and (b.) energy spectrum of the collected particles, for $I_e = 100\mu A$ with a collector barrier height of 170 meV.

Results

We present here some of the results of the simulation, compared with experimental data. Fig. 5b shows the collected electron spectrum for an emitter current of $100\mu A$ and a collector barrier height of 170 meV. A comparison with the injected particle spectrum reported in Fig. 5a shows that the width of the distribution at half maximum (of about 65 meV) is of the same order of the injected electron distribution, and this is a clear indication of the ballistic behaviour of the injected electrons.

For the analysis of the characteristics curve a few preliminary comments are needed. For a graded collector barrier, as in the THETA device, the dependence of the barrier height ΔE_c on the collector bias ΔV_c is linear only for negative bias values and, in this case, is given by:

$$\Delta E_c = \Phi_c - \Delta V_c \qquad (25)$$

where Φ_c is the collector barrier height for zero bias. For $\Delta V_c > 0$ this relation is no longer valid and is strictly connected with the grading profile which is, at present, not very well known. For Φ_c we used the value commonly quoted in the literature of $215^{+35}_{-15}meV$, where the errors are related to the uncertainty in the Al mole fraction in AlGaAs. Fig. 6 shows the theoretical energy distribution obtained using the linear approximation given by Eq. (25) (circles), compared with the experimental result (solid line). In this case one has to consider that the experimental curve is obtained indirectly, by taking the derivative of the collector current with respect to the collector bias. The two points in the interval (-1.0,0.0) V are due to an enhancement originating from the phonon interaction and are not present in the experimental curve. Because of the real injected particle distribution is probably affected by interbarrier diffusion, and inelastic tunneling should also be considered. Examining the (I,V) curve in more detail, one can see the small tail in the interval (-0.3,-0.2) V due to the corresponding tail in the electron distribution. The degeneracy of the electrons is now important, preventing about 85 percent of the scatterings via Pauli exclusion principle.

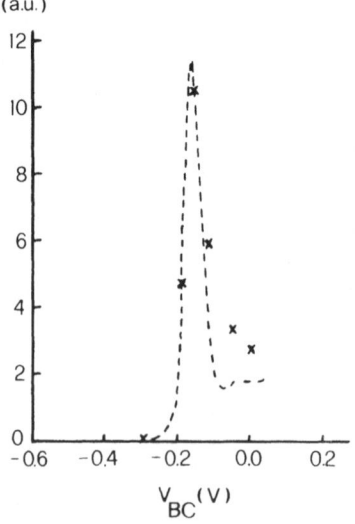

Fig. 6. Theoretical (crosses) and experimental (dashed curve) energy distribution for the THETA device at 4 K. The dotted lines are theoretical statistical errors.

As previously mentioned the comparison of the data is not straightforward for the positive collector bias because of the presence of the grading at the collector barrier. In this case it is not possible to use the linear relation (25). In fact the region between 0.0 and 1 V is actually still under investigation, in particular to understand the effect of the grading profile. Using the value for the collector barrier of 165 meV reported in the literature at 1 V bias, we obtain values for the transfer ratio in good agreement with experimental results. Figure 7, where we report all Monte Carlo results for $I_e = 100\mu A$ (circles) and $I_e = 60\mu A$ (crosses), shows that the point obtained for $1.0V$ base-collector bias is in good agreement with experiment (solid line). Then, assuming that the value of 165 meV is correct, this point is a good check of the model because in this case the effect of the resonant tunneling is knowwn to be irrelevant.

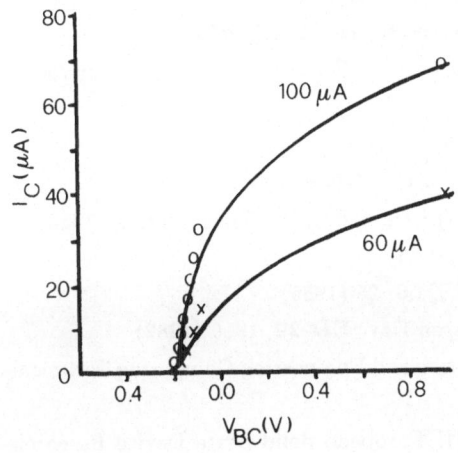

Fig. 7. Collector current for $I_e = 60\mu A$ and $100\mu A$. The solid line shows the experimental curves, circles and crosses the Monte Carlo results.

CONCLUSIONS

The Monte Carlo investigation of hot electron injection has shown the importance of electron-electron and electron-plasmon scattering when high doping is used. In order to achieve ballistic transport in devices based on the injection of fast electrons, the width of the base regions cannot be larger than few hundred anstrongs. Furthermore, devices have to operate at very low temperatures, in order to suppress or at least reduces all the scattering processes. It has been shown that it is possible to study ballistic transport using a semiclassical approximation, including a very careful and sophisticated treatment of quantum effects. The data reported, deprived from arbitrary derivation from the energy distribution show that the state of art is skillful enough to obtain a good description of the dynamics of the THETA device. The model used is capable also of successfully describe other ballistic devices. The problem of resonant tunneling , which seems to play an important role in the dynamics of HEIBD, needs further investigation.

ACKNOWLEDGMENTS

I would like to thank Dr. F. Antonelli and Dr. D. K. Ferry for their contribution to part of the work presented here. The Computer Center of the Modena University is gratefully acknowledged.

REFERENCES

1. J.R. Hayes, A.F.J. Levi and W. Wiegmann, *El.Lett.* **20**, 851 (1984)

2. M. Heilblum, *Sol. State Electr.* **24**, 343 (1981)

3. M. Heiblum, M. I. Nathan, D. C. Thomas, and C. M. Knoedler, Phys. Rev. Lett. **55**, 2200 (1985)

4. M. A. Hollis, S. C. Palmateer, L. F. Eastmann, H. V. Dandekar, and P. M. Smith, IEEE Electron Dev. Lett., **EDL4**, 440 (1983)

5. B.B. Varga, Phys. Rev. **137**. 1896 (1965)

6. M.E. Kim, A. Das, and S.D. Centuria, Phys. Rev. **B18**, 6890 (1978)

7. D. Bohm, and D. Pines, Phys. Rev. **92**, 609 (1953)

8. O. Madelung, *Introduction to Solid State Physics*, (Springer, Berlin), 1978

9. P. Lugli, Ph. D. Dissertation, Colorado State University (1985)

10. S.E. Kumenov, and V.I. Perel, Sov. Phys. Semicon. **16**, 1982 (1982)

11. B.I. Davidov, in *Plasma Physics and the Problem of Controlled Thermonuclear Reactions*, vol. 1, Izd. AN SSR (1958)

12. H. Ehrenreich, J. Phys. Chem. Solids **8**, 1996 (1965)

13. A. Mooradian, and G.B. Wright, Phys. Rev. Lett. **16**,999 (1966)

14. P.A. Wolf, in *Proc. Int. Conf. Light Scatt. Spectra Solids*, (Speinger, Berlin), 1969

15. P.Lugli and D.K.Ferry, Physica B **129**, 532 (1985)

16. P.Lugli and D.K.Ferry, IEEE El.Dev.Lett. **EDL6**, 25 (1985)

17. J. Y. Tang, and K. Hess, IEEE Trans. Electron Dev. **ED-29**, 1906 (1982)

18. F. Antonelli, IBM European Center for Scientific and Engineering Computing Technical Report (1987)

19. F. Antonelli, and P. Lugli, in Proc. of XVII European Solid State Device Research Conference ESSDERC87, Eds. P. Calzolari and G. Soncini, p. 177, Tecnoprint, Bologna (1987)

20. S. Bosi, and C. Jacoboni, J. Phys. C. **9**, 315 (1976)

21. P. Lugli, and D. K. Ferry, IEEE Trans. Electron Dev. **ED-6**, 25 (1985)

MODELS FOR SCATTERING AND VERTICAL TRANSPORT IN

MICROSTRUCTURES AND SUPERLATTICES

D. C. Herbert

Royal Signals and Radar Establishment
Malvern, UK

ABSTRACT

Novel techniques for solving hot carrier transport problems are presented. These techniques are combined with calculations of scattering rates in superlattices to discuss vertical transport.

1. INTRODUCTION

The ability to fabricate microstructures with feature sizes of 30A and less has revitalised the subject of semiconductor physics and is forcing device physicists to rethink some of the basic concepts and approximations used in device modelling. As device dimensions continue to shrink, various physical effects which can often be neglected for large devices start to become important and may dominate in some of the ultra-small structures of current interest. These effects may be classified into four main subject types, namely hot carrier transport, many body theory, electronic structure and scattering.

Heterojunctions or very high doping levels can lead to large built in electric fields which inevitably drive the carriers far from equilibrium. If the spacial and temporal scales are sufficiently short, then a quantum description of hot carrier transport is required. This implies that hot electrons with associated quasi-ballistic overshoot effects are likely to form a central theme for modelling next generation devices. High carrier concentrations are necessary to localise space charge in small regions so that electron-electron interactions and associated many-body effects become important. In this context we may note that it is believed that plasmon emission in the heavily doped base of conventional hot electron transistors, is at present limiting the achievable current gain in these devices[1]. When potentials from hetero-junction structures vary coherently on length scales which are short compared to carrier mean free paths, then novel band structure effects become important. Superlattices provide the most important examples, but we note that all aspects of transport and scattering are ultimately tied to the electronic structure. The ability to engineer the band structure of superlattices therefore offers very exciting possibilities for next generation devices. Finally, scattering rates are required for any realistic model of non-equilibrium effects and these still have some uncertainty, particularly for hot carriers, even in bulk crystalline semiconductors. In the microstructures of current interest, the lack of knowledge of scattering rates is likely to be a major limitation in simulating device behaviour.

The possibility of exploiting fast tunnelling and quasi-ballistic transport for high speed electronics has led to considerable interest in vertical transport in semiconductor microstructures. So far most research has focussed on transport associated with the lowest sub-band of a superlattice, with low-lying resonant states in double barrier structures or with the various hot electron transistor structures. In a recent paper[2] however, we have proposed that transport associated with the higher mini-bands of a superlattice, which has been neglected to date, may in fact have great potential for next generation devices. It is argued that a band structure engineering approach may be employed to taylor both scattering and transport for beneficial effects in hot electron devices. Some aspects of this proposal will be developed further in this paper.

The complexity of the new physics of microstructures makes it highly desirable to develop simple models of transport. Traditional Monte-Carlo methods involve excessive computer time, particularly when inhomogeneity in more than one spatial dimension is important, and in view of the many uncertainties in physical parameters and scattering, it seems that the numerical accuracy achieved by this method is not essential in many cases. The variety of structure available in microstructures is enormous, and ideally one would like fast computer models which can be used in an interactive manner to uncover trends in device related physics, using monte carlo simulation for final tuning of optimised structures.

In this paper we outline such an approach to microstructure physics that we are currently developing at RSRE. The emphasis is on obtaining simple numerical models which can be used in an interactive manner for device modelling purposes. In section (2), as a first step we map the Boltzmann equation (or weak coupling quantum kinetic equation[3,4]) onto a one dimensional system which allows analytic solutions. In section (3), models for scattering in superlattices are described and in section (4) results of transport calculations, which use the models for scattering and transport, are presented. Throughout the paper, the theory is illustrated using calculations for three superlattice structures with parameters shown in table (I). Effective mass theory is used for the electronic structure and wave functions are calculated numerically using the technique described by Vigneron and Lambin[5]. We have justified this approximate treatment by direct comparison with full pseudo-potential calculations[12].

Table (I)

Superlattice Parameters for GaAs/AℓGaAs Examples

Example	Unit Cell Parameters			Band edges (meV)					
	Well Width(A°)	Barrier Width(A°)	Barrier Height(eV)	Band 1		Band 2		Band 3	
I	100	10	0.25	17	44	81	166	206	344
II	80	40	0.2	36	43	127	164	242	335
III	40	30	0.2	68	118	232	411	458	761

2. MODELS FOR TRANSPORT

To motivate our approach to transport we first consider the semi-classical

Boltzmann equation. It is convenient to split the full one–electron distribution function into terms which are symmetric (f) and anti–symmetric (g) with respect to k–space inversion on surfaces of constant total energy. The steady state Boltzmann equation may then be written in the form

$$F \frac{\partial}{\partial k_x} f_t(\underline{k}, x) + v \frac{\partial}{\partial x} f_t(\underline{k}, x) = - g_t(\underline{k}, x) S_0 + \int S(\underline{k}, \underline{k}^1) g(\underline{k}^1, x) d^3 k' \quad (1)$$

$$F \frac{\partial}{\partial k_x} g_t(\underline{k}, x) + v \frac{\partial}{\partial x} g_t(\underline{k}, x) = - f_t(\underline{k}, x) S_0 + \int S(\underline{k}, \underline{k}^1) f(\underline{k}^1, x) d^3 k' \quad (2)$$

where subscript t labels surfaces with total energy E_t, v is a velocity, F is an electric field, S is the scattering matrix and $S_0 = \int d^3 k^1 S(\underline{k}, \underline{k}^1)$. Atomic units are used throughout this paper. Later, we shall map these equations onto a one–dimensional system when t will label trajectories. In this paper we consider systems with the dominant inhomogeneity in the x direction and assume that the distributions have axial symmetry.

As a preliminary we consider solving for the steady state in a homogenious superlattice. In this case the distributions f_t, g_t, for a given sub–band, can be expanded in low order polynomials:

$$f_t = \sum_{j=1}^{N} f_{t,j} (k_x/K)^{2j-2}, \quad g_t = \sum_{j=1}^{N} g_{t,j} (k_x/K)^{2j-1} \quad (3)$$

where K is the maximum allowed value of k_x on the energy surface E_t. For energies greater than the bandwidth, $K = \pi/L$ where L is the superlattice period. If these forms are substituted into the Boltzmann equation, the electric field dependent term gives a coupling to higher order polynomials and to obtain a closed set of equations we replace this coupling with an equivalent coupling to terms within the polynomial subspace. This procedure has some analogy with the self consistent decoupling schemes employed for chains of Green functions in many–body theory. For energies less than the bandwidth the method is similar to expansions in Legendre polynomials[14] With a closed set of equations on a finite polynomial sub–space, the Boltzmann equation can be solved by standard matrix methods. In practice, with fairly coarse energy meshes, care is needed to avoid numerical instability and we add a weak 'stiffness' parameter roughly proportional to the second derivaties of the distribution function. The method gives a very large saving in computer time when compared with Monte–Carlo methods and allows the complex anisotropic scattering processes encountered in microstructures, to be included with ease .

In fig (1) we show a comparison with ensemble Monte–Carlo simulation for example I of Table (1), described in more detail elsewhere[6]. The calculation used bulk scattering rates, adjusted to allow for the mini–gap. It is found that quite high order polynomials are required to achieve close agreement with Monte Carlo, but the lower order approximations can also give adequate order of magnitude results. These particular calculations are performed at $100°K$ for a superlattice designed to have a bandwidth less than the phonon energy, taken to be 36 meV. The first band gap also exceeds the phonon energy so that scattering into the second sub–band is negligible. In such a structure, the scattering at low temperature is dominated by acoustic phonon scattering and the very weak phonon absorption. For increasing applied fields, the transport is rapidly dominated by Bloch oscillations. These show up explicitly as a

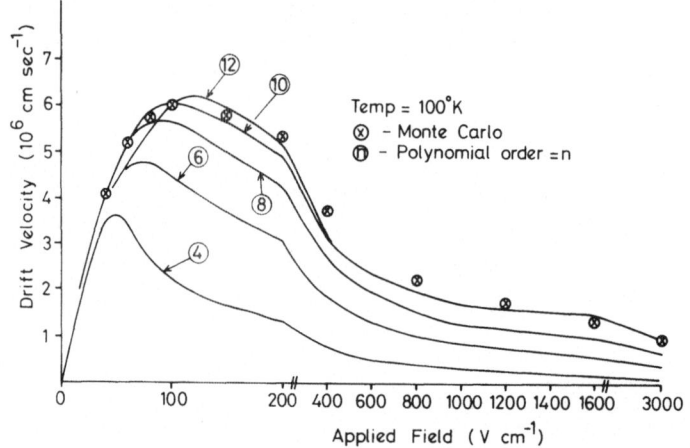

Fig. 1.   Comparison of various polynomial approximations with
Monte Carlo simulation for superlattice (I).

transient in the Monte-Carlo simulation[6], and a very large negative differential mobility is predicted.   These calculations were terminated at F = 3 kV/cm as tunneling is expected to become important above this and has not yet been included in the models.   At higher electric fields, the potential drop across the superlattice unit cell can also become comparable with the superlattice mini-band widths and a field dependent band structure will have to be considered.

The problem is more complicated for inhomogenious structures and to obtain simplified models of transport we take averages of the Boltzmann equations (1) and (2) on constant energy shells for positive k_x and map these averaged equations onto a one-dimensional system.   For E_t < bandwidth the mapping is given by

$$\rho_t^- = \langle j_t \rangle, \quad \rho_t^+ = E_t \langle f_t \rangle / \alpha(E_t) \tag{4}$$

where $\langle j_t \rangle$ denotes the mean current density on surface 't' and $\alpha(E_t)$ is a parameter depending on the shape of the full distribution. E_t is the energy of the surface.   In the one dimensional system current (J) and charge (Q) are given by

$$Q = N_1 \int \alpha(E_t) \rho_t^+ E_t^{-\frac{1}{2}} dE_t, \quad J = N_2 \int \rho_t^- dE_t \tag{5}$$

where N_1, N_2 are normalisations.

The scattering terms in (1) and (2) are simplified using relaxation approximations that retain the correct gain loss structure.   The kinetic equations in the one-dimensional representation then take the form

$$\frac{\partial^2 \rho_t^+}{\partial x^2} = \frac{1}{\ell_1[t]} \left\{ \rho_t^+ \left[\frac{1+b}{\ell_2[t]} + \frac{b}{\ell_2[t-1]} \right] - \rho_{t-1}^+ \frac{(1+b)}{\ell_2[t-1]} - \rho_{t+1}^+ \frac{b}{\ell_2[t+1]} \right\} \tag{6}$$

$$\rho_t^- = \ell_1[t] \frac{\partial}{\partial x} \rho_t^+ \tag{7}$$

where $\ell_1[t]$, $\ell_2[t]$ are momentum and energy relaxing path lengths respectively and b is a boson population appropriate to the inelastic scattering mechanism. Equations (6) and (7) have analytic solutions[3] but we find that it is a very good approximation to simplify further to obtain

$$\frac{\partial^2 \rho_t^+}{\partial x^2} = \frac{1}{\ell_1[t]} \left\{ \frac{\rho_t^+}{\ell_2[t]} - \frac{\rho_{t-1}^+}{\ell_2[t-1]} \right\} \tag{8}$$

Analytic solutions may now be obtained in the form

$$\rho_t^+ = \sum_{j=0}^{t} \left[\gamma_t^j \, e^{-x/\ell j} + \delta_t^j \, e^{x/\ell j} \right], \quad \ell_j^2 = \ell_1[j] \, \ell_2[j] \tag{9}$$

where

$$\gamma_t^j = \sum_{r=0}^{t-j} a_{t,r}^j \left[\frac{x}{\ell_j} \right]^r, \quad \delta_t^j = \sum_{r=0}^{t-j} b_{t,r}^j \left[\frac{x}{\ell j} \right]^r \tag{10}$$

Here trajectories 't' are labelled such that '0' denotes the highest total energy (normally the energy at which electrons are injected from the device cathode). The coefficients in the solutions are readily obtained from recursion relations.

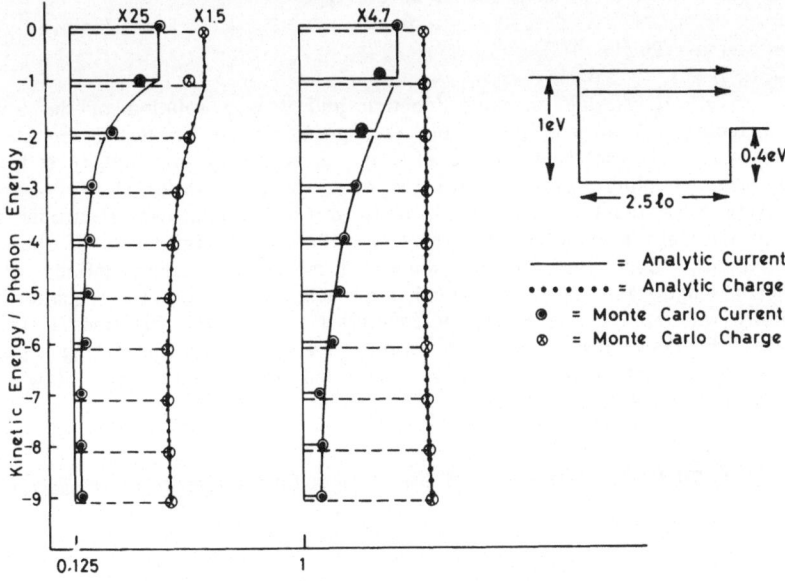

Fig. 2.   Comparison of results from the analytic model derived from the one-dimensional mapping, with Monte Carlo simulation for a silicon hot electron transistor at zero temperature at two positions in the base. Horizontal lines indicate magnitudes on energy trajectories, ℓ_0 is the ballistic path length for optic phonon emission.

To allow for electric fields either built in or applied, the space is divided into regions where each region contains a fixed number of trajectories and an average scattering length is associated with each trajectory within a region. By matching ρ^+_t at the boundaries of regions, using the contact boundary condition that growing exponential solutions vanish, and requiring that $\rho^-_t = 0$ at points of contact with band edges where the current density vanishes, all coefficients are determined. The method gives excellent agreement with Monte Carlo simulation for silicon monolithic hot electron structures [3] and also agrees very closely with the method recently introduced by Bertz[7]. Some results for a silicon hot electron transistor are shown in fig (2), discussed in more detail in ref (3). The analytic results differ slightly from the previous reported results due to the different approximations used for the path lengths. Due to the randomising character of scattering in Si, the analytic solution assumed that the symmetric part of the distribution becomes isotropic after a single scattering event and the small discrepancy with Monte-Carlo on trajectory 'l' may be related to this assumption. When f_t is isotropic and g_t is dominated by the linear term in k_x, the one-dimensional mapping yields $\alpha(E_t) = \sqrt{3}$, $\ell_1[t] = \ell_1[0]/\sqrt{3}$, $\ell_2[t] = \ell_2[0]/\sqrt{3}$. For ballistic motion on trajectory '0', $\alpha(E_0) = 1$. It is interesting to note that the current density has a strong quasi-ballistic peak on the highest trajectory, but that the charge density has a much flatter distribution over the trajectories. This reflects the fact that true ballistic transport is unattainable in dissipative systems. The distributions cannot be characterised by a temperature so that many approximations commonly used in device modelling[8] e.g drifted maxwellians, become inappropriate and the hot carrier aspects must be treated in detail.

The inclusion of quantum transport effects raises serious difficulties (for a recent review see for example Reggiani[9]). However, consistently with our philosophy of constructing simple models for device modelling applications, we are currently developing one-dimensional models which can be solved with the same algorithms outlined above for the Boltzmann equation. Our basic model is obtained by considering the motion of a wave packet of the type[4]

$$\Phi = \varphi(x - R)\psi(x - R) \qquad (9)$$

where φ is a slowly varying envelope function and ψ is a solution of the Shrödinger equation including any electric fields, but setting irreversible scattering terms to zero. This approximation is analagous to the effective mass theory of impurity states and is limited to weak coupling systems where the envelope term varies slowly compared to the underlying wave function. In microstructures with complicated electronic structure we think of electrons existing in wave packets with spacial dimensions of the order of scattering lengths. Such electrons will sense a local band structure defined within the electron path length and in a numerical approach, the Shrödinger equation can be solved locally with appropriate boundary conditions. The basis functions ψ in (9) are then obtained by diagonalising the local current operator. By considering equations of motion for the model wave packet, we find that for a sufficiently slowly varying envelope the average envelope density (ρ) obeys the equation

$$\frac{\partial}{\partial t}\rho + J\frac{\partial}{\partial x}\rho = 0 \qquad (10)$$

where J is the current associated with the function ψ. The term $J \, \partial/\partial x \, \rho$ is completely analagous with the term $v \, \partial/\partial x \, f$ in the Boltzmann equation (l) and consequently model kinetic equations of identical form to (6,7) are obtained. For superlattices it is possible for scattering lengths to become comparable with unit cell dimensions. In this case the process of optimising wave functions to allow for local band structure introduces quantum reflections, due to wave function missmatch and the electrons sense the 'grainyness' of the superlattice unit cell. Deviations of the wave function from the assymptotic W.K.B form also gives corrections both to scattering rates and to the density of states factor $(E_t^{-\frac{1}{2}})$ in the one-dimensional model of charge. Scattering rate corrections can be included in the path lengths and density of states corrections can formally be included in the shape function $\alpha(E_t)$. We are currently using these transport models to study hot carrier effects in superlattices. For realistic calculations a detailed understanding of scattering is required to obtain numerical values for the path lengths ℓ_1, ℓ_2.

3. SCATTERING RATES IN SUPERLATTICES

Hot carrier transport in superlattices appears to hold great potential for next generation devices. Current–voltage characteristics can show strong negative differential mobility due to Bragg reflection from Brillouin zone edges (see fig (1)) with possibilities for amplification and power generation. Fast Zener tunnelling between mini–bands in high electric fields suggests possibilities for switching. The very low electron mass in the higher sub–bands would also lead to very high sub–band mobility if carriers could be maintained in these sub–bands for sufficiently long times[2]. In addition the III–V semiconductors also have interesting optical properties which one can envisage combining with the transport features. All of these aspects are sensitive to electronic band structure and can be engineered towards desirable values. Scattering rates in the superlattices turn out to have a great deal of interesting structure. For possible high speed three terminal devices based on superlattices it will probably be necessary to consider quite high doping levels to minimise access resistance and RC couplings. For hot carriers we are therefore involved with plasmon scattering in addition to the usual phonon and impurity scattering.

To calculate the plasmon emission cross sections we consider a model in which a hot electron in one of the higher mini–bands interacts with a thermal gas of electrons in the lowest sub–band. In this case we can work with the one–electron Green function for the hot carrier and use thermal Green function techniques[10] to calculate the imaginary self energy in the standard manner. The Green function is calculated from the screened exchange diagram (Fig 3) where the plasmon loss terms come from the bubble diagrams in the screened coulomb interaction. We have to consider both inter–sub–band and intra sub–band processes. Details may be found in ref (2) where it is shown that the imaginary self energy for an inter–band process between states 'f' and 'i' can be approximated to the form

$$\text{Im} \sum = \frac{nm^*}{L} \left| V_k \right|^2 \left| M^{fi} \right|^2 \tag{11}$$

where L is the superlattice period, V_K is the coulomb interaction at wave vector K, where $E_B = \hbar^2 K^2/(2m^*)$ and E_B is the energy of the initial state band relative to the minimum of the final state band. $|M^{fi}|^2$ is essentially a matrix element between the initial state band and final state band averaged over all final states with corrections for band dispersion and Umklapp, n is the thermal electron density. Detailed calculation shows that this inter–sub–band scattering can be very weak (Figs 4,5). This is explained by noting that conservation of energy and momentum force a large wave vector transfer in the plane of the layers for inter–sub–band scattering so that the coulomb term V_K in (ll) can be small. Orthogonality of wave functions also act to make the term M^{fi} small. We expect that this suppression of inter–sub–band scattering is a general feature for coulombic type (small angle) scattering processes when the mini–gaps are sufficiently large, with interesting implications for hot electron devices.

Fig. 3. Dyson equation for the one-body propagator using the screened exchange approximation

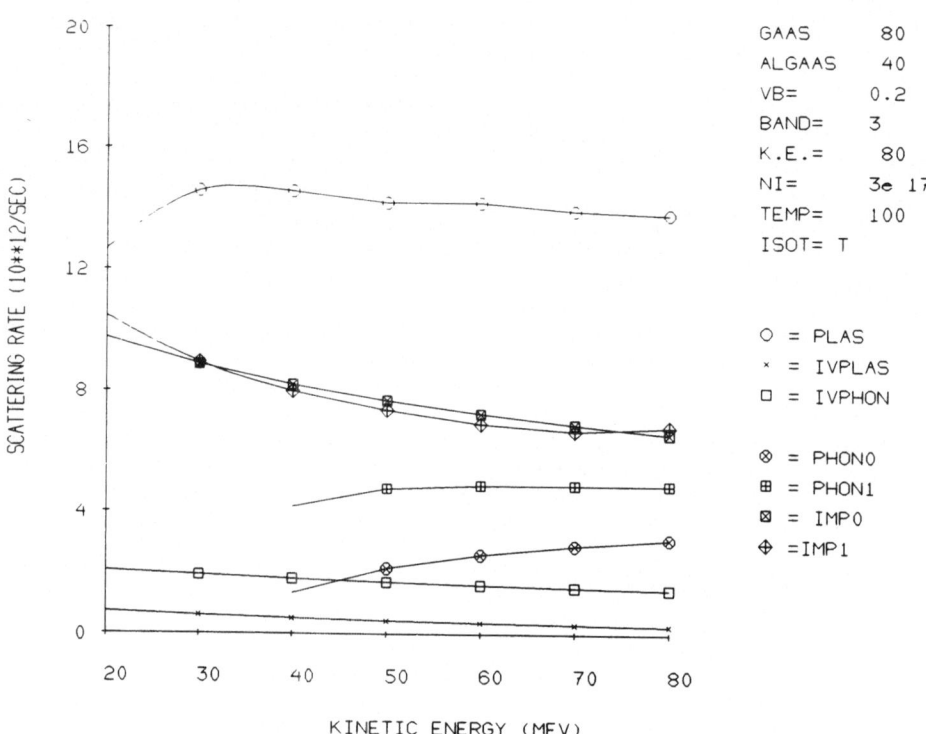

Fig. 4.   Scattering rates for superlattice (II) with initial
state in band 3. NI denotes doping density. Plas,
phon, imp refer to plasmon, phonon and impurity
scattering. iv indicates an inter-band process, 0
and 1 indicate that the initial state k_x value takes
the minimum and maximum value for the given kinetic
energy. Isot = T indicates that the factor $E_{qs}/2$
has been neglected in the screening function.   VB is
the barrier height in eV, KE is the maximum initial
kinetic energy within the band and the widths of the
GaAs and AℓGaAs layers are given in A°. Temp refers
to lattice temperature in °K.

The intra–sub–band scattering is much stronger and can be calculated from Fig (3) within the plasmon pole approximation. We approximate the plasmon dispersion to the form

$$\omega_\rho^2(q) \;=\; E_q^2 \;+\; \omega_{\rho\ell}^2(q_s^2 + \gamma\, q_x^2)/(q_s^2 + q_x^2) \tag{12}$$

where q_s is the wave vector in the plane of the layers, x is the superlattice growth direction and γ is the mass ratio m_s/m_x. The term containing γ is included to ensure that the plasma frequency approaches the correct value for isotropic bands. In many superlattices of interest, γ is small for the lowest sub–band and the plasmons have the form appropriate for a two–dimensional electron gas. $\omega_{\rho\ell}^2 = 4\pi n e^2 (\epsilon_0 m^*_s)^{-1}$, the bulk crystalline plasma frequency. To calculate scattering rates it is also necessary to allow for band non–parabolicity, wave function overlap and umklapp terms. From Figs (4,5) it is clear that for doping levels $\gtrsim 3 \times 10^{17}$/cc, the intra–band plasmon scattering tends to dominate.

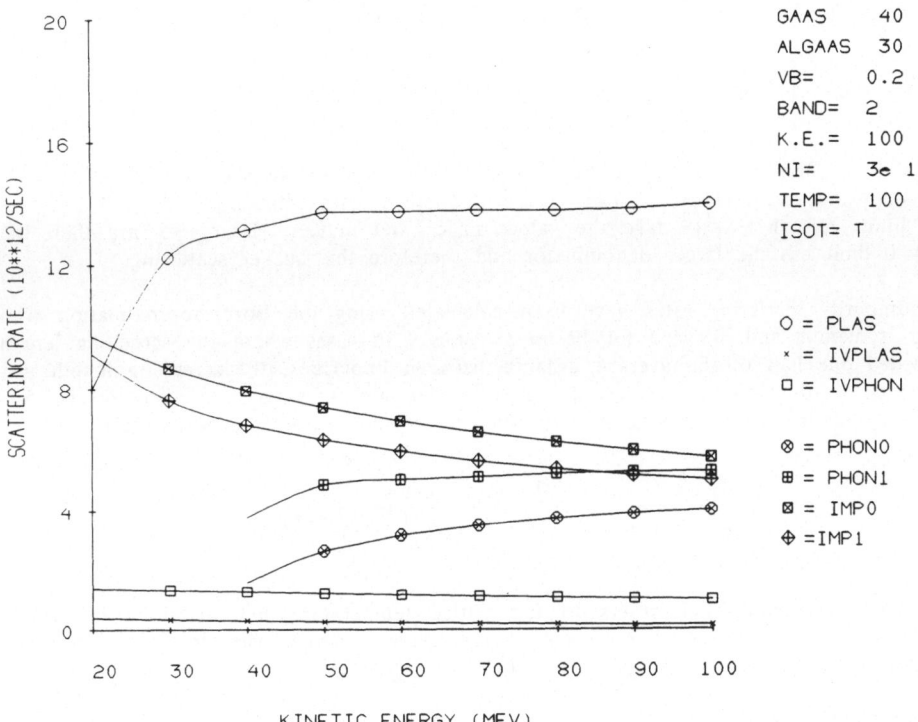

Fig. 5.   Scattering rates for superlattice (III) with initial state in band 2. Notation is the same as for Fig (4).

To calculate scattering from polar optic phonons we write the Fröhlich interaction in the form

$$H_I(q) \;=\; V \sum_G \left| q + G \right|^{-1} \exp\!\left[i(q + G)\cdot r \right]\!\left[a_{q+G} + a^+_{-q-G} \right] \tag{13}$$

$$|V|^2 \;=\; 2\pi\,\omega_0\!\left[\frac{1}{\epsilon_\infty} - \frac{1}{\epsilon_0}\right]$$

Here G are superlattice reciprocal lattice vectors, ω_0 is the phonon energy and ϵ_0, ϵ_∞ are the usual low and high frequency dielectric constants. a, a^+ are phonon annihilation and creation operators. As with inter-sub-band plasmon scattering, we find a suppression of inter-sub-band phonon scattering (Figs 4,5) wave function overlap and umklapp also significantly affect the intra sub-band scattering. A similar suppression has been reported for calculations of electron phonon scattering rates in quantum wells[13]. In the higher subbands, the scattering rates are highly anisotropic due to the large mass anisotropy. This is expected from the analytic form of the scattering integral which contains the term $|\underline{k} - \underline{k}^1|^{-2}$ where $\underline{k}$ is the initial state wave vector. If we transform to new coordinates in k-space to make the energy surfaces isotropic $k_s \to \sqrt{m_s} k_s$, $k_x \to \sqrt{m_x}\, k_x$, then the coulomb factor has the form

$$\left|\underline{k} - \underline{k}^1\right|^2 = m_x\left[k_x - k_x^1\right]^2 + m_s\left[k_s - k_s^1\right]^2 \tag{14}$$

as $k_x \to 0$, $|\underline{k} - \underline{k}^1|^2 \to m_s\,(K - k_s^1)^2 + m_x(k_x^1)^2$

as $k_s \to 0$, $|k - k^1|^2 \to m_x(K - k_x^1)^2 + m_s(k_s^1)^2$

where $K = (2E)^{\frac{1}{2}}$ and E is the initial state energy. Here we are assuming that E is less than the band width. Close to the phonon emission threshold $k^1 \sim 0$ and the two limits for $|k - k^1|^2$ take the values $m_s K^2$ and $m_x K^2$. If $m_s \gg m_x$ then the $k_x = 0$ limit has the larger denominator and therefore the weaker scattering.

Impurity scattering rates have been calculated using the Born approximation with r.p.a. screening and allowing for Fermi statistics. In cases where the screening length exceeded one half of the average distance between impurities, the screening length was set equal to this average. Wave function overlap and umklapp was also allowed for. Some calculations were also performed using a screening parameter of the form

$$V_I(q) = \frac{4\pi e^2}{\epsilon_0(q^2 + \kappa^2)} \quad , \quad \kappa^2 = \frac{4\pi n e^2}{\epsilon_0(kT + E_{qs}/2)} \tag{15}$$

For Debye screening, kT refers to the lattice temperature but to allow for Fermi statistics, kT was corrected to make this expression agree with the more accurate numerical value. The term $E_{qs}/2$ is included to force the screening parameter to take the correct assymptotic value for large values of q_s[(2)] and E_{qs} is the energy transfer in the plane. In many cases the results are sensitive to the E_{qs} correction indicating that a full wave-vector dependent screening should be used in an accurate theory.

In Fig (6) we note that for energies greater than the band width, the intra sub-band impurity scattering becomes very anisotropic. This is also expected from the analytic form of the scattering integral. We note that the scattering integral is dominated by the pole of $V_I(q)$ and for an initial k_x wave vector on the boundary of the integration range (K), ie on the Brillouin zone boundary, only one side of the complex pole contributes so that the scattering rate is reduced by about one half

4. VERTICAL TRANSPORT IN SUPERLATTICES

Both low field and high field transport shows interesting structure for superlattices. To calculate the low field mobilities, the distributions f_t, g_t are expanded in low order polynomials (3). For bulk semiconductors, it is only necessary to keep terms up to linear in k_x, when the method becomes equivalent to the technique described by Rhode[11]. For many superlattices, however, the bandwidth of the lowest sub-band is comparable with kT so that a significant part of the electron distribution lies at energies greater than the bandwidth. In this case the higher order terms are necessary to calculate mobilities accurately. In table (2) we show results for superlattices I and III which demonstrate this effect very clearly. Mobilities are also calculated neglecting the overlap and umklapp terms and we see that the results are significantly lower.

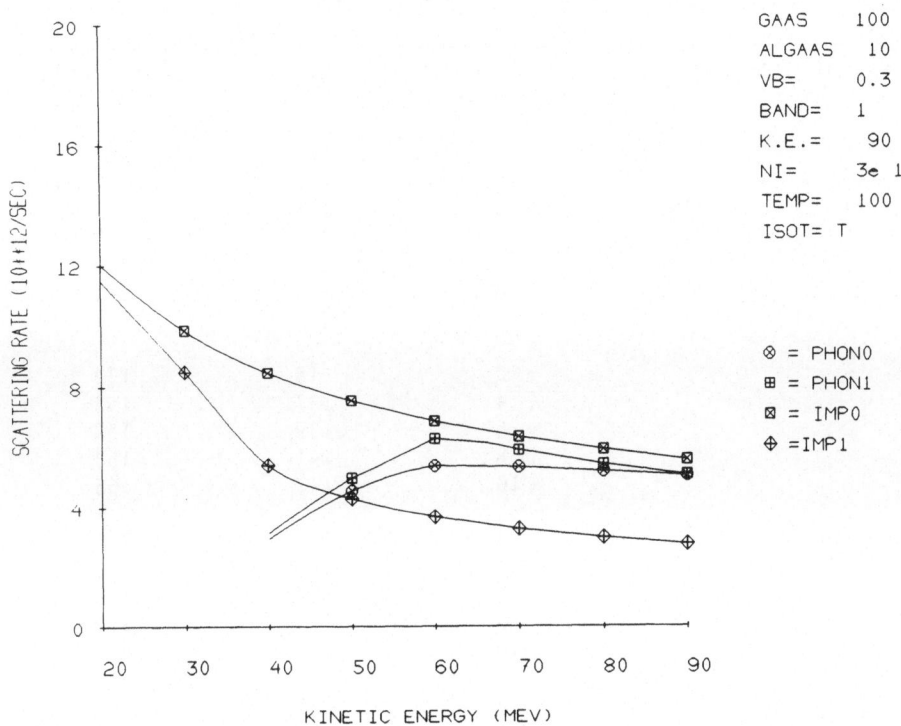

Fig. 6.   Scattering rates for superlattice (I) for initial state in band 1. Notation is the same as for Fig (4).

This is explained by a suppression of scattering from overlap terms, particularly for the doped superlattice. The relatively large effect of overlap on the impurity scattering for example (I) is related to an interference from umklapp terms. For example (III) in table (2), the lowest sub-band width is much greater than kT at 100°K and the effect of using higher order polynomials is much smaller. At these temperatures the material behaves very much like bulk GaAs. It is also noticeable that the phonon limited mobilities are larger for example (I). This is due to the first mini-gap suppressing the phonon absorption scattering in this case.

We have also performed calculations in which electrons are injected ballistically at various energies in the higher mini bands and computed the decay of the current in the sub-band with distance in the superlattice. Some results are shown in Figs (7,8). We note that the decay rate varies considerably between the superlattice types again demonstrating that the transport properties can be engineered towards desirable values.

Table (2)

Mobilities for lowest sub-band at 100°K

Example	Doping(cm^{-3})	Polynomial order	Mobility($10^3 cm^2 v^{-1} sec^{-1}$)	
			Overlap Neglected	Overlap Included
I	3×10^{17}	2	1.8	2.3
		4	2.1	2.9
		6	2.2	3.0
		8	2.3	3.1
		10	2.3	3.1
	0	2	118	132
		4	137	153
		6	139	158
		8	141	163
		10	142	165
III	3×10^{17}	2	2.6	2.8
		4	2.75	3.0
		6	2.72	3.0
		8	2.73	3.0
		10	2.73	3.0
	0	2	95	102
		4	100	112
		6	102	113
		8	103	114
		10	103	114

Finally we have used the model for quantum transport to look at hot carrier effects in the lowest sub-band for example (I). To perform these calculations we have allowed for the distribution of wave packet sizes, assuming that only wave packets which can be localised within the Brillouin zone carry current. This leads us to scale the current J(3) by the factor $(1 - e^{-T/\tau})$ where T is the period of Block oscillation and τ is the momentum relaxation time. The results in Fig (9) indicate quite a large effect at high fields, and it would be valuable to test this prediction using more rigorous formulations of quantum transport.

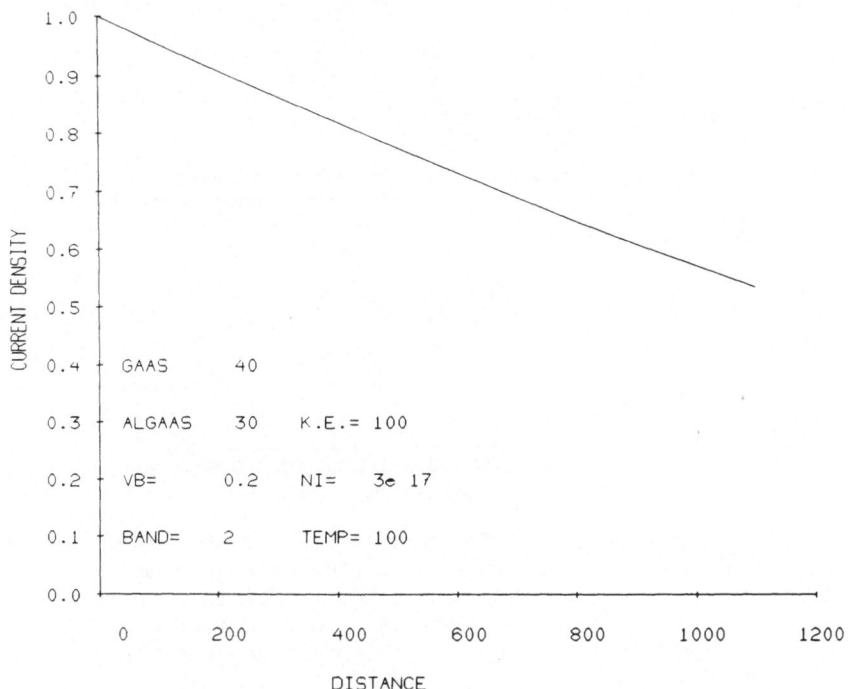

Fig. 7.   Variation of current density within band 2 with distance
          (A°) for ballistic injection for superlattice (III).
          Band refers to the subband for injection and KE refers
          to initial kinetic energy within this band. NI is the
          the doping density and Temp is the Lattice temperature °K.

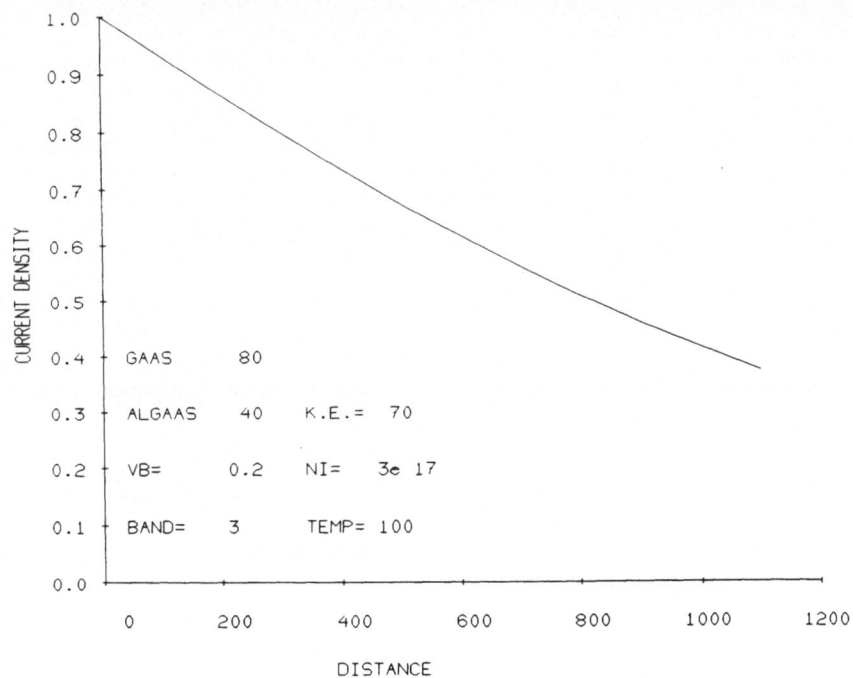

Fig. 8.   Variation of current density within band 3 with dstance
          (A°) for ballistic injection for superlattice (II).
          Notation is the same as for Fig (7).

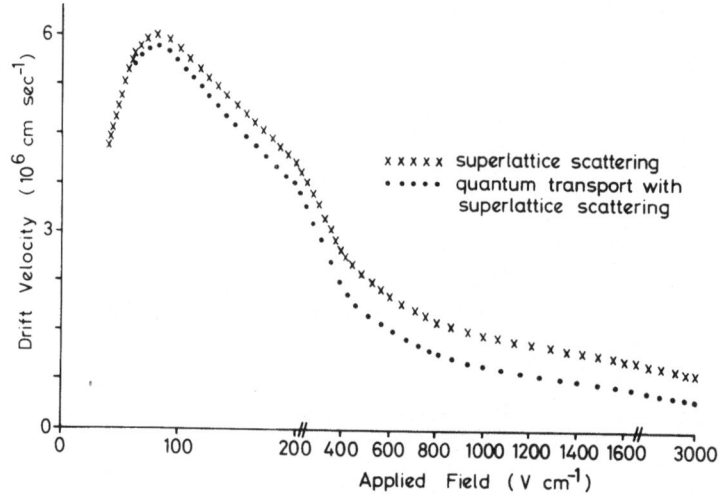

Fig. 9.   High field transport for superlattice (I) at 100°K
showing effects of using the approximate quantum
transport theory. The results differ slightly from
the values previously reported[6] due to some numerical
inaccuracies in the first version of the scattering rate
computer code.

CONCLUSION

In this paper we have presented some theoretical results for scattering rates and
transport in semiconductor superlattices. The techniques for modelling transport give
analytic or semi-analytic solutions and allow very fast computation of hot carrier
transport properties. For the superlattice of example (I) in table (I) in which the
band structure is engineered to suppress phonon emission scattering, very long electron
path lengths are predicted for ideal structures so that Bloch oscillation effects should
be readily observable experimentally at low temperature. The calculation of scattering
rates in superlattices reveal a great deal of interesting structure. The strong
suppression of inter-sub-band scattering is of greatest interest for applications but the
strong anisotropies in intra sub-band scattering are also of interest. These would be
rather complicated to handle with monte-carlo methods, but are readily allowed for
with the transport models described in this work. The computed decay of current
density associated with ballistic injection into the higher mini-bands of a superlattice
looks particularly interesting for applications by demonstrating that transport properties
can be tuned by engineering the superlattice band structure.

Having established reliable and numerically fast models for computing hot carrier
transport, the next step will be to consider high field transport in which the field
dependence of the superlattice band structure is also calculated self consistently with
the poisson equation for space charge. It will also be necessary to include quantum
transport aspects in these studies. In view of the rich structure for scattering and
transport revealed by the preliminary calculations outlined in this paper, we may
anticipate that many new and interesting non-linear phenomena remain to be
discovered in these structures.

ACKNOWLEDGEMENT

The monte-carlo results in Figs (1) and (2) were obtained by J. H. Jefferson. I would like to thank J. H. Jefferson and P.Lugli for helpful comments.

REFERENCES

(i)   J. M. Rorison, D. C. Herbert (1986) J. Phys C 19, 3991; 6357.

(2)   D. C. Herbert (1988) Semicond Sci. Technology 3, 101.

(3)   D. C. Herbert, J. H. Jefferson, M. J. Kirton (1985) Physica 134B, 82.

(4)   D. C. Herbert (1984) J. Phys C 17, 6749.

(5)   J. P. Vigneron Ph Lambin (1979) J. Phys A 12, 1961.

(6)   D. C. Herbert, J. H. Jefferson, M. Gell (1987) Proc Int Symposium on GaAs and related compounds. To be published.

(7)   F. Bertz (1986) Solid State Electronics 29 No 12, 1213-22.

(8)   'Semiconductor device modelling' Proceedings of the 1987 short course on semiconductor device modelling and applications. Edited by C. M. Snowdon and M Aℓ-Mudares.

(9)   L. Reggiani (1985) Physica 134 B/C, 123.

(10)   A. A. Abrikosov, L. P. Gorkov and I. E. Dzyaloshinski, 1963, 'Methods of quantum field theory in statistical physics' published Prentice-Hall.

(11)   D. L. Rhode (1972) Phys. Rev B 2, 1012.

(12)   M. Gell, D C Herbert (1987) Phys. Rev. B 35, 9591.

(13)   F. A. Riddock and B. K. Ridley (1983) J. Phys C 16, 6971

(14)   W. Fawcett (1973) 'Electrons in crystalline solids' (Vienna: IAEA) 531.

ELECTRON BEAM SOURCE MOLECULAR BEAM EPITAXY OF

$Al_x Ga_{1-x} As$ GRADED BAND GAP DEVICE STRUCTURES

R. J. Malik, A. F. J. Levi, B. F. Levine
R. C. Miller, D. V. Lang, L. C. Hopkins, and R. W. Ryan

AT&T Bell Laboratories
Murray Hill, NJ 07974

ABSTRACT

A new method has been developed for the growth of graded band-gap $Al_x Ga_{1-x} As$ alloys by molecular beam epitaxy which is based upon electron beam evaporation of the Group III elements. The metal evaporation rates are measured real-time and feedback controlled using beam flux sensors. The system is computer controlled which allows precise programming of the Ga and Al evaporation rates. The large dynamic response of the metal sources enables for the first time the synthesis of variable band-gap $Al_x Ga_{1-x} As$ with arbitrary composition profiles. This new technique has been demonstrated in the growth of unipolar hot electron transistors, graded base bipolar transistors, and M-shaped barrier superlattices.

The optical and electronic properties of $Al_x Ga_{1-x} As / GaAs$ heterojunctions and quantum wells have been extensively studied in materials grown by molecular beam epitaxy (MBE). There are, however, a number of interesting structures which have been proposed which require precise alloy grading over atomic dimensions. The ability to tailor band structure to obtain novel electrical and optical properties is termed "band-gap engineering".[1] Previous attempts to obtain alloy grading by varying the temperature of MBE effusion cells have been only marginally successful and have serious limitations.[2-4] This is related to the large thermal inertia and restrictions in heating and cooling rates of the MBE effusion cells which leads to slow modulation of the beam flux along with a time lag in response. In addition, variations in the growth rates determined by the Group III molecular beam fluxes seriously complicates the problem of obtaining arbitrary alloy grading. Thus Group III sources with large dynamic response are required for band gap engineering of $Al_x Ga_{1-x} As$. Several different approaches could be used which include electron beam heated sources which are typically used to evaporate Si and refractory metals,[5] gas sources,[6] and ion beam sources.[7]

We recently reported on the use of electron beam source MBE growth of III-V compounds.[8] In this paper, we discuss the use of this technique for precise analog grading of $Al_xGa_{1-x}As$ and its application in the growth of unipolar hot electron transistors, graded base bipolar transistors, and M-shaped barrier superlattices.

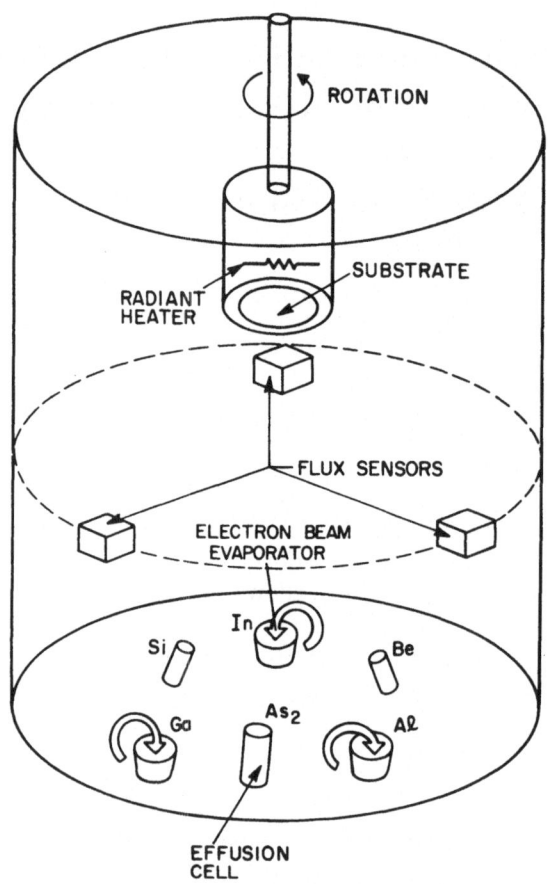

Fig. 1.   Schematic diagram illustrating the geometry of the electron-beam source III-V molecular beam epitaxy system.

The geometry of the e-beam source MBE system is shown schematically in Figure 1. The MBE system is a vertical design which is necessitated by the use of the electron beam evaporation sources and is similar in design to systems used for silicon or metals epitaxy. Three electron beam sources are used to evaporate Ga, Al, and In since these elements have unity sticking coefficients and determine the alloy composition in III-V compound semiconductors grown by MBE. A large capacity, 300 cm^3 arsenic source with a high temperature cracker is used to generate an As_2 molecular beam. A Si strip heater and small 2 cm^3 Be effusion cell are used as n- and p-type doping sources, respectively. All the molecular beam sources have separate mechanical shutters which are computer controlled. A substrate manipulator which allows substrate heating and rotation is located 250 mm above the source flange. Indium-free wafer holders for 50 mm and 75 mm substrates can be fitted on the manipulator.

A block diagram of the feedback control loop used to measure and program the evaporation rates for the Ga and Al e-beam sources is shown in Figure 2. Both a modulated ion gauge and a commercially available Inficon Sentinel III flux sensor[9] have been used successfully to measure the molecular beam fluxes. They both have adequate sensitivities to measure and control evaporation rates as low as 0.01 Å/sec. The Sentinel III flux sensor which uses a transducer based upon electron impact emission spectroscopy is superior from the standpoint of materials selectivity and its immunity from charged particles causing measurement error. An amplified analog signal which is directly proportional to the molecular beam fluxes or evaporation rates is input to a digital controller which interfaces with the e-gun power supply. The bandwidth for the feedback control loop of the Sentinel III flux sensor is 8 Hz. A computer communicates with the digital controller on a RS 232 bus and updates the evaporation rate setpoints every second. An absolute calibration of the molecular beam fluxes is made in situ by use of reflection electron diffraction oscillations.[10]

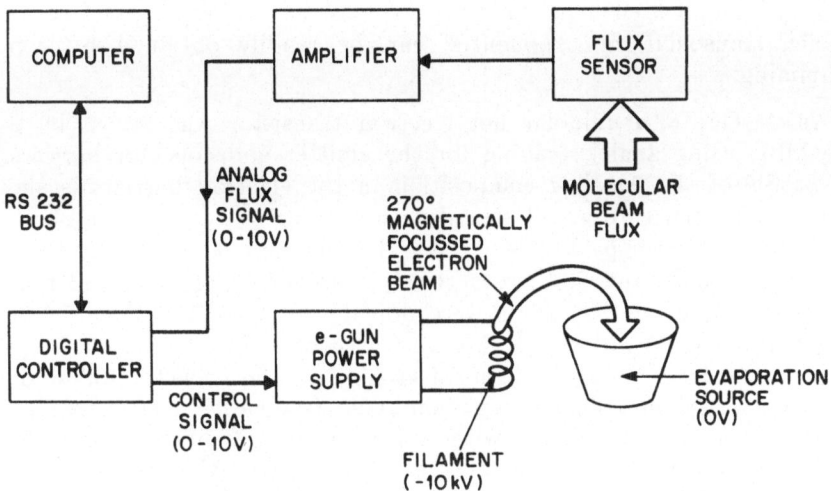

Fig. 2.   Block diagram of the feedback control loop for the electron-beam evaporation sources.

Precise alloy grading of $Al_xGa_{1-x}As$ is accomplished by the following method. The $Al_xGa_{1-x}As$ growth rate is kept constant during epitaxy and is just the sum of the individual growth rates for AlAs and GaAs. The normalized growth rate for $Al_xGa_{1-x}As$ can be expressed as

$$r_{AlGaAs} = r_{AlAs} + r_{GaAs} = 1$$

The Al mole fraction of the alloy is given by the ratio of the AlAs growth rate to the total growth rate

$$x = \frac{r_{AlAs}}{r_{AlAs} + r_{GaAs}}$$

Now if the individual growth rates are varied such that

$$r_{AlAs}(t) = f(t)$$

$$r_{GaAs}(t) = 1 - f(t)$$

then the time dependent alloy composition during growth is simply given by

$$x(t) = f(t).$$

Also since r_{AlGaAs} remains constant, then the position dependence of Al alloy composition can also be expressed as

$$x(d) = f(t)$$

where $d = r_{AlGaAs} \cdot t$ is the position within the epilayer. Thus it is evident that even very complicated mathematical expressions for alloy grading such as parabolic, sinusoidal, or exponential can be readily obtained by computer programming.

An $Al_x Ga_{1-x} As$ unipolar hot electron transistor was grown by e-beam source MBE using analog grading for the emitter and collector barriers.[11] A SIMS profile of the Al alloy composition in the electron barriers is shown in Figure 3. The structure was designed with a collector barrier peak of $x = 0.35$ and an emitter barrier peak of $x = 0.3$. The long and short arms of the graded barrier were 1000Å and 90Å, respectively with a barrier peak width of 50Å. The emitter, base, and collector were doped n-type to $1 \times 10^{18} cm^{-3}$ and the graded barriers to $1 \times 10^{16} cm^{-3}$. Mesa transistors were fabricated using anodic etching to the base and wet chemical etching to the collector followed by an alloyed Au/Ge/Ni metallization for contacts. Hot electron spectroscopy[12-13] was performed on a transistor with a base width of 400Å at 4.2° K. The derivative spectrum of collector current shows clear evidence of a "ballistic" electron peak. This is the first time that ballistic transport has been seen in analog graded $Al_x Ga_{1-x} As$ structures. It should be noted that in a number of attempts to simulate alloy grading by using variable duty cycle chopped $Al_x Ga_{1-x} As/GaAs$ superlattice,[14] no ballistic peak was observed. We infer from this that the multiple interfaces or trapped impurities at the interfaces in the chopped superlattices leads to enhanced electron scattering suppressing ballistic transport.

The Al beam flux signal in the growth of a graded band-gap base $Al_x Ga_{1-x} As/GaAs$ heterojunction bipolar transistor is shown in Figure 4. The Al composition is graded from $x = 0.04$ to $x = 0.2$ in the 1000 Å base region. The emitter junction is also graded from $x = 0.2$ to $x = 0.33$ over 300 Å and then graded back down to $x = 0$ over 500 Å for the emitter contact region. It has been previously demonstrated that alloy grading in the base results in a quasi-electric field in the base which reduces the base transit time.[15-16] The precise grading control is clearly illustrated here.

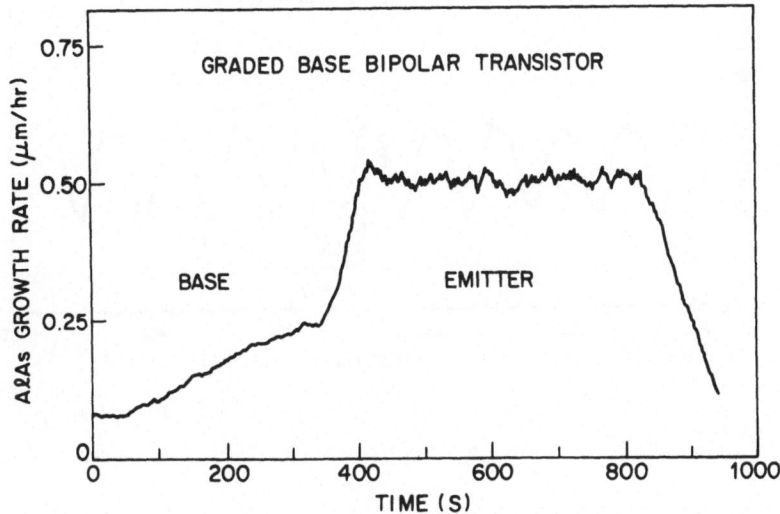

Fig. 3.   Secondary-ion-mass-spectrometer measurement of the Al alloy composition in the analog-graded, unipolar, hot electron electron transistor.

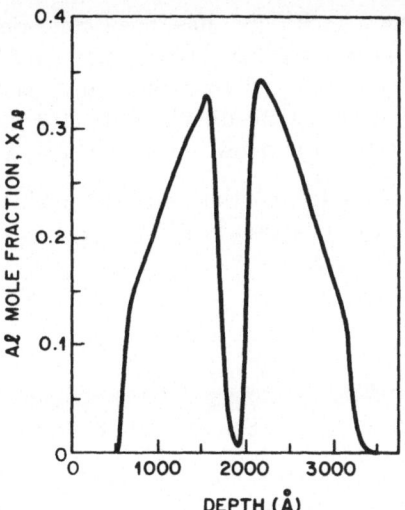

Fig. 4.   Al beam flux signal in the growth of a graded base heterojunction bipolar transistor.

221

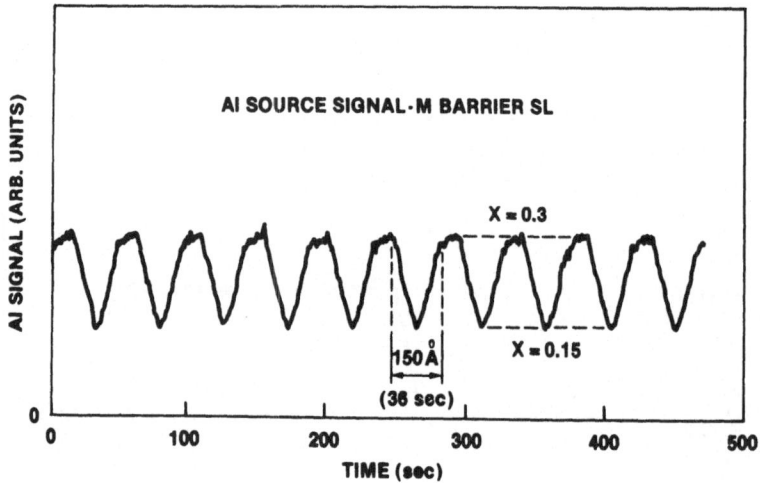

Fig. 5.   Al beam flux signal in the growth of an M-shaped barrier superlattice.

The Al beam flux signal in the growth of an M-shaped barrier superlattice is shown in Figure 5. This superlattice is used in an intersubband absorption quantum well photodetector.[17–18] The Al composition was linearly modulated from x = 0.3 to x = 0.15 and back to x = 0.3 over a distance of 150 Å. The sample contained a 50 period superlattice and was grown continuously without interruption. The excellent linear grading profile is apparent, and it is obvious that such a structure would be impossible to grow using conventional effusion cell sources.

In conclusion, a new technique has been developed for the growth of graded $Al_xGa_{1-x}As$ structures with arbitrary alloy profiles by molecular beam epitaxy. The system uses Ga and Al electron-beam heated evaporation sources which are measured real-time and feedback controlled using beam flux sensors. The MBE system has been used to grow novel $Al_xGa_{1-x}As$ graded alloy devices.

We gratefully acknowledge partial support of this work through NASA Goddard Space Flight Center contract NAS5-28602.

REFERENCES

[1]   F. Capasso, Ann. Rev. Mater. Sci., *16*, 263-91 (1986).

[2]   J. P. Harbison, L. D. Peterson, J. Leskoff, Proc. Fourth Int. Conf. MBE, U. of York, September 1986 (to be pub. J. Cryst. Growth).

[3]   N. J. Sauer, T. Y. Chang, A. H. Dayem, E. H. Westerwick, J. Vac. Sci. Tech. B, *5*, 718 (1987).

[4]   K. Alavi, A. Y. Cho, F. Capasso, and J. Allam, J. Vac. Sci. Tech. B, *5*, 802 (1987).

[5]   J. C. Bean, Proc. Fourth Int. Conf. MBE, U. of York, September 1986 (to be pub. J. Cryst. Growth).

[6] M. B. Panish, H. Temkin, and S. Sumski, J. Vac. Sci. Tech. B, *3*, 657 (1985).

[7] S. Shimizu, O. Tsukakoshi, S. Komiya, and Y. Makita, Jap. J. Appl. Phys., *24*, 1130-40 (1986).

[8] R. J. Malik, J. Vac. Sci. Tech. B, *5*, 722 (1987).

[9] Inficon Leybold-Heraeus Co., Syracuse, NY 13057.

[10] J. H. Weave, B. A. Joyce, P. J. Dobson, and N. Norton, Appl. Phys. A, *31*, 1 (1983).

[11] R. J. Malik and A. F. J. Levi, Appl. Phys. Lett., *52* (1988).

[12] J. R. Hayes and A. F. J. Levi, IEEE J. Quantum Electron., *QE-22*, 1744 (1986).

[13] M. Heiblum, M. I. Nathan, D. C. Thomas, and C. M. Knoedler, Phys. Rev. Lett., *55*, 2200 (1985).

[14] M. Kawabe, M. Kondo, N. Matsuura, and K. Yamamoto, Jpn. J. Appl. Phys., *22*, L64 (1983).

[15] B. F. Levine, C. G. Bethea, W. T. Tsang, F. Capasso, K. K. Thornber, R. C. Fulton, and D. A. Kleinman, Appl. Phys. Lett., *42*, 769 (1983).

[16] R. J. Malik, F. Capasso, R. A. Stall, R. A. Kiehl, R. W. Ryan, R. Wunder, and C. G. Bethea, Appl. Phys. Lett., *46*, 600 (1985).

[17] B. F. Levine, K. K. Choi, C. G. Bethea, J. Walker, and R. J. Malik, Appl. Phys. Lett., *50*, 1092 (1987).

[18] K. K. Choi, B. F. Levine, C. G. Bethea, J. Walker, and R. J. Malik, Appl. Phys. Lett., *50*, 1814 (1987).

FUTURE TRENDS IN QUANTUM SEMICONDUCTOR DEVICES

M. J. Kelly

GEC Hirst Research Centre
East Lane
Wembley
Middlesex   HA9 7PP
United Kingdom

ABSTRACT

Between physics discoveries and commercial products there is a large attrition rate. The status of some heterojunction devices is reviewed, and some indications are given of the evolution of all-heterojunction devices.

INTRODUCTION

Only in the last year has it been announced[1] that a major supercomputer (the Cray-3) is to be partially implemented in GaAs. For all the advantages that GaAs has over Si in theory, such as intrinsic speed, it has taken a long time to rival the ever-resilient Si technology. Even now $1\mu m$ long by $50\mu m$ wide MESFET technology is being used only for the processor and not for the memory. The characteristic internal propagation delays of order 80 ps are four times smaller than in the Si technology of Cray-2, and issues such as yield, reliability etc have been settled. Were it not for the fact that GaAs can do two things that silicon does not, i.e. lase and give negative differential resistance, and further that GaAs substrates can retain very high resistivity levels after processing, it is doubtful that GaAs LSI would ever have been achieved.

Bipolar and field - effect devices involve potential barriers due to junctions with metals and/or dielectrics, or rapid changes in doping profiles. These barriers are all limited by defects, materials integrity etc. Semiconductors heterojunctions permit the construction of perfect potential barriers at the atomic layer scale, improving on standard device designs and allowing the possibility of "new" devices. If we leap from the $1\mu m$ gate MESFET above to the state-of-the-art compound semiconductor transistors, and extrapolate realistic technology developments, we achieve impressive performances that leave silicon behind (c.f. Table 1 courtesy of L.F. Eastman, Cornell University). The state of the art is often with factors of 2-5 of Table 1.

TABLE 1

COMPOUND SEMICONDUCTOR TRANSISTOR PREDICTED PERFORMANCE

SMODFET - Al,GaAs/In,GaAs/Al,GaAs/GaAs
$1\,\mu m$ M gate and p doped buffer
1000 mS/mm, .8 A/mm
f_T = 200 GHz, f_{max} = 400 GHz
Switching time < 4 ps

MODFET - Al,InAs/GaInAs/AlInAs/Inp
$.1\,\mu m$ M gate and P doped buffer
1200 mS/mm, > 1A/mm
f_T = 250 GHz, f_{max} = 500 GHz

HBT - Al,GaAs/In$_y$Ga$_{1-y}$As/GaAs
Submicron Emitter, fast electron transit,
1-5 x 10^5 A/cm^2,
f_T = 200 GHz, f_{max} = 400 GHz
Switching time < 4ps

PBT - GaAs with Tungsten Control Electrodes
f_T = 100 GHz, f_{max} = 400 GHz
High operating voltage for power

Several points can be made:

(i) for devices whose operation depends on a transit time, such as the field effect and bipolar transistors, a common set of performance limits apply in terms of the voltage - frequency and power - impedance - frequency squared products. One sees this in the ultimate projection of performances[2].

(ii) the HEMT has advantages over the MESFET, but the higher mobility reduces access resistances, and noise levels rather than improves any fundamental transit time.

(ii) the choice between field effect and bipolar is made on issues such as noise, current drive capability, input impedance, thermal stability etc rather than just speed.

In the context of the above evolution, what is the role for the quantum and other phenomena on the 20 nm length scale of the de Broglie wavelength of room temperature carriers in semiconductors? At present we have ballistic motion, resonant tunnelling and quantum size effects at our disposal. We can see some further improvements using these new effects, but will they survive through to production?

HOT-ELECTRON EFFECTS

The merits advanced for the heterojunction bipolar transistor over the homojunction bipolar version are well documented[3]. They include (i) higher emitter efficiency, because holes are prevented by a valence band barrier from flowing from the base to emitter, (ii) higher base doping that leads to decreased base resistance without compromising the emitter efficiency, (iii) improved frequency response because of higher gain and lower base resistance, and (iv) wider range of operating temperatures.

In principle a majority carrier device with a comparable base resistance might have several further advantages, such as rapid switching with a low emitter-base capacitance. Suggestions for a high speed transistor with a metal base go back to the 1960's, and failure was a matter of materials integrity in the first instance. The revival of a monolithic semiconductor hot electron transistor with a heavily doped semiconductor replacing the metal base has followed the mastery of MBE and MOCVD in fabricating them, and several teams have been active[4]. The AlGaAs/GaAs system has again led the way because of materials maturity, as was the case in the HEMT and QW lasers. This system now looks rather less encouraging because the two conditions for a useful transistor cannot be met simultaneously: high gain and high speed. The high gain is based on a highly efficient transfer of electrons from the emitter to the collector. This has been achieved with structures where the base has low doping, which implies an intolerably large base resistance. With an acceptable base resistance (i.e. $>10^{18}$ cm^{-3} Si doping in the base) the inelastic scattering length for hot electrons is only of order 40 nm and after two or more energy losses (to coupled plasmon-phonon modes and to single electron scatterings) the collector barrier cannot distinguish them from thermally excited carriers in the base[5]. Two solutions have been proposed, and both so far have given encouraging results: (i) the use of other materials that allow higher injection energy (limited by intervalley scattering in the GaAs system) e.g. InGaAs/InAlGaAs[6], AlSbAs/InAs[7] etc, so that after several energy losses, the carriers are still hot, and (iv) the use of a two-dimensional electron gas base, either from modulation doping the collector barrier or δ- doping the base near the collection barrier[8]. In this latter case, the coupling between the plasmons and phonons is weakened on dimensional grounds (cf the dispersion of the plasmon mode in 2D) and high transfer efficiency ($>90\%$) is achieved. The collector barrier is still marginally too small, so that while high gain can be achieved, any leakage in the collector barrier would limit the load on the collector.

At this point the hot-electron-transistor becomes a contender with the other class of transistor in Table 1. Our own study shows that it can achieve some of the performance of the high-speed field effect devices without resort to sub-micron lithography. Further work, to achieve current densities of order 10^5 A/cm^2, is required. To date, however, there are no reports of performance of micron-scale devices or their circuit properties, even though their equivalent circuit does look encouraging.

In conventional n$^+$n-n$^+$ Gunn devices, an appreciable part of the device is used to heat the electrons to make intervalley transfer possible at ˜0.33 eV. This heating process is stochastic and noisy with optic phonon losses acting against the heating effect of the applied field. We have found[9] that judicious design of heterojunction layers and doping profiles could prove useful in improving the efficiency and reducing the noise levels in Gunn diodes at 50 GHz and above. The key point is to give electrons 95% or so of the energy they require for intervalley transfer when they reach the drift region, generally of relatively low field, doped n-. A considerable folk lore has been built up on the use of a metal contact directly on to the n- region. In practice this acts as a leaky Schottky barrier providing the electrons with the required 0.3 eV energy[10]. We have found that a layer of AlGaAs with the Al ratio increasing from 0 to 30%-40%, followed by a thin region of n+ spike doping has the same effect. The spike is essential as under bias it depletes to allow a reduced field in the drift region and a greater control over the subsequent field instabilities. The depletion field at the spike adds to the energy step without degrading the electron energy through electron collisions (c.f. the results of hot-electron spectroscopy). We have tested this idea in increasing the fundamental frequency of operation which now stands at over 80 GHz.

RESONANT TUNNELLING EFFECTS

Much of the effort devoted to 2D physics and the HEMT system 5-10 years ago seems now devoted to resonant tunnelling in double barrier diodes. The peak-to-valley ratio at room temperature has improved with both material quality and material system with over 4:1 at room temperature in the GaAs/AlGaAs and 17:1 with the InAlAs/GaInAs system[11]. For such impressive static characteristics, the doping has been kept well away from the barriers.

In order to produce a high frequency or high speed device the doping profile is of paramount importance. In such thin devices the large capacitance between contacts together with the various series resistances places an RC-type ceiling on the speed of operation which limits all such tunnel oscillators to date. It is important to note that by comparison with the Esaki diode, the principle advantages here are the independent control over doping and tunnelling. Thin tunnel barriers in the Esaki diode were related to hyper- abrupt p-n junctions in which the dopant positions were unstable in high fields. In resonant tunnel structures the effects occur in thin undoped layers while depletion and other effects can be tailored to improve voltage swings etc. In our own work, heavy doping of the cathode superlattice and light doping of the anode superlattice help to increase the peak current density and the depletion voltage swing over that achieved with a symmetrically doped structure[14].

In all the double barrier diode work to date, there seems to be a correlation between peak-to-valley ratio and peak current density to the extent that their products is approximately constant[13]. This is of concern in all the applications considered, both two-terminal devices for microwave and three-terminal devices for digital. In the former one is wanting increased power and high efficiency, while in the latter, in the case of the RHET and other such devices, one wants high current density for fast switching while still maintaining a significant difference in the internal state of the transistor.

On a on-off basis the two-terminal resonant tunnelling device performances are highly creditable: 200 GHz+ operation[14], or 25% efficiency of microwave production at 2 GHz, but in terms of the competition from existing devices, a combination of high speed, high efficiency and low noise are required. Resonant tunnel barriers in the RHET[15] and in other configurations[16] have been used to demonstrate three-terminal devices that have sufficiently complex internal structure to be able to perform the functions of several conventional transistors. So far the demonstrations have been confined to the GaAs/AlGaAs system, and to devices operating at slow speed and low temperature. A considerable further effort will be required to produce performances that approach those achieved with devices in Table 1. Room temperature operation and high current densities are required.

QUANTUM SIZE EFFECTS

It could be argued that the heterojunction represents a natural extension of enhancing potential barriers and so improving device performance. The all heterojunction resonant tunnelling device in its multifunction applicaton represents potentially radical advance. It is really in optical applications where quantum size effects have been used to improve the performance of lasers that the new physics is making its mark. In practice the precise advantages of quasi-two-dimensional electron states in a laser are quite complex. The reduced density of states means that population inversion is achieved more easily, and threshold currents reduce. Further, for a given number of injected carriers, stronger inversion is achieved and a better volume gain is achieved. By contrast the overlap of the optical wave and the active region is reduced, and this offsets much of the advantage. The tailoring of the optical confinement layers have some bearing on the reduced temperature dependence of operation[17]. In optical modulaton at wavelengths away from the principal band-gap, quantum size effects make device functions feasible where otherwise the losses would be intolerable[18]. In both the HEMT and the QW laser, the advantages of the specifically two-dimensional nature of the carriers is often offset by complex compromises with the rest of the circuit. If one progresses from quantum wells to ribbons and dots, further compromises will be needed. If one considers the quantum dot or quantum rod laser the constraints on density of states, volume gain, sensitivity to fluctuations in the fabricated microstructures, etc, give rise to further scientific and technological challenges[18].

SUMMARY

The mainstream heterojunction devices have been refined to produce very high performances. Quantum semiconductor devices will find a place if they offer clear improvements over the operation of mainstream devices. Radical new devices will have to do something that conventional devices cannot, or provide a leap in performance to justify the investment. Otherwise the new ideas will be victims of attrition.

ACKNOWLEDGEMENTS

I would like to thank my co-workers at GEC for their advice.

REFERENCES

1. D. Kiefer and J Heightley, "Cray-3: A GaAs Implemented Supercomputer System" IEEE GaAs IC Symposium Technical Digest, pp 3-6 (1987)

2. P H Ladbrooke "Comparison of Transistors for Monolithic Microwave and Millimetre Wave Integrated Circuits", GEC Journal of Research 4 114-123 (1986), and references therein.

3. S M Sze "Physics of Semiconductor Devices" (Wiley, Second Edition 1981)

4. M Heiblum, I M Anderson and C M Knoedler" D.C. Performance of Ballistic Tunnelling Hot-Electron Transfer Amplifiers", Appl. Phys. Lett. 49 207-9 (1986)

   S Muto, K Imamura, N Yokoyama, S Hiyamizu and H Nishi "Sub-picosecond Base Transit Time Observed in a Hot-Electron Transistor", Electronics Letters 21 555-6 (1985)

   A P Long, P H Beton and M J Kelly, "Hot Electron Transport in Heavily Doped GaAs", Semiconductor Science and Technology 1 63-70 (1986)

   J R Hayes, A F J Levi and W Wiegmann "Hot Electron Spectroscopy of GaAs", Phys. Rev. Lett 54 1570-2 (1985)

5. P H Beton, A P Long and M J Kelly "Monte-Carlo Simulation of Hot Electron Spectra" Solid State Electronics 31 637-40 (1988)

   A F J Levi, J R Hayes and R Bhat "Ballistic injection devices in semiconductors", Appl. Phys. Lett 48 1609-11 (1986)

6. K Imamura, S Muto, T Fujii, N Yokoyama, S Hiyamizu and A Shibatomi "InGaAs/InAlGaAs Hot-Electron Transistor with a current gain of 15", Electronics Letts 22 1148-50 (1986)

7. A F J Levi and T H Chiu "Room-temperature operation of hot electron transistors" Appl. Phys. Lett. 51 984-6 (1987)

8. A P Long, M J Kelly and T M Kerr (unpublished)

9. N R Couch, P H Beton, M J Kelly, T M Kerr, D J Knight and J Ondria "The use of linearly graded composition AlGaAs Injectors for Intervalley Transfer in GaAs: Theory and Experiment", Solid State Electronics 31 613-6 (1988)

10. B Fank "Indium Phosphile mm-Wave devices and components" Microwave Journal 27 #4 p 95-101 (1984)

11. The GaAs/AlGaAs results have been achieved in several laboratories including our own. See also
    T Inata, S Muto, Y Nakata, T Fujii, H Ohnishi and S Hiyamizu, "Excellent Negative Differential Resistance of InAlAs/InGaAs Resonant Tunnelling Barrier Structures Grown by MBE" Jap. J. Appl. Phys. 25 L983-5 (1986)

12. R A Davies to be published, and
    M J Kelly, S R Andrews, N R Couch, R A Davies and T M Kerr "Novel Tunnelling Phenomena in Quantum Wells and Superlattices" to appear in Physica Scripta

13. S Muto, S Hiyamizu and N Yokoyama
    "Transport Characteristics in heterojunction devices" in "High Speed Electronics", (edited by B Kallback and H Beneking) (Springer-Verlag 1986) pp 72-8

14. T C L G Sollner, E R Brown and H Q Le "Microwave and Millimetre - Wave Resonant - Tunnelling Devices" to appear in "Physics of Quantum Electron Devices" editor F Capasso (Springer-Verlag 1988)

15. N Yokoyama and K Imamura "Flip-Flop Circuit using a Resonant Tunnelling Hot Electron Transistor (RHET)", Electronics Letters $\underline{22}$ 1228-9 (1986)

16. S Sen, F Capasso, A Y Cho and D Sivco "Resonant Tunnelling Device with Multiple Negative Differential Resistance: Digital and Signal Processing Applications with Reduced Circuit Complexity", IEEE Electron Devices $\underline{34}$ 2185-91 (1987)

    T K Woodward, T C McGill, H F Chung and R D Burnham "Applications of Resonant - Tunnelling Field-Effect Transistors", IEEE Electron Device Letters 9 122-4 (1988)

17. D A B Miller, J S Weiner and D S Chemla "Electric-Field Dependence of Linear Optical Properties in Quantum Well Structures: Waveguide Electro- absorption and sum rules" IEEE Journal of Quantum Electronics $\underline{22}$ 1816-30, 1986 and references therein

18. C Weisbuch "Novel Heterostructure Devices for Electronics and Optoelectronics" to appear in the (SPIE 869) Proceedings of the International Conference on Technologies for Optoelectronics, Cannes, November 1987

NOVEL OPTICAL PROPERTIES OF InGaAs-InP QUANTUM WELLS

M. S. Skolnick, K. J. Nash, S. J. Bass, L. L. Taylor,
A. D. Pitt, L. J. Reed and M. K. Saker

Royal Signals and Radar Establishment
St Andrews Road, Malvern, Worcestershire, WR14 3PS, UK

ABSTRACT

Novel phenomena in the low temperature optical properties of InGaAs–InP quantum wells, not previously reported for the GaAs–GaAlAs system, are discussed. These include the observation of strongly enhanced exciton–LO phonon coupling, screening of the Fröhlich interaction as a function of free carrier density, the observation of a Fermi energy edge singularity and the deduction of a value for the density of states between Landau levels in high magnetic field by a spectroscopic technique.

1. Introduction

Since the pioneering studies of Dingle and co–workers[1] in 1974, there has been a great deal of work on the optical properties of GaAs–GaAlAs quantum wells (QWs). By contrast, investigations of the InGaAs–InP materials system are at a much less mature stage. The first reports of the study of the photoluminescence (PL) from InGaAs–InP QWs were published by Kodama et al[2] and Razeghi et al[3] in 1983. Since then high quality InGaAs–InP quantum wells have been produced by a variety of different growth techniques (see eg References 4 to 11) with very good low temperature PL linewidths as low as $\sim$ 10 meV at 10 Å being reported.[4,5,7,8] The recent observation of well resolved PL lines from monolayer fluctuations in well width[4,10] is another indication of the high degree of structural perfection and electronic quality that can now be obtained for QWs in this materials system.

The purpose of the present paper is to demonstrate that a number of new physical phenomena can also be observed in InGaAs–InP QWs, by contrast with the well studied GaAs–GaAlAs system. Recent work on the optical properties of InGaAs–InP QWs grown by atmospheric pressure metal–organic chemical vapour deposition (AP–MOCVD) will be reviewed. Attention will be concentrated on novel results obtained on both nominally undoped and modulation doped QWs, with particular emphasis being placed on those phenomena which have not been reported previously for GaAs–GaAlAs QWs. It will be seen that the key factor which allows the observation of most of the new phenomena is the existence of hole localisation, probably in alloy fluctuations, in the AP–MOCVD grown, InGaAs alloy QWs.

The results which will be discussed include the observation of strongly enhanced exciton–LO phonon coupling in QWs, the screening of the Fröhlich interaction as a function of free carrier density, the observation of a Fermi energy edge singularity in the PL spectra of modulation doped QWs and the effects of high magnetic fields on the PL spectra, where a new method to determine the density of states between Landau levels in a quasi–two dimensional system will be presented.

2.   Observation of LO phonon satellites of exciton recombination lines and screening of the Fröhlich interaction

Low temperature (2K) PL spectra for a series of nominally undoped AP–MOCVD grown InGaAs–InP QWs are shown in Figure 1 (1a, b, c for ∼ 100 Å well width, 1d for 75 Å well width). Details of the samples and a review of the crystal growth techniques can be found elsewhere [11,12]. The PL spectrum for each sample is

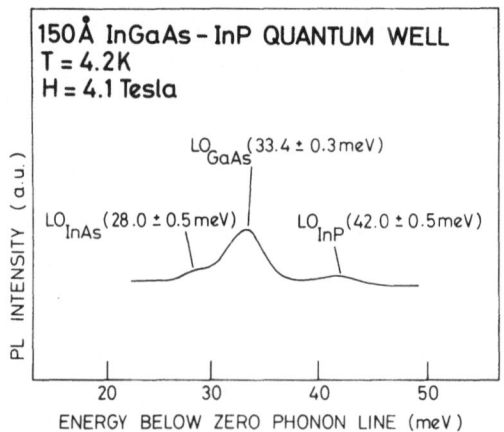

FIGURE 1

Series of low temperature (2K) PL spectra for nominally undoped InGaAs–InP quantum wells. Each spectrum is composed of an excitonic recombination line (X) together with an LO phonon satellite (X_{LO}) of the exciton line at 33 meV to lower energy. The well widths are indicated on the figure. The asymmetric lineshapes (a to c) with a low energy tail and a sharp cut-off at high energy (the electron Fermi energy) are indicative of the presence of free carriers in the wells.

composed of a dominant exciton recombination line (labelled X) and a weaker subsidiary line at ~ 33 meV to lower energy (X_{LO}). The X_{LO} peaks are ascribed to LO phonon satellites of the main exciton recombination lines. There are a number of observations which support this attribution of X_{LO}, as opposed to other possible processes, such as electron to acceptor recombination, which could give rise to features ~ 30 meV below the main exciton line. The strongest piece of experimental evidence in favour of the LO phonon satellite attribution is shown in Figure 2. An expanded version of the spectrum in the X_{LO} satellite region, for our highest quality sample (150 Å well width), is presented in this figure. Replication of the exciton PL line by the three LO modes of the system, the InAs and GaAs LO modes of the InGaAs QW

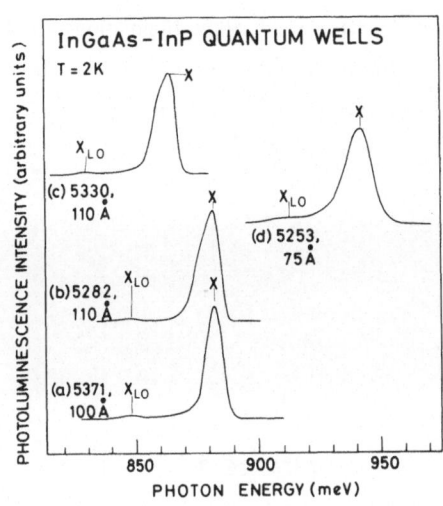

FIGURE 2

Expanded spectrum of X_{LO} satellite region for a high quality sample of 150 Å well width. The spectrum is taken in a magnetic field of 4.1 Tesla to remove any free carrier broadening effects. Coupling to the InAs and GaAs–like LO modes of the InGaAs quantum well, and to the InP LO mode of the barrier material is observed demonstrating that X_{LO} is an LO phonon satellite of the exciton PL line.

and the LO mode of the InP barrier material is observed, thus providing very strong evidence in favour of the LO phonon satellite attribution. Further support to this identification is given by the observation of both one and two LO phonon satellites of X, and the independence of the X, X_{LO} separation in magnetic fields up to 10 Tesla.[13]

The ratio of the intensities of X_{LO} to X is of the order 1 to 5% for most of the AP–MOCVD grown InGaAs–InP samples we have investigated. This contrasts with the situation for direct gap GaAs–GaAlAs QWs where evidence for LO phonon replication has not so far been reported.[14,15] We attribute the strongly enhanced LO phonon coupling in the AP–MOCVD grown InGaAs–InP QWs to the existence of hole localisation, at low temperature, in this system. The occurrence of carrier localisation leads to a large change in charge density during recombination. This change in charge density gives rise to a large change in lattice polarisation via the Fröhlich interaction and hence to enhanced LO phonon coupling of the exciton recombination. Hole rather than electron localisation is proposed to explain the enhanced phonon coupling since the significant diamagnetic shifts of the exciton line which are observed (~ 5 meV in a magnetic field of 10 Tesla), indicate weak electron binding in the exciton. Also it is much easier to localise the hole than the electron because of its larger effective mass (mass ratio m_h/m_e ~ 10). In order to explain the magnitude of the observed phonon coupling, hole localisation on a length scale of 10 to 30 Å is required.[16]

It is suggested that the carrier localisation occurs in alloy fluctuations in the InGaAs QWs principally because comparable phenomena are not observed in GaAs–GaAlAs QWs where the QW is the binary material. Transmission electron microscope images of thick layers of InGaAs grown under similar conditions to the InGaAs–InP QWs show clear evidence for the occurrence of alloy fluctuations on length scales from 10 to 200 Å.[11] In solid source MBE (SSMBE) grown InGaAs–InP QWs the LO phonon coupling is much weaker (X_{LO}/X ~ 10^{-4})[17] probably indicative of the more uniform alloy growth on a microscopic scale in this case.

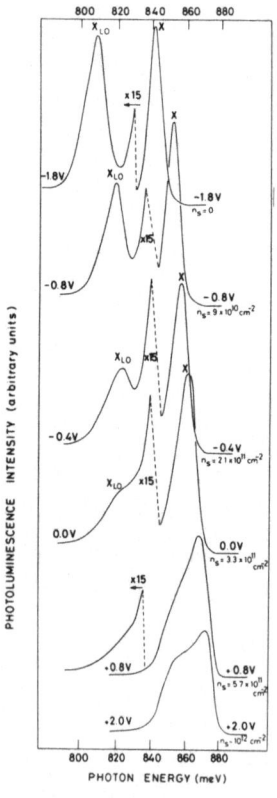

FIGURE 3

Series of PL spectra (10K) for Schottky-gated, single sided modulation doped quantum well, as a function of applied bias to Schottky barrier. The carrier density (n_s) values on each trace are obtained from C–V measurements. When the well is depleted the exciton line X and its GaAs LO phonon satellite, X_{LO}, are observed. With increasing n_s, X_{LO} decreases relative to X due to screening of the Fröhlich interaction, and is no longer observable for n_s > 5 x 10^{11} cm^{-2}.

The spectra shown in Figures 1, 2 are obtained for nominally undoped QWs. Any free carriers (n_s ~ 2 x 10^{11} cm^{-2}) in the QWs arise due to the occurrence of persistent photoconductivity effects in these structures at low temperature.[18,19] Phonon coupling in the PL spectra can also be observed in modulation doped structures where the free carrier density can be varied from 0 to 10^{12} cm^{-2} by changing the applied bias across the structure using a semi–transparent Schottky electrode. The use of such Schottky gated modulation–doped structures provides a new method to study the screening of the carrier–LO phonon (Fröhlich) interaction in quasi–two dimensional structures as a function of n_s.[15] A series of PL spectra are shown in Figure 3 for a single sided modulation doped structure as a function of n_s from 0 to 10^{12} cm^{-2}. At $n_s = 0$ at the top of the figure the exciton PL line X and its LO phonon satellite X_{LO}, with $\sim$ 5% of the intensity of X, are observed. As n_s is increased the ratio of X_{LO} to X decreases approximately linearly with n_s up to 3×10^{11} cm^{-2} where X_{LO} is very weak, and for $n_s > 5 \times 10^{11}$ cm^{-2}, X_{LO} is no longer observable. The variation of the ratio X_{LO}/X versus n_s is shown in graphical form in Figure 4.

The strong decrease of the intensity of the phonon satellite intensity with increasing n_s could arise either from screening of the Fröhlich interaction which gives rise to the occurrence of LO phonon replication, or to delocalisation of the hole (since the hole localisation leads to the enhanced phonon coupling) with increasing n_s. However, comparison of photoconductivity (PC) (where only free holes participate) and PL spectra (where localised holes dominate due to thermalisation at low temperature) in high magnetic fields[20,16] shows that the holes remain localised even at $n_s = 10^{12}$ cm^{-2}, and so it can be concluded with confidence that the decrease of X_{LO}/X with n_s shown in Figures 3, 4 is due to screening of the carrier (hole)–LO phonon interaction.

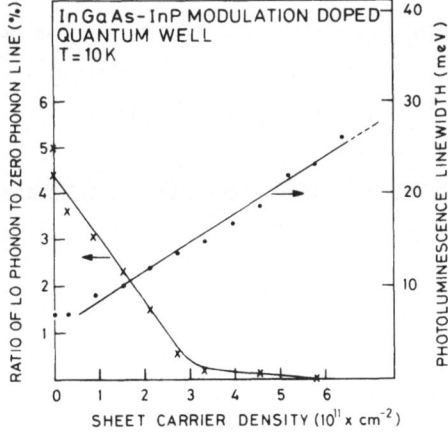

FIGURE 4

Variation of ratio of intensity of LO phonon satellite (X_{LO}) to zero phonon PL line (X) with carrier density (n_s) (left hand axis). The ratio X_{LO}/X decreases approximately linearly with n_s up to $\sim 3 \times 10^{11}$ cm^{-2} and is unobservable beyond 5×10^{11} cm^{-2}. The linewidth of the main PL line against n_s is also plotted on the figure (right hand axis). The linear variation of linewidth with n_s is as expected for a two dimensional electron gas where PL from all carriers from E=0 to E_F is observed.

237

The nearly complete screening of X_{LO}/X at $n_s \approx 4 \times 10^{11}$ cm^{-2} contrasts with that predicted for other polaron-related phenomena in two dimensions. For example Das Sarma and Mason[21] predict only a factor of 2 reduction in the polaron mass enhancement at $n_s \sim 6 \times 10^{11}$ cm^{-2}. However, direct comparison with such calculations is difficult since they treat the interaction of a free carrier with a cloud of virtual LO phonons, whereas the present experiments probe the Fröhlich interaction via the change in polarisation of the lattice which occurs during electron-hole recombination. Measurement of the polaron mass enhancement by cyclotron resonance experiments is also not straightforward, since it is difficult to separate such contributions from mass increases due to band non-parabolicity. However, Hopkins et al[22] have concluded recently that the Fröhlich interaction is strongly screened in two dimensions, in agreement with the present conclusions, since they could account for the measured dependence of effective mass on energy from the band edge solely in terms of band non-parabolicity as calculated by five band k.p theory.

To close this section, it should be emphasised that the observation of LO phonon satellites is an unambiguous manifestation of the Fröhlich interaction. Furthermore, the ability to vary n_s in a Schottky gated QW structure provides a new means to study the screening of the carrier-LO phonon interaction by optical spectroscopy. In three dimensions the same experiment cannot be carried out since the carrier density cannot be varied uniformly over the spatial region from which PL is obtained, as opposed to the present quasi-two dimensional situation where the signal originates only from the 100 Å QW where n_s is uniform.

3. Observation of a Fermi energy edge singularity in low temperature photoluminescence spectra

In the previous section the screening of the LO phonon satellite of the electron-hole recombination line in Figure 3 was emphasised. However, another notable feature of the spectra in Figure 3 (taken at 10K) as a function of n_s is the change in linewidth with increasing n_s from 7 meV (full width half maximum 5.5 meV at 4.2K) at $n_s = 0$ to 25 meV at $n_s \approx 6 \times 10^{11}$ cm^{-2}, reasonably close to the value of the two dimensional electron Fermi energy of 29.9 meV for this carrier density. This suggests that PL from all the electrons from the bottom of the band (E=0) up to the Fermi energy (E_F) is observed in the recombination. This is further demonstrated by the approximately linear variation of linewidth against n_s shown in Figure 4, as expected in two dimensions where $E_F = h^2 n_s/4\pi m$, if PL from E=0 to E_F is observed.

The situation is even clearer for the PL spectrum in Figure 5a, obtained from a separate, slightly higher-quality 100 Å modulation doped sample (double-sided doped), with $n_s = 9.2 \times 10^{11}$ cm^{-2} as determined from low temperature Shubnikov-de Haas

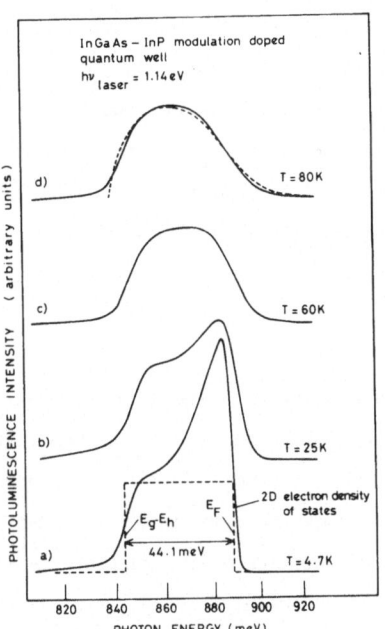

FIGURE 5

Series of spectra from 4.7K to 80K for double-sided modulation doped InGaAs–InP quantum well. The width of the spectrum in (a) is 44.1 meV, very close to the electron Fermi energy of 45.0 meV. The dashed curve in (a) indicates the form of the spectrum expected for recombination without k–restriction in two dimensions without inclusion of many body effects. The enhancement of the experimental curve relative to the 2D density of states is attributed to the Fermi energy edge singularity. With increasing temperature, (b) to (d), the enhancement is suppressed until at 80K a good fit to the experimental curve is obtained from the convolution of the electron and hole densities of states alone.

measurements. The form of the spectrum (taken at 2K) is similar to that of the highest n_s spectrum in Figure 3. It is composed of a low energy onset at $\sim$ 840 meV of width $\sim$ 5 meV, a nearly energy independent region over the next 10 meV, and then a rapid increase towards the sharp high energy cut–off at $\sim$ 885 meV. The overall width of the spectrum is 44.1 meV (see Figure 5a) very close to the calculated electron Fermi energy of 45.0 meV. As for the sample of Figure 3, it is seen that all electrons from E=0 up to E_F participate in the recombination, even though the Fermi wavevector (k_F) for electrons at E_F is 2.4 x 10^6 cm^{-1}. Efficient k–conserving recombination of electrons at E_F could only be explained for a highly improbable hole temperature of $\sim$ 200K. However, the participation of all electrons in the Fermi sea (from E=0 to E_F) in the PL spectrum can be understood if the photocreated holes thermalise to localised states with wavevector components $\sim$ 2.4 x 10^6 cm^{-1}, of the order of k_F. Using the uncertainty relation $\Delta r \Delta k = \frac{1}{2}$ this corresponds to hole localisation on a length scale of $\sim$ 20 Å, in the same range as the values of 10 to 30 Å required to explain the strength of the exciton–LO phonon coupling in these structures, discussed in section 2.

In the absence of many–body interactions, if the recombination proceeds without k–restriction, a PL spectrum of the form of the energy independent electron density of states in two dimensions, of total width E_F, would be expected, as shown by the dashed curve in Figure 5a. The main difference between this dashed curve, and the experimental spectrum is the strong enhancement of the experimental curve towards E_F. This enhancement is attributed[23] to a "Fermi energy edge singularity" and arises from the enhanced multiple electron–hole (e–h) scattering rate for electrons which recombine close to E_F. For electrons far below E_F such interactions have very low probability since the only empty states into which the electrons can scatter are above E_F. This increased (Coulomb) e–h interaction for electrons near to E_F leads to a many body "excitonic" enhancement of the oscillator strength for recombination near to E_F, as first discussed for degenerate semiconductors and metals by Mahan[24,25] and Nozieres and di Domenicis[26] nearly twenty years ago. The only previous observation before the present work was in the soft X–ray emission and absorption spectra of metals.[27]

The present InGaAs–InP QWs are particularly suited to the observation of this many body effect. Firstly, the occurrence of hole localisation permits recombination of all electrons in the Fermi sea without k–restriction, and thus allows recombination at E_F to be observed. The analogy with the X–ray emission case, where an inner shell localised core hole recombines with the electrons in the conduction band of the metal, is close since in the X–ray case k–restriction for electrons at E_F ($k_F \sim 10^8$ cm^{-1}) is again lifted by the hole localisation. In GaAs–GaAlAs QWs by contrast, the recombination is dominated by electrons close to the bottom of the band[28,29] since the photocreated holes thermalise to free or very weakly localised states, thus precluding the observation of any Fermi energy anomaly. A second important feature in the present work is the use of modulation doping, unachievable in three dimensions, which mimimizes impurity broadening of the spectra. Thirdly Coulomb interactions are enhanced and screening is reduced in two relative to three dimensions, thus increasing the size of the Fermi energy enhancement in the two dimensional QW compared to bulk semiconductors.

Further support to the many body interpretation of the form of the spectrum in Figure 5a is obtained from the temperature dependence of the PL shown in Figures 5b to 5d. It is seen that the enhancement at E_F becomes weak at a temperature of ~ 60K, of the order of the quasi–two dimensional excitonic Rydberg ($E_x \sim 6$ meV = 70K) in InGaAs. Such a suppression of the Fermi energy enhancement has been predicted to occur at temperatures such that $kT/E_x \sim 1$ by Schmitt–Rink et al[30] for recombination of an electron–hole (e–h) plasma of equal e–h density, in good qualitative agreement with the results of Figures 5b–d. The observation of the Fermi energy edge singularity depends on the existence of a sharp Fermi cut–off. When the width of this cut–off is of the order of the Coulomb interaction energy (E_x), the enhancement at E_F will be strongly attenuated and broadened. In Figure 5d it is shown that a good fit to the PL lineshape at 80K can be obtained from a convolution of the electron and hole densities of states with no E_F enhancement at a temperature where $kT \gtrsim E_x$.

A theoretical many–body treatment of the PL lineshape expected for recombination between a degenerate two dimensional electron gas and a low density of photocreated, localised holes has been carried out by Rorison.[31,23] The effects of multiple e–h scattering and static screening were included in the Bethe–Salpeter equation for the e–h Green's function including only ladder diagrams. Good qualitative agreement was obtained between the calculated lineshapes, showing a strong enhancement at E_F superimposed on the two dimensional density of states, and the experimental spectrum of Figure 5a.

4.   Magneto–spectroscopy of InGaAs–InP modulation doped quantum wells

In this section the results of Zeeman spectroscopic studies of the modulation doped InGaAs–InP quantum wells are discussed. It is shown that such investigations in magnetic fields up to 10 Tesla provide a very informative means for the study of the nature of the electron and hole states participating in the recombination, and lead to a new method for the study of the density of states between Landau levels (LLs) in high magnetic field.

A series of PL spectra in fields from zero to 9.6 Tesla is shown in Figures 6a to e. The zero field spectrum is seen to break up into a series of LLs, labelled

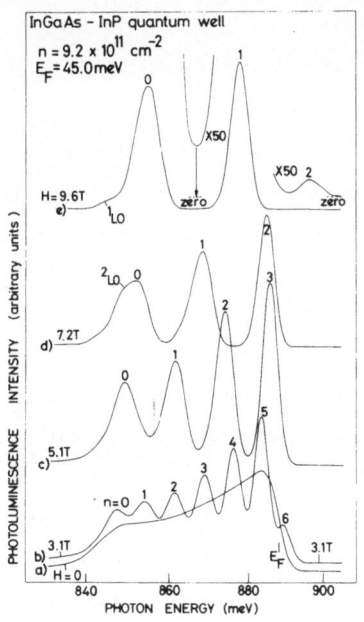

FIGURE 6

PL spectra at 2K for InGaAs–InP modulation doped, 100 A wide quantum well with n_s = 9.2 x 10^{11} cm^{-2}, E_F = 45.0 meV in magnetic fields (H) from O to 9.6T (same sample as Figure 5). The H=0 spectrum breaks up into a series of electron Landau levels (LLs). The LL intensities correlate well with filling factors deduced for n = 9.2 x 10^{11} cm^{-2}. The very weak PL intensity between the n=0 and 1 LLs at 9.6T enables an upper limit for the background density of states of 0.036 D_O to be deduced. E_F indicates the electron Fermi energy at H=0. The gain is decreased by a factor of 1.5 for d,e relative to a,b,c.

n = 0 to 6, with spacing given by the electron cyclotron energy $\hbar\omega_c$ ($\hbar eH/m^*c$), for an effective mass of 0.049 m_O, expected for electrons in InGaAs. A PL signal is observed from each populated electron LL. With increasing field the LLs move through the electron Fermi energy, and at 9.6T only two LLs remain populated. The transition energies are plotted against magnetic field in Figure 7. The straight lines through the points are drawn for LL slopes of $(n+\tfrac{1}{2})\hbar\omega_c$ with $m^* = 0.049\ m_0$. There is no evidence for any hole LL splittings as opposed to observations for GaAs–GaAlAs QWs,[32] providing further support for the participation of localised holes in PL at low temperature, in the present case. The existence of hole localisation lifts the $\Delta n = 0$

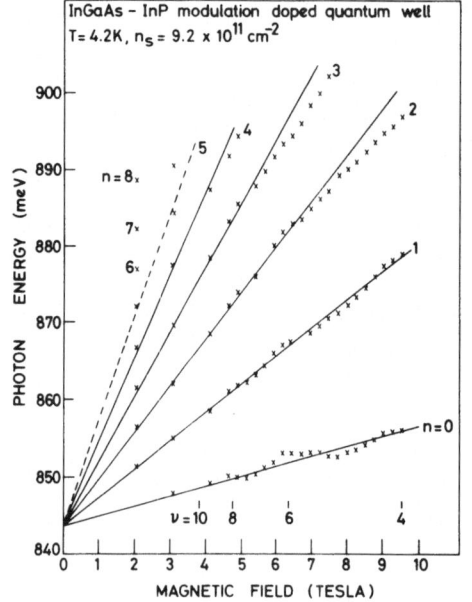

FIGURE 7

PL transition energies (x), for spectra of Figure 6 plotted against magnetic field. The full lines correspond to $(n+\tfrac{1}{2})\hbar\omega_c$ variations with field for the various LLs. Anomalies in the energies of filled LLs occur at $\upsilon = 8, 6, 4$. In addition, the energies of the LLs at E_F deviate from the $(n+\tfrac{1}{2})\hbar\omega_c$ slope as the level empties ($n = 3$ and 2 begin to empty at $\upsilon = 8$ and 6 respectively), as discussed in Ref 33.

selection rule for interband LL transitions, and thus permits recombination from all populated electron LLs to be observed. Furthermore, the absence of any hole LL splittings allows the behaviour of the electron LLs to be studied without complications from superimposed hole splittings.

The lines drawn through the observed LL positions on Figure 7 all extrapolate to the same zero field energy (the renormalised quantum well bandgap in the presence of a carrier density of $9.2 \times 10^{11}\ \mathrm{cm}^{-2}$). This demonstrates that the unusual shape of the spectrum of Figures 5a and 6a is not due to the superposition of more than one band. In addition the observation of equally spaced LLs shows that the electron states are not perturbed by disorder or localisation effects. Anomalies in the slopes of filled LLs at even integer filling factors and in the slopes of partially filled LLs on Figure 7 are discussed in Reference 33.

One of the most striking features of the spectra in Figure 6 is that at high fields the Landau levels are very well resolved. In an ideal 2D system the LL density of states is a series of delta functions, each of which is then broadened by the potential fluctuations (ionised impurity, alloy broadening etc) in any real sample. The spectrum in Figure 6e is seen to approximate closely to this situation, although it should be borne in mind that in a PL experiment the LL peaks are also subject to inhomogeneous bandgap broadening which does not affect experiments sensitive to the electron states alone. Indeed the width of the n=1 LL at 9.6T of 5.3 meV is close to that found in similar samples with $n_s = 0$ at zero magnetic field, where the linewidth is determined by inhomogeneous processes such as well width or alloy broadening.

For this reason the widths of the electron Landau levels cannot be obtained directly from the spectra of Figure 6. However, a value for the DOS between LLs (D_b) can be deduced from the spectra at 9.6T. A knowledge of the form of the DOS in a 2D system is essential for a detailed interpretation of experiments in the Quantum Hall system. Large values for D_b, of 10 to 50% of the zero field value D_0, have been deduced from a variety of experiments (eg thermal activation of ρ_{xx}, specific heat, de Haas–van Alphen) which are sensitive only to the form of the DOS at E_F.[34-37] Such large values of D_b are not predicted by perturbation theories which treat the interaction of the 2D electron gas with short range scatterers,[38] nor by exact theories for electrons in a random potential.[39]

The ratio of the PL intensity between the n=0 and n=1 LLs at 9.6T and the PL intensity at the same energy below E_F at zero field is only 0.024. In order to relate this ratio to the ratio of D_b at 9.6T to D_0, it needs to be determined whether there is any change in oscillator strength for electrons in localised LL states compared to those in states near the centres of LLs. There is no variation of the integrated PL intensity between 0 and 9.6T, thus providing normalisation in the comparison of intensities between the spectra.

Simultaneous Shubnikov–de Haas (SdH) and PL experiments were carried out to investigate whether the oscillator strength for recombination for electrons in LL tail states at high field was modified relative to electrons in states near the centres of the LLs. The SdH measurements allow the LL populations to be accurately determined. Comparison of the LL PL intensity, when the Fermi energy is in the LL tail states, with that predicted from the populations obtained from SdH, allows an upper limit of 1.5 to be placed on any reduction of oscillator strength for electrons in the tail states relative to states near the centres of the LLs. This permits an upper limit for D_b of $0.036\ D_0$ (1.5 times $0.024\ D_0$ from the ratio of the PL intensities at 0 and 9.6T) to be deduced.

This spectroscopically determined value is a factor of three to ten times less than that deduced from the swept field (or swept n_s) experiments sensitive to the form of the DOS at E_F, but is in agreement with theoretical expectation.[38,39] The large values for D_b deduced in References 34–37 may arise from the effects of spatial inhomogeneity of n_s or chemical potential on these experiments, as discussed in the statistical model of Gerhardts and Gudmundsson.[40] In a swept field experiment spatial variation of n_s will lead to the integer filling factor condition being fulfilled at different fields for different spatial positions, and thus to the deduction of a large value for the DOS between LLs.

5.   Conclusions

A number of key features of the optical properties of InGaAs–InP quantum wells grown by AP–MOCVD have been discussed. Localisation of the photocreated holes in the alloy QWs has been shown to be crucial in leading to the strong phonon coupling, permitting the observation of the Fermi energy edge singularity, and allowing the investigation of electron Landau level populations and density of states in high magnetic field.

References

1.   R. Dingle, W. Wiegmann and C.H. Henry, Phys Rev Lett 33, 827, 1974 and reviewed by R. Dingle in Festkorperprobleme (Advances in Solid State Physics), edited by H.J. Queisser (Pergamon–Vieweg, Braunschweig 1975), Vol 15, p21.

2.   K. Kodama, M. Ozeki and J. Komeno, J Vac Sci Tech B1, 696, 1983.

3.   M. Razeghi, J.P. Hirtz, U.O. Ziemelis, C. Delalande, B. Etienne and M. Voos, Appl Phys Lett 43, 585, 1983.

4.   T.Y. Wang, K.L. Fry, A. Persson, E.H. Reihlen and G.B. Stringfellow, Appl Phys Lett 52, 290, 1988.

5.   W.T. Tsang and E.F. Schubert, Appl Phys Lett 49, 220, 1986.

6.   B.I. Miller, E.F. Schubert, U. Koren, A. Ourmazd, A.H. Dayem and R.J. Capik, Appl Phys Lett 49, 1384, 1986.

7.   P.A. Claxton, J.S. Roberts, J.P.R. David, C.M. Sotomayor–Torres, M.S. Skolnick, P.R. Tapster and K.J. Nash, J. Cryst Growth 81, 288, 1987.

8.   M.S. Skolnick, K.J. Nash, M.K. Saker, S.J. Bass, P.A. Claxton and J.S. Roberts, Appl Phys Lett 50, 1885, 1987.

9.   M.B. Panish, H. Temkin, R.A. Hamm and S.N.G. Chu, Appl Phys Lett 49, 164, 1986.

10.   Y. Kawaguchi and H. Asahi, Appl Phys Lett 50, 1243, 1987.

11.   S.J. Bass, S.J. Barnett, G.T. Brown, N.G. Chew, A.G. Cullis, M.S. Skolnick and L.L. Taylor in Thin Film Growth Techniques for Low Dimensional Structures edited by R.F.C. Farrow, S.S.P. Parkin, P.J. Dobson, J.H. Neave and A.S. Arrott (Plenum 1987), p137.

12.   M.S. Skolnick, P.R. Tapster, S.J. Bass, N. Apsley, A.D. Pitt, N.G. Chew, A.G. Cullis, S.P. Aldred and C.A. Warwick, Appl Phys Lett 48, 1455, 1986.

13. M.S. Skolnick, K.J. Nash and S.J. Bass, Appl Phys Lett $\underline{52}$, 674, 1988.

14. Acoustic and optic phonon satellites phonon satellites of exciton recombination have been observed recently for indirect gap GaAs-AlAs short period superlattices by E. Finkman, M.D. Sturge, M.H. Meynadier, R.E. Nahory, T.C. Tamargo, D.M. Hwang and C.C. Chang, J Lumin $\underline{39}$, 57, 1987. However, in this case, at least the acoustic phonon (27 meV) is momentum conserving in the indirect gap system.

15. Similar observations to those of ref 14 have been reported by K. J. Moore, P. Dawson and C. T. Foxon, J. Physique $\underline{48}$, C5-525, 1987.

16. M.S. Skolnick, K.J. Nash, P.R. Tapster, D.J. Mowbray, S.J. Bass and A.D. Pitt, Phys Rev B$\underline{35}$, 5925, 1987.

17. K.J. Nash, M.S. Skolnick, P.A. Claxton and J.S. Roberts, unpublished.

18. M.J. Kane, D.A. Anderson, L.L. Taylor and S.J. Bass, J Appl Phys $\underline{60}$, 657, 1986.

19. D. A. Anderson, S. J. Bass, M. J. Kane and L. L. Taylor, Appl Phys Lett $\underline{49}$, 1360, 1986.

20. D.J. Mowbray, J. Singleton, M.S. Skolnick, S.J. Bass, L.L. Taylor, R.J. Nicholas and W. Hayes, Superlatt and Microstruct $\underline{3}$, 471, 1987.

21. S. Das Sarma and B.A. Mason, Phys Rev B$\underline{31}$, 5536, 1985.

22. M.A. Hopkins, R.J. Nicholas, M.A. Brummell, J.J. Harris and C.T. Foxon, Phys Rev B$\underline{36}$, 4789, 1987.

23. M.S. Skolnick, J.M. Rorison, K.J. Nash, D.J. Mowbray, P.R. Tapster, S.J. Bass and A.D. Pitt, Phys Rev Lett $\underline{58}$, 2130, 1987.

24. G.D. Mahan, Phys Rev $\underline{153}$, 882, 1967.

25. G.D. Mahan, Phys Rev Lett $\underline{18}$, 448, 1967.

26. P. Nozieres and C.T. Dominicis, Phys Rev $\underline{178}$, 1097, 1969.

27. see eg T.A. Callcott, E.T. Arakawa and D.L. Ederer, Phys Rev B$\underline{18}$, 6622, 1978.

28. A. Pinczuk, J. Shah, R.C. Miller, A.C. Gossard and W. Wiegmann, Solid State Commun $\underline{50}$, 735, 1984.

29. M. H. Meynadier, J. Orgonasi, C. Delalande, J. A. Brum, G. Bastard, M. Voos, G. Weimann and W. Schlapp, Phys Rev B$\underline{34}$, 2482, 1986.

30. S. Schmitt-Rink, C. Ell and H. Haug, Phys Rev B$\underline{33}$, 1183, 1986.

31. J.M. Rorison, J Phys C$\underline{20}$, L311, 1987.

32. M.C. Smith, A. Petrou, C.H. Perry, J.M. Worlock and R.L. Aggarwal, Proceedings of 17th Int Conf on the Phys of Semicond, edited by J.D. Chadi and W.A. Harrison (Springer, New York, 1984), p547.

33. M.S. Skolnick, K.J. Nash, S.J. Bass, P.E. Simmonds and M.J. Kane, Solid State Commun, 1988.

34. D. Weiss, K. von Klitzing and V. Mosser in Springer Series in Solid State Sciences, Vol 67, edited by G. Bauer, F. Kuchar and H. Heinrich, Springer-Verlag (Berlin, Heidelberg, 1986), p240.

35. M.G. Gavrilov and I.V. Kukushkin, Pis'ma Zh. Eksp. Teor. Fiz $\underline{43}$, 79, 1986 (Sov Phys JETP Lett $\underline{43}$, 103, 1986).

36. E. Gornik, R. Lassnig, G. Strasser, H.L. Stormer, A.C. Gossard and W. Weigmann, Phys Rev Lett $\underline{54}$, 1820, 1985.

37. J.P. Eisenstein, H.L. Stormer, V. Narayanamurti, A.C. Gossard and W. Weigmann, Phys Rev Lett $\underline{55}$, 875, 1985.

38. T. Ando and Y. Uemura, J Phys Soc Japan $\underline{36}$, 959, 1974.

39. F. Wegner, Z Phys B$\underline{51}$, 279, 1983.

40. V. Gudmundsson and R.R. Gerhardts, Phys Rev B$\underline{35}$, 8005, 1987.

TIME RESOLVED SPECTROSCOPY OF GaAs/AlGaAs QUANTUM WELL STRUCTURES

Ernst O. Göbel

Philipps-Universität, Fachbereich Physik, Renthof 5

3550 Marburg, Fed. Rep. Germany

INTRODUCTION

Optical spectroscopic techniques like absorption, photoluminescence, and photoluminescence excitation spectroscopy are widely used to characterize and investigate the properties of quantum wells (QW) and provide significant information on the electronic structure of the respective samples. Conclusions on the dynamics of nonequilibrium carriers, however, generally cannot be drawn from stationary experiments but require time resolved spectroscopy techniques. The dependence of the fundamental relaxation and recombination processes on the dimensionality, i.e. the thickness L_z and L_b of the quantum well and barrier layers, respectively, is one of the key questions as far as QW are considered. In addition, new relaxation processes come into play in QW, which are not present in bulk material, like carrier trapping from the barriers into the QW, intersubband scattering between states with different quantum number, and intervalley scattering including spatially seperated bands (real space transfer).

In this paper some of our recent results on time resolved luminescence spectroscopy of GaAs/AlGaAs QW are briefly summarized. In the first part representative data on the efficiency and dynamics of carrier trapping in single QW will be described. The second part deals with the exciton recombination dynamics at low temperatures (T < 60K) including electric field effects.

CARRIER TRAPPING

The efficiency and dynamics of carrier trapping into single QW with L_z below 10 nm is of particular importance for QW laser devices. It is established that for these lasers the threshold current is significantly dependent on the band structure profile of the confinement layers. In particular, SQW lasers with graded index separate confinement heterostructures (GRINSCH) exhibit lower threshold current than conventional single QW lasers[1].

We have investigated the trapping behaviour of molecular beam epitaxy grown single QWs with L_z = 5 nm and 1.2 nm with different configuration of the confinement layers[2],[3]. The trapping efficiency is determined from photoluminescence excitation (PLE) spectra and conclusions on the trapping dynamics can be drawn from time resolved luminescence. PLE spectra in the spectral range corresponding to the onset of the AlGaAs absorption of a 5 nm SQW with a linear and a constant band gap profile of the cladding layers are shown in the left and right hand side of Fig. 1, respectively. The full lines represent the experimental data. The dashed lines are calculated assuming that all the photoexcited carriers get trapped into the QW[11]. The ratio of the experimental and calculated curves yields the trapping efficiency as a function of the excitation photon energy. The trapping is equal to one per definition for direct excitation of the QW (i.e. excitation photon energy smaller than the AlGaAs band gap). For indirect excitation of the QW the trapping efficiency differs significantly for the different structures and is close to one in the graded index structure, but is only about 0.5 in the conventional seperate confinement heterostructure with $Al_{0.3}Ga_{0.7}As$ barrier layers with 0.1 μm thickness. This low trapping efficiency reveals that about half of the carriers recombine or get trapped within the AlGaAs.

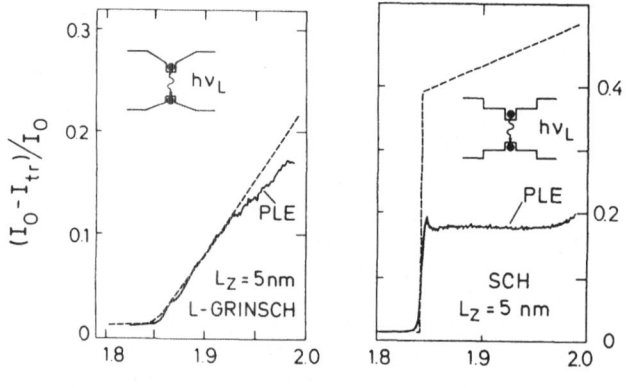

PHOTON ENERGY (eV)

Figure 1
PLE spectra of GaAs/AlGaAs single QW with L_z = 5 nm for a linear graded band gap of the AlGaAs cladding layers (left) and a conventional seperate confinement herterostructure with constant band gap in the barrier (right). The spectra are calibrated in terms of the relative absorbed light intensity and compared to calculated curves taking into account the actual band gap profile[3].

The time behaviour of the lowest subband exciton luminescence of the same samples is compared in Fig. 2. The rise of the luminescence is almost the same for indirect (a) and direct (b) excitation, which reveals that the actual trapping times are faster than the luminescence risetime, which is mainly determined by the thermalization of the carrier system[4].

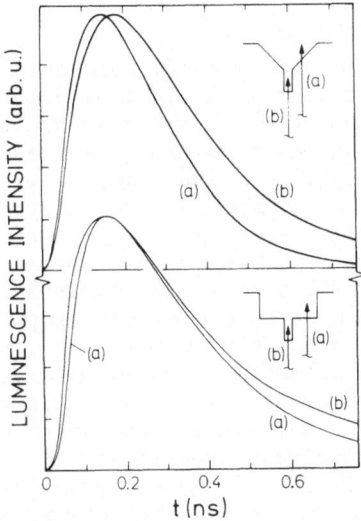

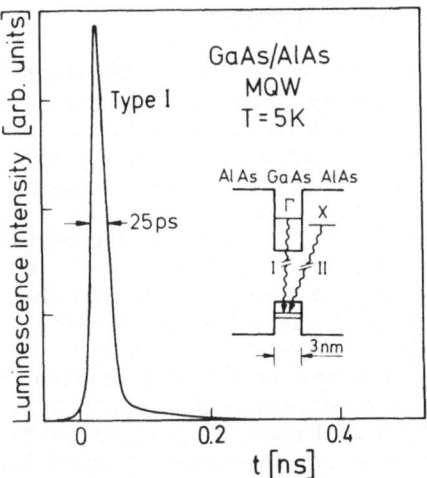

Figure 2
Time dependence of the spectrally integrated photoluminescence intensity of a 5 nm QW with a linear graded barrier (upper) and a separate confinement heterostructure (lower part) for indirect (a) and direct excitation (6).

Figure 3
Time dependence of the type I luminescence in a GaAs/AlAs QW with L_z = 3 nm. The inset depicts schematically the recombination processes. The decay of the type II luminescence (not shown in the Figure) is much slower with decay times in the order of ns to μs[8].

Surprisingly, the luminescence decay is slightly different for direct and indirect excitation. This behaviour is not completely understood at present but it could be related to some built in electric fields, which can be screened for the indirect excitation, where carriers can move within the AlGaAs before being trapped in the QW.

It should be mentioned that the trapping efficiency of SQW with L_z = 1.2 nm is comparable to samples with L_z = 5 nm for the same confinement layer profile[3]. This indicates that the electron wavefunctions of the 1.2 nm QW are not completely confined within the well and their spatial extension is expected to be comparable to the 5 nm wells[5].

In very thin QW with indirect AlGaAs barriers (x > 0.43) a new relaxation process occurs as soon as the confined electron state in the QW is higher in energy than the lowest indirect conduction band state in the barrier material. Electrons then will relax from the QW into the barrier material. For a QW with binary AlAs barrier material this crossover from a type I to a type II superlattice occurs for $L_z \le$ 30 nm. The optical properties of these type I/type II quantum wells have been studied in great detail[6],[7],[8]. An upper bound for the relaxation times of electrons from the Γ-state in the GaAs to the energetically lower X states in the AlAs barrier material can be obtained from time resolved photoluminescence. In Fig. 3 the temporal variation of the type I luminescence is depicted for a GaAs/AlAs multi QW structure with L_b = 7 nm. The luminescence decays within the temporal resolution of the experiment ($\sim$ 20 ps), which is much faster than for a L_z = 3 nm QW with $Al_{0.3}Ga_{0.7}As$ barrier layers. This demonstrates that the transfer of electrons from the Γ-state in the GaAs to the X-state in AlAs barriers takes place with time constants below 20 ps.

The properties of excitons in QW will be discussed in more detail in the contributions by Bajaj[9] and Chang[10] within this book. Most important for the subject of this chapter is the increase of exciton binding energy with decreasing QW thickness L_z and the corresponding increase in oscillator strength for optical transitions. The increase in exciton oscillator strength has been verified experimentally by absorption measurements[11]. Exciton recombination strength does not directly reveal the oscillator strength, which is generally normalized to the volume of a unit cell but also depends on the coherent extension of the exciton[12]. The change of the radiative exciton lifetime in QW thus does not exhibit an explicit dependence on L_z but is directly proportional to the square of the in plane exciton Bohr radius. Experimental data for the decay times of the exciton luminescence for GaAs/AlGaAs multi QW samples with different thickness L_z are summarized in Fig. 4. The photoluminescence decay times decrease continously with decreasing well thickness from about 1.6 ns for L_z = 15 nm to about 0.3 ns for L_z = 2 nm. The decay is exponential over 1 to 2 orders of magnitude followed by a slower decay. This slower decay could be related to exciton localization or bound exciton recombination.

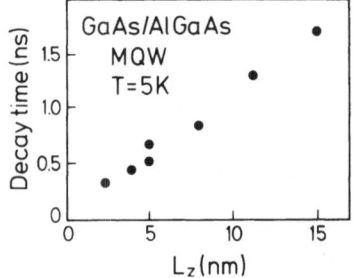

Figure 4
Photoluminescene decay times of the exciton recombination in GaAs/AlGaAs multi Qws with different thickness L_z.

One of the particularly exciting features of excitons in QWs is the possibility of tuning the spectral position of exciton absorption[13] and luminescence[14] by external electric fields applied perpendicular to the QW layers. This energy shift of the exciton resonance due to the quantum confined Stark effect is accompanied by a respective change of the exciton oscillator strength[15,16,17,28]. This in turn is again reflected in the radiative exciton decay. Results for the dependence of the exciton luminescence decay times on the external electric field strength are summarized in Fig. 5 for QW with different thickness[19]. The electric fields are applied via a semitransparent Schottky contact (built in voltage ~ 1V; 1V corresponds to about 15 kV/cm). A pronounced increase of the photoluminescence lifetime with increasing electric field strength is

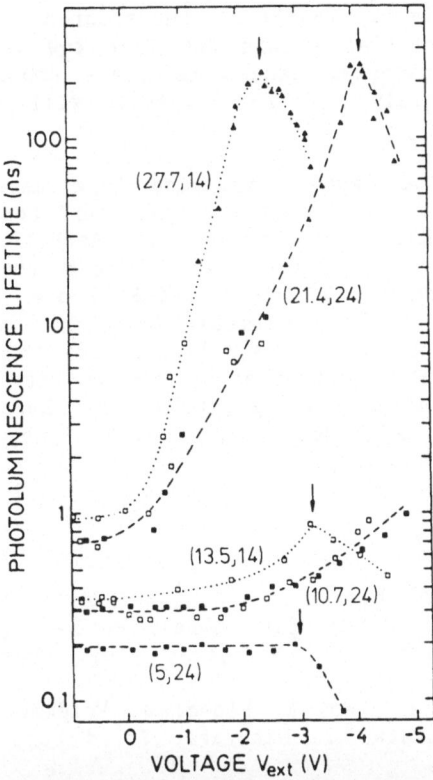

Figure 5
Photoluminescence decay times
of the exciton recombination in
GaAs/AlGaAs QWs (x = 0.3) with
different QW thickness (L_z) and
barrier layer thickness (L_b)
labelled by (L_z,L_b) in nm as a
function of the voltage applied
to a Schottky contact.

observed for QW with $L_z > 5$ nm. In accordance with the results for the
spectral shift this increase of the decay times in most pronounced for
thicker wells due to the fact that the field induced spatial seperation of
the electron and hole wavefunction is larger in thick wells. At high
electric fields the decay times decrease again due to tunneling ofcarriers
into the AlGaAs barriers. The critical field at which this turn over of
the decay times is observed depends on the one hand on L_z which
determines the effective barrier height for a given composition, and on
the other hand on the barrier layer thickness L_b as verified
quantitatively by the results shown in Fig.5.

In summary, the examples which we have briefly described may
demonstrate the potential of optical spectroscopy, in particular, time
resolved luminescence spectroscopy to investigate the dynamical electronic
properties of quantum wells with "engineered" band structure. The time
resolution of the experiments described here is limited to about 20 ps and
thus is not able to provide information on the very fast initial

relaxation processes in QW. The tremendous progress in ultrashort optical pulse generation techniques within the last few years[20], however, now provides direct experimental access to these subpicosecond relaxation processes[21] and certainly further results will come up in the near future.

Acknowledgement: The results reported here have been obtained in cooperation with the Max-Planck-Institut für Festkörperforschung in Stuttgart, Philips Res. Laboratories in Redhill, England and the University of Oxford. I particularly would like to thank H.-J. Polland, J. Kuhl, K. Ploog (Stuttgart). P. Dawson and K. Moore (Philips). R. Elliott (Oxford) and G. Peter and J. Feldmann (Marburg) for their significant contributions. The high quality samples used for the investigations have been grown in different laboratories and we are very much indebted to C. Foxon (Philips, Redhill), K. Fujiwara, T. Nakayama and Y. Ohta (Mitsubishi), C.W. Tu (AT&T Bell Labs), and K. Ploog (MPI Stuttgart) for their continuous support.

REFERENCES

1   W.T.Tsang, Appl. Phys. Lett. 39, 134 (1981)

2   J. Feldmann, G. Peter, E.O. Göbel, K. Leo, H.-J. Polland, K. Ploog, K. Fujiwara, T. Nakayama, Appl. Phys. Lett. 51, 226 (1987)

3   H.-J. Polland, K. Leo, K. Ploog, J. Feldmann, G. Peter, E.O. Göbel, K. Fujiwara, T. Nakayama, Solid Sate Electr. 31, 341 (1988)

4   Recent results on hot carrier cooling in GaAs QW can be found e.g. in: K. Leo, W.W. Rühle, H.-J. Queisser, K. Ploog, Appl. Phys. A 45, 35 (1988)

5   R.L. Greene, K.K. Bajaj, Sol. State Commun. 45, 831 (1983)

6   P. Dawson, B.A. Wilson, C.W. Tu, R.C. Miller, Appl. Phys. Lett. 48, 541 (198)

7   P. Dawson, K.J. Moore, C.T. Foxon, SPIE, 782 Quantum Well and Superlattice Physics, 208 (1987)

8   E. Finkmann, M.O. Sturge, M.-H. Meynadier, R.E. Nahory, M.C. Tamargo, D.M. Hwang, C.C. Chang, Journ. Luminesc. 39, 57 (1987)

9   K.K. Bajaj, same issue

10   V.C. Chang, same issue

11   Y. Masumoto, M. Matsuura, S. Tarucha, H. Okamoto, Phys. Rev. B52, 4275 (1985)

12   J. Feldmann, G. Peter, E.O. Göbel, P. Dawson, K. Moore, C. Foxon, R.J. Elliott, Phys. Rev. Lett. 59, 2337 (1987)

13   D.A.B. Miller, D.S. Chemla, T.C. Damen, A.C. Gossard, W. Wiegman, T.H. Wood, C.A. Burrus, Phys. Rev. Lett. 53, 2173 (1984)

14   H.-J. Polland, L. Schultheis, J. Kuhl, E.O. Göbel, C.W. Tu, Phys. Rev. Lett. 55 2610 (1985)

15   D.A.B. Miller, D.S. Chemla, T.C. Damen, A.C. Gossard, W. Wiegmann, T.H. Wood, C.A. Burrus, Phys. Rev. B32, 1043 (1985)

16   G. Bastard, E.E. Mendez, L.L. Chang, L. Esaki, Phys. Rev. B28, 3241 (1983)

17   J.A. Brum, G. Bastard, Phys. Rev. B31, 3893 (1985)

18   G.O. Sanders, K.K. Bajaj, Phys. Rev. B35, 2308 (1987)

19   H.-J. Polland, K.Köhler, L. Schultheis, J. Kuhl, E.O. Göbel, C.W. Tu, Superlattices & Microstructures 2, 309 (1986)

20   see e.g. R.L. Fork, C.H. Brito Cruz, P.C. Becker, C.V. Shank, Optics Lett. 12, 4403 (1987)

21   see e.g. W.H. Knox, D.S. Chemla,G. Livescu, Solid State Electr. 31, 425 (1988)

RECOMBINATION MECHANISMS IN A TYPE II GaAs/AlGaAs SUPERLATTICE[+]

T.W Steiner[*], D.J.Wolford, S.W. Tozer and T.F. Kuech

IBM T.J. Watson Research Center
P.O. Box 218, Yorktown Heights, NY, 10598

M. Jaros

Dept. of Theoretical Physics
The University, Newcastle upon Tyne, NE1 7RU, UK

A variety of experiments[1,2,3] have shown that the GaAs/GaAlAs band offset can be determined optically, with potentially meV resolution, provided type II recombination can be observed. Type II, or cross-interface recombination, can be induced in a GaAs/AlGaAs superlattice of arbitrary composition and periodicity by the application of hydrostatic pressure[1]. Figure 1 depicts schematically the band structure of such a superlattice at ambient pressure and at a non-zero pressure above the crossing point to type II behaviour. A typical above-crossing spectrum is displayed in Fig. 2. For pressures larger than the composition and periodicity dependent crossing point the X valley of the AlGaAs, rather than the Γ valley of the GaAs, is the lowest energy conduction-band minimum. The electrons will thus transfer to the AlGaAs while the holes remain in the GaAs. The oscillator strength of the optical recombination of these electrons and holes is very small since the k-space and real space overlaps of these electron and hole wavefunctions are both small. This qualitative arguement is in agreement with the results of a pseudopotential calculation[4] which predicts a small oscillator strength for the optical type II recombination. Experiments investigating the time-decay and recombination mechanisms of these processes are described below.

A GaAs/AlGaAs superlattice was loaded into a diamond-anvil-cell with liquid He as the pressure medium and a ruby chip as the manometer[5]. This cell was then introduced into an optical varitemp cryostat to allow spectroscopy at low temperatures. The sample was excited using an Ar-ion laser chopped by a fast acousto-optic modulator with a few nanosecond switching times. The luminescence was dispersed by a 0.85m, double-grating spectrometer and detected with a cooled GaAs photomultiplier. The time resolution was obtained using a time-to-amplitude converter in the usual time-correlated photon counting scheme. The photon detection rate was kept at least 2 orders of magnitude below the laser repetition rate to avoid distortion of the data.

Type II recombination, with its associated small oscillator strength and correspondingly long lifetime, has been observed in a MOCVD-prepared superlattice with 70-Å GaAs wells and barriers (0.24 mole-fraction Al), at very

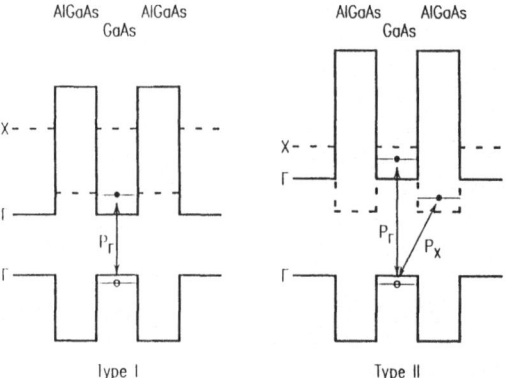

Fig. 1. Band edge picture of the ambient pressure type I
GaAs/AlGaAs superlattice and the pressure induced
type II superlattice. The direct-gap recombination
is labelled P_Γ while the cross-interface recombination is labelled Px.

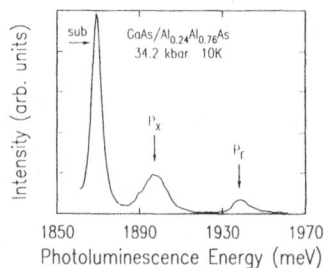

Fig. 2. A typical photoluminescence
spectrum of an GaAs/AlGaAs
superlattice above cross-over.
The direct-gap recombination
is labelled P_Γ while the
cross interface recombination
peak is labelled Px. The peak
labelled sub is due to substrate emission.

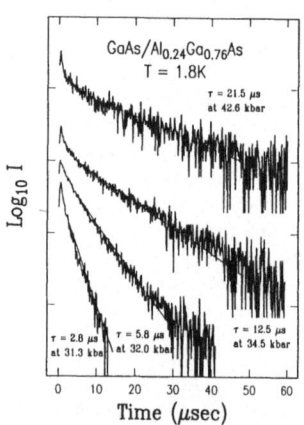

Fig. 3. Luminescence decays at
various pressures above
cross-over. The initial
non-linearities are due
to unavoidable sample
heating.

low temperature and excitation densities[6]. The luminescence intensity decay as a function of time after pulsed excitation for a variety of pressures above the crossing pressure to type II recombination are shown in Fig. 3. Well above crossing the lifetime is 5 orders of magnitude longer than the type I, direct-gap recombination observed at ambient pressure, yet the recombination still has a high quantum efficiency, thus attesting to its dominantly radiative origin. The observed dependence of the lifetime of this exponential decay, on the applied pressure, is in qualitative agreement with the predictions of the pseudopotential calculations and general principles. As the pressure is reduced towards cross-over the X minimum of the AlGaAs and the Γ minimum of the GaAs come closer together in energy and band mixing occurs. The admixture of components from the Γ minimum, which is solely responsible for the oscillator strength, increases as the GaAs Γ minimum approaches the AlGaAs X minimum. Consequently, the radiative lifetime of the cross-interface recombination should decrease as the pressure is decreased towards cross-over, as is observed. This long lifetime, high-quantum-efficiency behaviour was only observed at temperatures near 1.8K, and at the lowest possible excitation densities. In all cases, the excitation density was reduced until the observed lifetime was independent of excitation density, although, in the case of the luminescence from the superlattice far above crossing, this limited the observable decay to approximately one decade of intensity.

At higher temperatures or excitation densities cross-interface recombination with radically different characteristics was observed[7]. This luminescecence was characterized by a lifetime of order 100 nsec, a very low quantum efficiency (order 0.01%) and with its intensity peak ≈ 4 meV higher in energy than the above mentioned, high-quantum-efficiency luminescence. Furthermore, the observed lifetime of the cross-interface recombination was

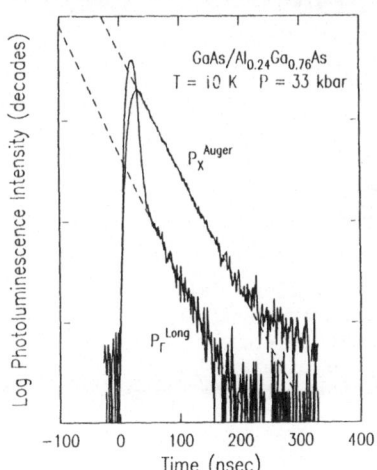

Fig. 4 The decay of the cross-interface Px and direct-gap P_Γ luminescecnce at higher (10K) temperature. Note the long-lived component of P_Γ whose lifetime is identical to the lifetime of Px.

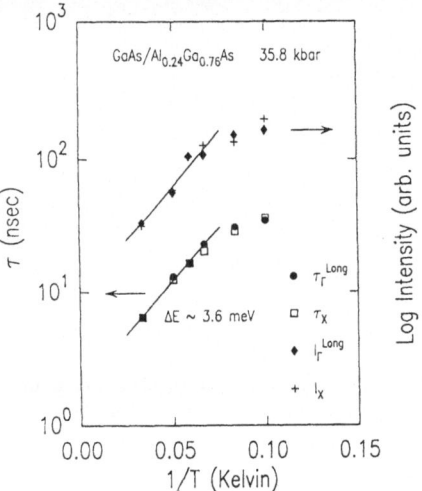

Fig. 5. The lifetimes τ and intensities I of the Px and the P_Γ long-lived components as a function of 1/T. A fit to the linear portion gives a localization energy of ≈ 3.6 meV.

replicated by a weak component in the higher energy, direct-gap recombination (Fig. 4). The observed lifetime and intensity of both the low-quantum efficiency, cross-interface recombination and the direct-gap replica were found to depend strongly and identically on temperature (Fig. 5), thus indicating that they are coupled together. As shown in Fig. 5 the simultaneous reduction of lifetime and intensity with an activation energy of 3.6 meV, indicates that this luminescence results from a center with a localization energy of 3.6 meV. The observation in Fig. 4 of a component in the Γ luminescence with identical lifetime as the X-gap recombination can be explained by a nonradiative Auger mechanism. The electrons responsible for the Γ luminescence are far from equilibrium and only the near unity direct-gap oscillator strength allows some of them to recombine radiatively before the rest are transferred to the X-minima. This process nevertheless, requires a substantial "feeding" in order for Γ luminescence to be observed. The proposed Auger mechanism involving two electrons and one hole as the dominant recombination channel for the low quantum efficiency cross-interface recombination explains the observed results. It acounts naturally for the obervation of the identical lifetime component in the Γ luminescence, since the electron-hole recombination energy is given to the remaining electron, which is then injected deep into the conduction band from where, after thermalizing, it will have a chance to contribute to the Γ luminescence. The number of such hot, Auger electrons depends on the cross-interface recombination rate which is proportional to the number of occupied centers giving rise to the cross-interface recombination. Hence, the Γ luminescence intensity will follow that of the cross-interface luminescence intensity.

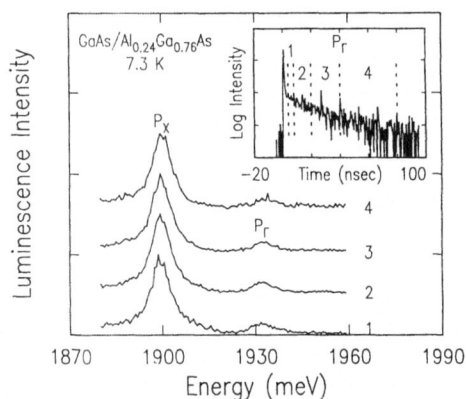

Fig. 6 Time-window spectra taken with various delays after pulsed excitation. The inset shows a luminescence decay curve at the P_Γ energy with the dashed lines indicating the time windows used for spectra 1 - 4. Note the definite long-lived peak at the P_Γ energy.

The low quantum efficiency of the type II recombination is also naturally explained in the current model since those elctrons and holes recombining with the Auger mechanism do not yield a photon, i.e. they recombine non-radiatively. The initial fast component of the Γ luminescecence, whose intrinsic decay rate cannot be deduced from these measure ments due to the limited turn off time of the acousto-optic modulator, results from the initial large number of hot electrons generated by the laser pulse. The purely exponential nature of these decays places restrictions on the possible recombination mechanism. The exponential decay implies fixed electron-electron and electron-hole overlap so that the decay rate does not depend on the carrier density. This requirement is satisfied by a bound exciton but not by free carriers. Since two electrons must be involved a possible center would be a neutral donor in the AlGaAs with a cross-interface exciton bound to it. That such a center would recombine with an Auger mechanism should not be surprising since, at pressures above cross-over to type II, the radiative recombination is quenched and the behaviour is more like that in indirect-gap semiconductors like Si where bound exciton recombination is known to be dominated by the Auger mechanism[8].

The long-lived Γ luminescence cannot be due to thermal activation since the barrier is of order 100 meV and kT is of order 1 meV. Furthermore, the integrated intensity of the long-lived component in the Γ luminescence decreases with increasing temperature. The long Γ luminescence decay can also not be due to a broad, structureless background decaying at the observed rate since time-window spectra displayed in Fig. 6 clearly show a long-lived peak at the energy of the Γ recombination.

An energy level diagram of the various processes discussed here has been drawn in Fig. 7. The pulsed laser excitation is represented by the upward pointing arrow on the right. This results in some direct-gap Γ recombination after thermalization and a predominant transfer to the lowest conduction band minimum, the X minimum of the AlGaAs, drawn on the left of the diagram. The neutral-donor-like level responsible for the low quantum efficiency lumin-escence is drawn below the X level representing the thermally deduced 3.6 meV binding energy. Two electrons have been drawn on this level to symbolize the Auger recombination with an arrow indicating the ejection of one of these electrons from the trap to deep within the conduction band upon electron-hole recombination. The luminescence from this level is only seen by virtue of the few pairs that nevertheless manage to recombine radiatively, but the process is dominated by the non-radiative Auger recombination as can be deduced from the extremely low quantum efficiency. The hot, Auger electrons act like delayed, laser-generated electrons and again have a chance to recombine radiatively across the direct-gap although most will again be transferred across the interface to the AlGaAs X minimum. This delayed direct-gap recombination results in the observed long component in the Γ luminescence.

The last level drawn $\simeq$ 4 meV below the neutral-donor-like level is the level resulting in the very long-lived recombination observed at the lowest temperatures and excitation densities. Since it is characterized by a very slow, high-quantum-efficiency recombination there can be no Auger process here and it must therefore be "iso-electronic" in origin. It has been drawn below the neutral-donor-like level since additional time-window spectroscopy (not shown here) during experimental conditions under which both processes are observable, indicate that the isoelectronic-like recombination peaks some 4 meV below the neutral-donor-like recombination. Consequently, this lumin-escence must originate from a center which has localized a cross-interface exciton with a localization energy of order 8 meV.

In conclusion, two distinctly different types of type II, cross-inter-face recombination have been observed in the MOCVD superlattice studied. Both originate from localized cross-interface excitons. One has a localization energy of order 8 meV, decays slowly and displays near-unity quantum effi-ciency, and hence has been labelled "iso-electronic like". The second lumin-escence process has an exciton localization energy of 3.6 meV, decays much

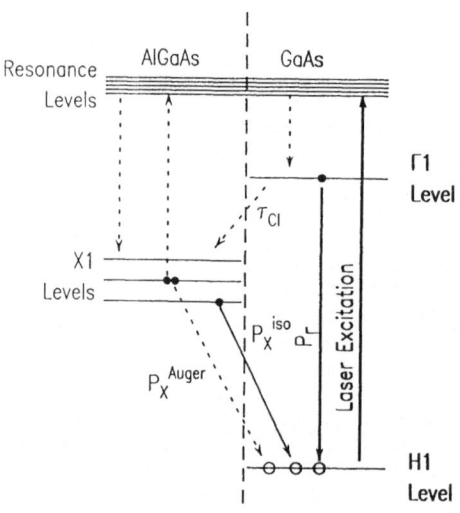

Fig. 7 An energy level diagram of the processes
discussed in the text. Radiative processes
are drawn as solid lines while predominantly
non-radiative processes are dashed. τ_{ci}
represents the cross-interface transfer of
electrons to the conduction band minimum in
the AlGaAs. The energy axis is not to scale.

faster and shows a very low quantum efficiency. By analogy to indirect-gap
donor-bound-excitons in Si and GaP and the experimental observation of delayed, hot, electrons it has been determined that this recombination is dominated by the non-radiative Auger mechanism and consequently labelled "neutral-
donor-like". The precise origin of the binding centers responsible for these
two processes remains, however, still unknown.

ACKNOWLEDGEMENT

[+]Supported in part by ONR under contract N00014-85-C-0868.
*Present address: Dept. of Physics, SFU, Burnaby, BC, V5A-1S6, Canada.

REFERENCES

1. D.J. Wolford, T.F. Keuch, J.A. Bradley, M.A. Gell, D. Ninno, M. Jaros,
   J. Vac. Sci. Technol. 4, 1043 (1986)
2. P. Dawson, B.A. Wilson, C.W. Tu and R.C. Miller, Appl. Phys. Lett. 48, 541
   (1986) and B.A. Wilson, P. Dawson C.W. Tu and R.C. Miller, J. Vac. Sci.
   Technol. 4, 1037 (1986)
3. G. Danan, B. Etienne, F. Mollot, R. Planel, A.M. Jean-Louis, F. Alexander,
   B. Jusserand, G. Le Roux, J.Y. Marzin, H. Savary and B. Sermage, Phys.
   Rev. B35, 6207 (1987)
4. M.A. Gell, D. Ninno, M. Jaros, D.J. Wolford, T.F. Keuch and J.A. Bradley,
   Phys. Rev. B35, 1196 (1987)
5. A. Jayaraman, Rev. Mod. Phys. 55, 65 (1983)
6. T.W. Steiner, D.J. Wolford, S.W. Tozer and T.F. Keuch, submitted to
   Physical Review B; D.J. Wolford, T.F. Kuech, T.W. Steiner M. Jaros, M.A.
   Gell and D. Ninno, Superlattices and Microstructures (1988), in press
7. T.W. Steiner, D.J. Wolford, T.F. Kuech and M. Jaros, Superlattices and
   Microstructures (1988), in press
8. T. Steiner and M.L.W. Thewalt, Solid State Commun. 49, 1121 (1984)

INTERFACE RECOMBINATION IN GaAs-GaAlAs QUANTUM WELLS

Bernard Sermage
Centre National d'Etudes des Télécommunications
Laboratoire de Bagneux
196 avenue Henri RAVERA - 92220 Bagneux, France

Interface recombination has been studied in MBE grown GaAs-$Ga_{1-x}Al_xAs$ undoped double heterostructures with GaAs thicknesses varying between 15 Å and 1 μm. We study the influence of quantum confinement and of a superlattice in the under confinement layer on interface recombination. We also show that radiative recombination at room temperature in quantum wells is bimolecular.

INTRODUCTION

Non radiative recombinations are prejudicial to minority carriers devices such as bipolar transistors, lasers, optical bistable devices, seeds, etc... With the appearance of quantum well devices, interface recombination has more and more importance on the proper work of these devices. For example, a reasonnable value like 300 cm/s for the interface recombination velocity S, leads to a non radiative lifetime τ_{nr} equal to 2 ns in a 60 Å wide quantum well and double the threshold current of quantum well lasers. Many papers report on carriers lifetime measurement in MBE grown multiple [1-6] or single [7-9] quantum well at room temperature. The large range of results (0.02 ns - 70 ns) means that non radiative recombination plays certainly a role.

METHOD

Our method for measuring non radiative lifetime and interface recombination velocity has already been described in ref. 8 and 9. Let us recall it briefly. This method can only been applied to undoped double heterostructures. They contain a Gallium Arsenide layer with thickness d surrounded by two $Ga_{1-x}Al_xAs$ layers with thicknesses d_1 and d_2. The composition of GaAlAs was generally 30 % aluminium and the confinement layers may contain superlattices. The samples are excited by the 3 ps pulses of a synchronously pumped dye laser working at 0.58 μm and the luminescence is spectrally dispersed by a weakly dispersing grating (100 groves/mm) and dispersed in time by a Hamamatsu synchroscan streak camera. The time resolution is about 10 ps. The two dimensions data acquisition system allows to select easily the luminescence wavelength

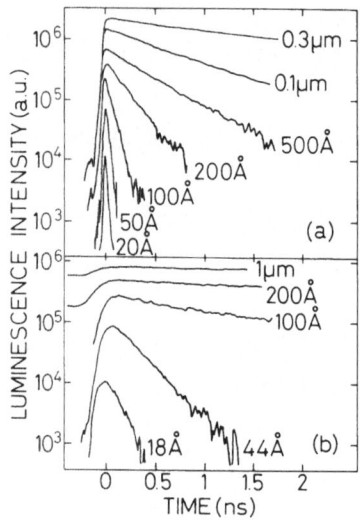

Fig. 1. Luminescence decay curves at
low excitation for two very
different series of MBE samples
showing the importance of
interface recombination and
the importance of the status
of the growth chamber.

range of interest. We can thus separate the luminescence of the quantum well from that of the buffer layer in the case of small d (< 200 Å). For large d, we have checked that there was no ambiguity by removing the substrate in some samples. We spectrally integer all the luminescence coming from the layer we study. Examples of luminescence decay are given on fig. 1. When the decay is fast enough (τ_L < 7 ns), the luminescence decay time τ_L is given by the slope of the curve Log (I_L) = f(t). When the decay is slow (τ_L > 7 ns), τ_L is obtained from the ratio of the luminescence intensities after and before excitation. Corrections due to the particular sweeping of the synchroscan streak camera are done.

The luminescence decay is studied as a function of excitation power over three orders of magnitude. Because the laser pulse and the time resolution of the streak camera (10 ps) are short compared to the carriers lifetime, carriers do not recombine during the laser pulse and the carriers density just at the end of the excitation pulse ($n(o)$) is proportionnal to the laser power P_{ex}. By exciting under the barrier gap in some samples, we have checked that this hypothesis was not modified by the carriers recombination and diffusion in the $Ga_{1-x}Al_xAs$ barrier. The variation of the luminescence intensity at the end of the excitation pulse ($I_L(o)$) as a function of the laser power P_{ex} gives the variation of the radiative lifetime τ_r :

$$\frac{1}{\tau_r} \sim I_L(o) / P_{ex} \qquad (1)$$

and allows to calculate the carriers lifetime from the luminescence decay time τ_L :

$$\tau = \tau_L \frac{\partial Log(I_L(o))}{\partial Log(n(o))} = \tau_L \frac{\partial Log(I_L(o))}{\partial Log(P_{ex})} \qquad (2)$$

Let us note that along a luminescence decay curve, the carriers density decreases and for this reason the different lifetimes : τ, τ_r, τ_L which depend on carriers density, vary. For this reason we just determine these lifetimes at the beginning of the decay where the carriers density is fixed by the excitation.

Fig. 2 shows the variation of $1/\tau$ and $1/\tau_r$ as a function of excitation in a 100 Å thick quantum well. The carriers recombination prolability $1/\tau$ is the sum of the radiative and the non radiative part which are supposed to be independant :

$$\frac{1}{\tau} = \frac{1}{\tau_r} + \frac{1}{\tau_{nr}} \qquad (3)$$

Eq. 1 gives the variation of $1/\tau_r$ but not its absolute value. Practically we fix the position of the $1/\tau_r$ curve (Fig. 2) so that $1/\tau_{nr} = 1/\tau - 1/\tau_r$ is as constant as possible.

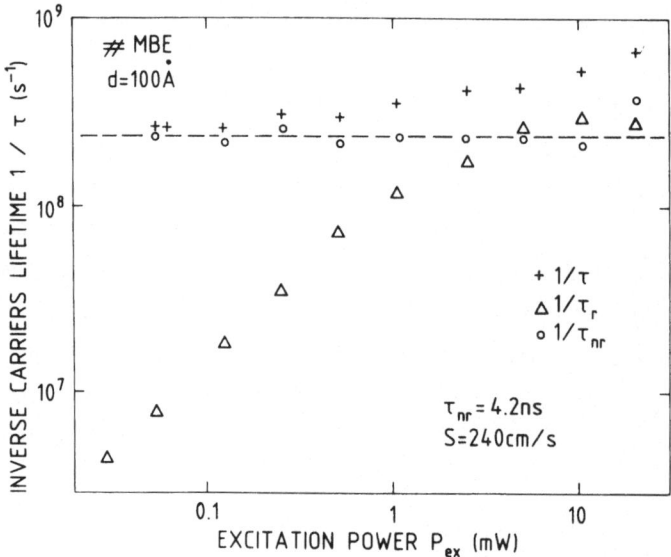

Fig. 2. Variation with excitation of the inverse
carriers lifetime $1/\tau$ (+), of the inverse
radiative lifetime ($\triangle$) and of the inverse
non radiative lifetime (o) for a 100 Å
thick quantum well. The under confinement
layer 0.5 μm thick contains a superlattice
with 43 wells 13 Å thick and $Ga_{0.7}Al_{0.3}As$
barriers 34 Å thick. The upper confinement
is obtained with a 1 μm thick $Ga_{0.7}Al_{0.3}As$
layer.

The variation of $1/\tau_{nr}$ in a series of samples grown in the same conditions with different thicknesses d of the GaAs layer allow to separate the
contribution of the interface from that of the bulk in the non radiative recombinations.

$$\frac{1}{\tau_{nr}} = \frac{1}{\tau_{nrb}} + \frac{S}{d} \qquad (4)$$

where S is the sum of the interface recombination velocities at the two GaAs-
GaAlAs interfaces, d is the thickness of the GaAs layer and τ_{nrb} is the bulk
non radiative life-time. Equation 4 is valid as long as d is small compared to
the carriers diffusion length L_D which is the case here since the maximum
thickness we have studied was 1 μm and L_D is larger than 3 μm. Actually in the
case of MBE grown GaAs-GaAlAs double heterostructures, Eq. 4 was always satis-

fied when d was larger than 200 Å i.e. whenever the quantum confinement effects were negligible.

RESULTS

1) First result concerns the radiative recombination in quantum wells. As can be seen on Fig. 2 and Fig. 3, the radiative recombination probability is proportionnal to the excitation power and then to the carriers density at low excitation which means that the radiative recombination at room temperature is dominated by band to band recombination and not excitonic recombination. If the radiative recombination was excitonic (monomolecular process), the radiative lifetime would be constant. In the 30 quantum well (d < 200 Å) samples we have studied, the slope of $1/\tau_r$ as a function of excitation power was always close to 1. There has been some controversy on this point[6,7], and we think

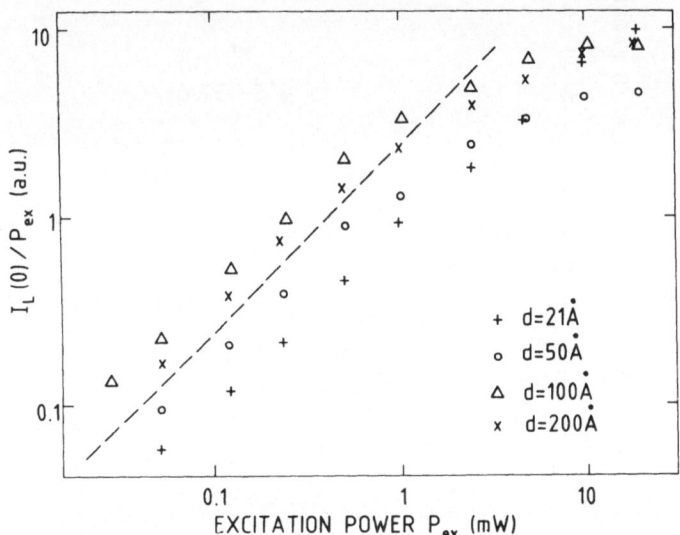

Fig. 3. Ratio between the maximum luminescence intensity and the excitation power (which is proportionnal to the inverse radiative lifetime) plotted as a function of the excitation power in different quantum well samples. This shows that the radiative recombination is bimolecular. The confinement layers are the same as in fig. 2. At low excitation, the carriers density is smaller than 10^{17} cm^{-3}.

that we have here some evidence that excitonic recombination plays a small role in radiative recombination at room temperature which agrees with the fact that kT is larger than the excitonic binding energy in GaAs quantum wells (E_b ~ 10 meV). Let us note that the variation of τ_r with excitation can not been explained by recombinations from different excitonic levels, each one having a constant τ_r and the relative population of these levels varying with excitation since at room temperature, the different excitonic levels are thermalised and decay at the same time.

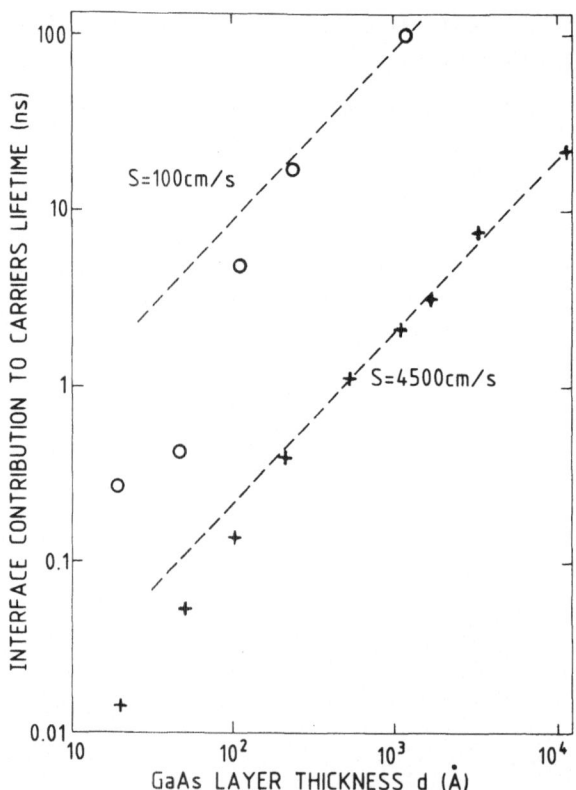

Fig. 4. Interface part of the non radiative carriers lifetime $1/(1/\tau_{nr} - 1/\tau_{nrb})$ as a function of the GaAs layer thickness d in two series of double heterostructures showing for large d the constancy of the interface recombination velocity S and for small d the increase of S due to quantum effects.

2) As we have already said, Eq. 4 is satisfied in the case of thick GaAs layers (d > 200 Å). For the thin layers, S should increase due to the increase of the probability of presence of the carriers in the barriers as foreseen by Quantum mechanics. This has been theoretically demonstrated by Duggan et al[10]. and observed experimentally[8,9]. Fig. 4 shows such a behaviour in two series of samples. The interface part of the non radiative lifetime : $1/(1/\tau_{nr} - 1/\tau_{nrb})$ decreases faster than S/d for the values of d smaller than 200 Å. However in two other series we have observed an other behaviour i.e. the interface recombination velocity is constant even for thin GaAs layers (Fig. 5). We don't know the exact reason of these different behaviours. It could be due to different natures of the recombination centers, to different residual doping or to different profiles of the recombination centers in the GaAs layer.

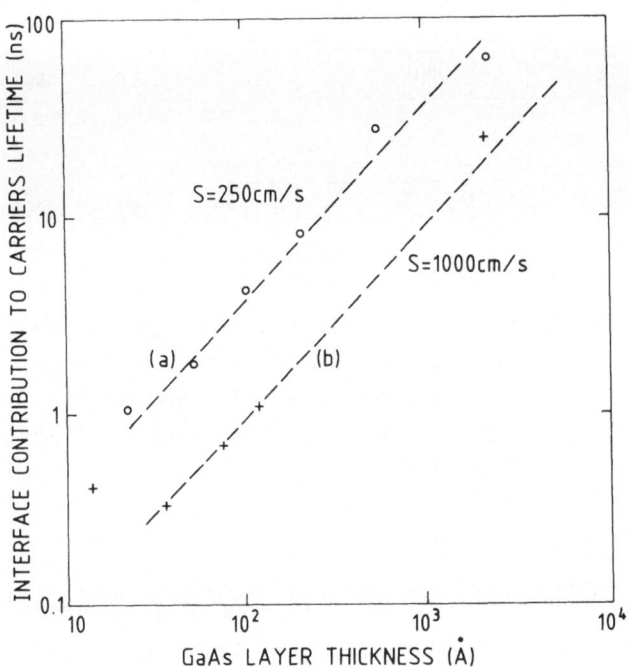

Fig. 5. Interface part of the non radiative
carriers lifetime as a function of
the GaAs layer thickness d in two
other series showing cases where the
increase of S in quantum wells is not
observed. Samples of series (a) have
the same structure as in Fig. 2.

Let us note that these different behaviours can not been explained by an improvement of the status of the growth chamber during a series as observed by Dawson et al.[1,2] since the series have been grown with different chronology without any correlation with the behaviour of the τ_{nr}(d) curves. Furthermore we have checked the constancy of the status of the growth chamber by growing after some series a sample with the same GaAs thickness as the first sample of the series. This sample had the same τ_{nr} as the first one.

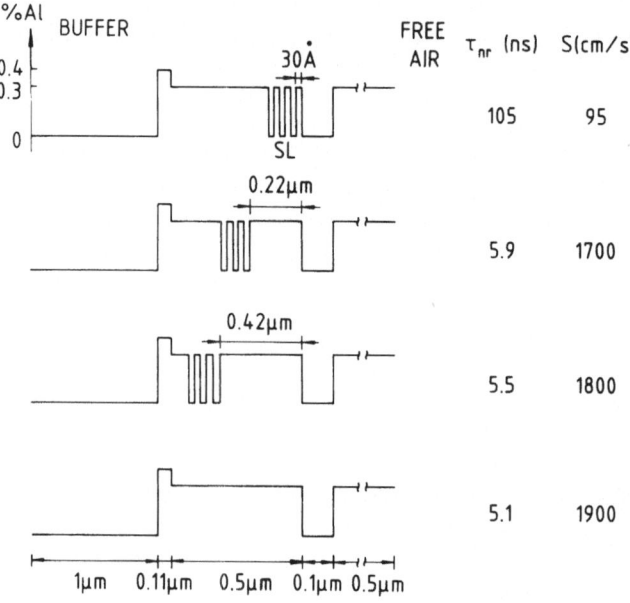

Fig. 6. Influence of the distance between the superlattices in the under confinement layer and the main GaAs layer on the non radiative carriers lifetime in the GaAs layer.

3) The last result concerns the influence of a superlattice in the under confinement layer. We have done about ten tests and in all the cases, the superlattice improves the non radiative lifetime in the main GaAs layer (quantum well or thick layer) as already observed by K. Fujiwara et al[7]. Results are a little puzzling because the improvement varies considerably from one case to an other and because there are many parameters. We will select one test : in the under confinement layer we have inserted a superlattice made of 6 wells 12 Å thick and five barriers 30 Å thick. The distance between the last well and the 0.1 μm thick GaAs layer has been made different in the three

samples we have studied : 30 Å, 0.22 μm and 0.42 μm. The non radiative carriers lifetime has been measured in these three samples and compared with a sample without superlattice. As shown in Fig. 6, the improvement due to the superlattice is important only when the superlattice is close to the GaAs layer. This means that the action of the superlattice is not to trap the impurities or dislocations coming from the substrate but more probably to trap impurities which lie at the surface of the GaAlAs during the growth of AlGaAs. An other explanation could be that when the aluminium oven is open, the shutter is cold and adsorb molecules of the residual vacuum. When we shut the aluminium oven, the temperature of the shutter is increased to nearly 900 degrees Celcius and outgas rapidly all the adsorbed molecules. Part of these molecules go to the sample in which they will be situated close to the AlGaAs-GaAs interface. The quantity of adsorbed molecules increases with the distance between the last GaAs well and the main GaAs layer.

ACKNOWLEDGEMENTS

We gratefully acknowledge F. Alexandre and J. Beerens who have collaborated in the major part of this work and G. Weimann from whom we have studied three samples.

REFERENCES

1. P. Dawson, G. Duggan, H.I. Ralph and K. Woodbridge, Photoluminescence decay times in multiple quantum well heterostructures prepared by molecular beam epitaxy, in the Proceeding of the 17th ICPS, Eds. D.J. Chadi and W.A. Harrison, Springer, New York, 1985, p. 551

2. P. Dawson, G. Duggan, H.I. Ralph and K. Woodbridge, Photoluminescence decay times in AlGaAs-GaAs multiple quantum well heterostructures, Superlattices and Microstructures, 1 : 173 (1985)

3. D. Bimberg, J. Christen, A. Werner, M. Kunst, G. Weiman and W. Schlapp, Evidence for excitonic decay of excess charge carriers in high quality GaAs quantum wells at room temperature, Appl. Phys. Lett. 49 : 76 (1986)

4. J.E. Fouquet and A.E. Siegman, Recombination times in GaAs/Al$_x$Ga$_{1-x}$As multiple quantum well structures, in the Proceedings of the 17th ICPS, Eds. D.J. Chadi and W.A. Harrison, Springer, New York, 1985, p. 583

5. J.E. Fouquet and R.D. Burnhour, Recombination dynamics in GaAs/Al$_x$Ga$_{1-x}$As quantum well structures, IEEE Journal of Quantum Electronics QE-22 : 1799 (1986)

6. E.H. Böttcher, K. Ketterer, D. Bimberg, G. Weiman and W. Schlopp, Excitonic and electron hole contributions to the spontaneous recombination rate of injected charge carriers in GaAs-GaAlAs multiple quantum well lasers at room temperature, Appl. Phys. Lett. 50 : 1074 (1987)

7. K. Fujiwara, A. Nakamura, Y. Tokuda, T. Nakayama and M. Hirai, Improved recombination lifetime of photoexcited carriers in GaAs single quantum well heterostructures confined by GaAs/AlAs short-period superlattices, Appl. Phys. Lett. 49 : 1193 (1986)

8. B. Sermage, M.F. Pereira, F. Alexandre, J. Beerens, R. Azoulay and N. Kobayashi, Interface recombination in GaAs-GaAlAs double heterostructures and quantum wells, in Proceedings of the 1987 Int. Symp. GaAs and related compounds, Eds. A. Christou and H.S. Rupprecht, Institute of Physics, Bristol, 1988, p. 605

9. B. Sermage, M.F. Pereira, F. Alexandre, J. Beerens, R. Azoulay, C. Tallot, A.M. Jean Louis and D. Meichenin, Interface recombination in GaAs-GaAlAs quantum wells, Journal de Physique, 48 : C5-135 (1987)

10. G. Duggan, H.J. Ralph and R.J. Elliott, Interface recombination in p type GaAs-GaAlAs quantum well heterostructures, Solid State Comm., 56 : 17 (1985)

THE INTERFACE AS A DESIGN TOOL FOR MODELLING OF OPTICAL AND ELECTRONIC PROPERTIES OF QUANTUM WELL DEVICES

J. Christen and D. Bimberg

Institut fur Festkorperphysik der Technischen Universitat
Hardenbergstrße 36, D-1000 Berlin 12, Germany

Abstract

The atomic scale crystallographic and chemical properties of interfaces between semiconductors are of decisive importance for the performance of novel generations of electronic and photonic devices and are in addition of large fundamental interest. Optical methods like luminescence and absorption have recently emerged to yield quantitative information on these properties, if the corresponding lineshape are carefully analyzed. We emphasize here luminescence. The natural lineshape of luminescence from a quantum well shows Gaussian broadening if its interfaces are not ideally abrupt. A detailed lineshape theory is outlined, allowing for a quantitative determination of the interface roughness distribution function. We find this function to depend in a delicate way on growth rates, temperature, interruption time and chemical compositon of the growth surface. The results of an experimental study of the model quantum well system AlGaAs/GaAs/AlGaAs grown by molecular beam epitaxy with and without interruption of the growth at the interfaces is presented. Roughness reduction upon growth interruption is analyzed in detail. For specific growth conditions and interruptions of 2 min at both interfaces formation of up to 7 μm large interface islands differing by a one monolayer step (2.8 Å) are observed. Consequently such quantum wells have a columnar structure, which can be directly visualized using cathodoluminescence imaging. Strong reduction of island size indicating transition from planar growth to three-dimensional growth is observed by CLI upon an increase of growth temperature from $T_g = 600°$ C to $660°$ C.

1. Introduction

Our present knowledge of structural, chemical and electronic properties of interfaces between semiconductors is inversely proportional to their importance for the design and the understanding of a whole generation of novel microstructured electronic and photonic devices. These devices such as double barrier diodes consist wholly or partly of interfaces and regions close to them. Two of the main difficulties which can be presently indentified to cause this situation are:

(i) Many properties of interfaces and of structures and devices consisting of multiples of interfaces depend on seemingly minor details of the technology used to fabricate them, as well as on fundamental physical properties of the materials used. Frequently it is extremely difficult to analyze and separate in an unambiguous manner the various causes giving rise to a certain property.

(ii) Clear-cut experimental information on interfaces is scarce, since some of the few existing methods are only qualitative or semiquantitative of nature, like reflection high-energy electron diffraction (RHEED), and for others, like some electrical methods, the evaluation procedures are dubious.

In addition, interfaces are only accessible to the presently existing surface characterization methods if the structure is grown by molecular-beam epitaxy (MBE). During the brief time interval before the second material is grown on top of the first one the future interface can be inspected in the form of an open surface e.g., by RHEED.

Detailed mathematical analysis of QW luminescence lineshapes is demonstrated to provide quantitative information on the statistics of structural/chemical disorder at semiconductor interfaces with sub-Angstrom depth resolution. A model, recently proposed by some of us /1/, established the qualitative relationship between quantum well luminescence lineshapes and interface disorder. An extension of this model is presented here, allowing for a quantification of the information gained from lineshape analysis. Our results /2/ are found to be in general agreement with recent predictions /3-5/ based upon observations of the dynamics of RHEED oscillations demonstrating the reduction of surface disorder upon interruption of MBE growth for a few seconds. In contrast to RHEED observations, however, the luminescence results can now be interpreted quantitatively. Thus, structural parameters of interfaces are correlated with optical properties of quantum wells. In addition we present a cathodoluminescence imaging system enabling us to visualize for the first time directly monoatomic steps at semiconductor interfaces and the columnar structure of single quantum wells (SQW's) grown by growth interrupted MBE. Shrinkage of the lateral extension of these islands for increasing growth temperature at constant growth rate is observed.

This paper is organized as follows: Section 2 contains a brief description of the samples used for our experiments and the cathodoluminescence imaging (CLI) and the photoluminescence set-ups are described. The theoretical background necessary for an understanding of our derivation of interface roughness data from luminescence lineshapes is given in Sect. 3. Photoluminescence results of quantum wells grown at various growth rates, temperatures and interruption times are presented and analysed in Sect. 4. Conditions of carrier thermalization and nonthermalization are discovered. Direct images of growth islands at interfaces are presented in Sect. 5. Sect. 6 contains a summary.

2. Experiment

a) Samples

In order to eliminate effects of interlayer well size fluctuations produced e.g. by fluctuations of the substrate or effusion cell temperature during MBE growth, predominantly single quantum wells are investigated. All samples are grown on GaAs substrates oriented within $\Delta \alpha \leqslant 0.1°$ to (100). The As_4/Ga flux ratio is about 4. Results from three series of samples grown at largely varying temperatures and rates will be presented here. For growth details see Ref. 2.

b) Photoluminescence

The main features are: Excitation wavelength 514 nm or 632 nm, typical excitation density 90 mW/cm², temperature range 1.5 K - 300 K spectral resolution 0.06 - 0.3 nm, being largest at low temperature. Details of the method were described recently /1/.

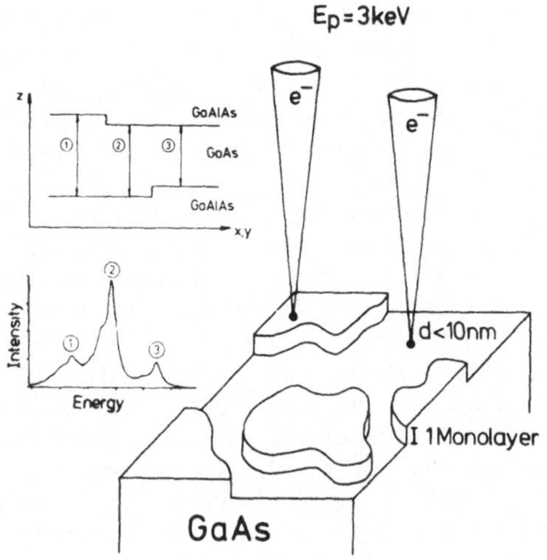

$E_p = 3keV$

Fig. 1 Principle of cathodoluminescence imaging: The lower half of an interface is shown. Triplet splitting of recombination results from single monolayer steps at each interface if the island area is larger than the projection of the recombination volume.

c) Cathodoluminescence Imaging - CLI

Features essential for the present system /2/ are the following: The miniature continous-flow cryostate is fitted to a standard specimen stage and allows a variation of the temperature in the range 5 K $\leqslant$ T $\leqslant$ 300 K. The kinetic energy of the electrons used for excitation in the CLI experiments is reduced to 3 keV, as compared to 30 keV in our standard CL work /6,7/. At 3 keV the center of excitation is 0.04 µm below the surface and the diameter of the generation volume of secondary electrons and final-state electron-hole pairs, the Bethe-range, is thus largely reduced to 0.2 µm. This value and not the much smaller spot size of the electron beam limits the lateral resolution if no deconvolution procedure is applied. The emitted light is dispersed by a 30 cm vacuum monochromator having a linear dispersion of 6 nm/mm. Spectral resolution in the temperature range 5 K -100 K is $\leqslant$ 0.6 nm. The electron beam is digitally scanned over an rectangular area of typical dimensions 32 µm x 18 µm. This area is equally divided into 128 x 100 pixels. (maximum value is 512 x 400 pixels). On each pixel the luminescence intensity is measured at a fixed wavelength with a dynamic range of 10^3. The emission wavelength from quantum well areas of different widths is different. Thus information on structural properties of a QW can be translated into information on optical properties. A CLI discloses directly the width distribution of the QW under investigation (its columnar structure) and thus the variation of the z-coordinate of the interface positions in the x-y-plane.

Figure 1 shows schematically the operating principle of CLI for an interface showing an island-like structure with single monolayer steps. Details of the point focus CL system have been described recently[7].

3. Correlation of Interface Roughness with QW-Luminescence Lineshapes: Theoretical Background

A complete model for the lineshape of the luminescence from a quantum well is established, extending considerably our previous approach /1/ of a simple Gaussian or Lorentzian shape. A detailed mathematical procedure is given, which demonstrates how to calculate interface roughness from the observed lineshape broadening.

The interface of a heterostructure consists of two adjacent lattice planes of different chemical composition. Growth of the second material usually commences on a plane which is not monoatomically flat but shows a structure of monolayer height steps due to incomplete coverage. The position of the interface in growth direction z thus depends on the x-y-coordinate. This effect is sometimes called interface roughness. Interdiffusion of atoms at the usually rather high growth temperatures additionally contributes to roughness.

A quantum well consists of a narrow, chemically uniform layer limited by two interfaces to the barrier material. The well is truely chemically uniform if it is made up out of an element or a binary compound. In all other cases there exists a mean composition which is defined as an average over an information volume /8/. The information volume might contain a statistical or a nonstatistical distribution (e.g. due to ordering effects /9/) of the compound. In what follows we will consider only wells of binary compounds like GaAs. The roughness of each of the two interfaces contributes to a x, y-dependence of L_z, the QW width. In general, the contributions from the top and the bottom interface are different, since the growth surface of the bottom interface consists of the barrier material, whereas the growth surface of the top interface consists of the well material. We have demonstrated recently /10/ that the surface kinetics of such rather similar materials as GaAs and AlGaAs are still far from identical.

The lineshape function I(E) of the luminescence of an electron-hole pair from a quantum well as a function of photon energy E is given by

$$I(E) \sim f(E) \cdot \int_0^\infty D(E,E') \cdot B(E',E_n) \, dE' \tag{1}$$

if momentum conservation is taken care of by a separate mechanism. D(E) is the two-dimensional density of states, B (E) is a broadening function, which, in the case of discrete quantization energies E_n, is a Lorentzian broadening due to final state interaction /11/. It is largest for small energies close to E_n and can be neglected if the statistical contribution to the lineshape broadening is much larger.

f (E) is the thermal distribution function. In general, Fermi functions with different Fermi energies for electrons and holes have to be inserted. Under the present circumstances of excitonic recombination after low to medium excitation a Maxwell-Boltzmann function presents an excellent approximation. The low energy onset of recombination occurs at:

$$E_n = E_g + E_e(L_z) + E_h(L_z) - E_x^{2D}(L_z) \tag{2}$$

with E_g the 3-dimensional gap, E_e the energy of the lowest two-dimensional electron subband, E_h the energy of the lowest two-dimensional hole subband and E_x the binding energy of the two-dimensional exciton, if the recombination is excitonic, as it is here /12/. Otherwise $E_x = 0$. E_e, E_h are the well known solutions of the two-dimensional "particle in the box" Schrödinger equation /13/ which are L_z-dependent. E_x also depends on L_z, but its absolute value is much smaller. If L_z is statistically distributed over a certain "information volume", then E_e, E_h and thus E_n are also statistically distributed. E_n in Eq.(1) and Eq.(2) has then to be replaced by its probability distribution $P(E_n, \langle E_n \rangle)$ and its expectation value $\langle E_n \rangle$ respectively. $B(E',E_n)$ is no longer given by a Lorentzian but its convolution with the probability distribution $P(E_n, \langle E_n \rangle)$. The information volume V_I is equal to the exciton volume if the recombination is excitonic. Otherwise it is equal to the recombination volume which can be calculated from the mean e-h distance.

In a luminescence experiment recombination of many electron-hole pairs from different, although closely neighbouring areas are detected. The experimentally observed lineshape I_{exp} thus presents a statistical average over the observation volume of all "single event" lineshapes Eq. (1).

$$I_{exp} = \int_V I \left\{ E\,(V) \right\} \, dV \qquad (3)$$

This additional broadening effect has to be taken into account for larger lateral inhomogenities. Then the variance $\overline{\sigma}$ of the continuous distribution of the mean values of the width $\overline{L_z}$ cannot be neglected as compared to the variance σ of the width L_z within one single recombination volume. Temperature dependent luminescence taken at varying positions of the samples shows that this additional broadening is observable, but small, and can be neglected for the present set of samples.

With increasing growth interruption time the lateral size of the islands at the interfaces becomes larger than the recombination volume. Then discrete mean width values L_z^ν exist, and Eq. (3) is replaced by

$$I_{exp.} = \sum_{\nu=1}^{n} I_\nu\,(E) \qquad (4)$$

with n = 3, if exactly one monolayer step occurs at each interface and the islands at both interfaces are larger than the projection of V_I onto them. Fig. 1 depicts in a schematic way this correlation of structural and optical properties.

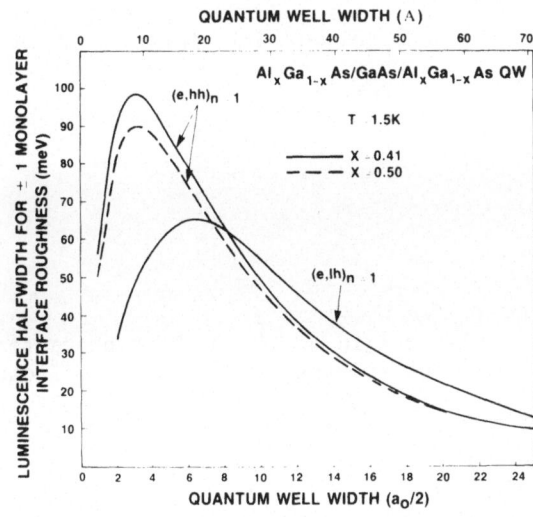

Fig. 2 L_z-dependence of energy broadening of the luminescence for a fixed FWHM of one lattice constant of the L_z-distribution function. Broadening values are given for the $(e, hh)_{n=1}$ and the $(e, lh)_{n=1}$ transitions for $[Al]$ = 41.2 %, and for the $(e, hh)_{n=1}$ transition for $[Al]$ = 50 %.

A practical example of the impact of quantum well width variation, and thus of interface roughness, on the luminescence broadening is given in Fig. 2. The L_z-dependence of the spectral broadening is given for a fixed halfwidth equal to a, the lattice constant, of the L_z-distribution function. Fig. 2 is calculated for a finite square well potential, ($[Al]$ = 41.2 % and 50 %, $\Delta E_c = 0.62\,E_g$), assuming continuity of the envelope function $\psi(z)$ and of $\psi'(z)/m^*$ at the interface and energy dependent masses[13]. The effective masses which we use for this calculation are

for GaAs:
$m_e^* = 0.0665$, $m_{lh}^* = 0.091$, $m_{hh}^* = 0.377$
for $Al_{0.41}Ga_{0.59}As$:
$m_e^* = 0.0895$, $m_{lh}^* = 0.137$, $m_{hh}^* = 0.424$
for $Al_{0.5}Ga_{0.5}As$:
$m_e^* = 0.0895$, $m_{lh}^* = 0.148$, $m_{hh}^* = 0.435$

Fig. 3 b presents a fit of Eq. (1) to an experimental luminescence line obtained from a 4.2 nm QW, grown according to Sec. 2 without interruption of the growth. The fit is perfect, although only one fitting parameter (the variance of the $P(E_n)$ distribution) is used. The temperature is directly taken from a semilogarithmic plot of the luminescence line.

The variance of the theoretical spectral broadening function of Fig. 4 b (σ_E = 1.3 meV) together with Fig. 3 yields a variance of the quantum well width distribution function σ = 0.33 Å, which is approximately 20 % less than the value we derived earlier /1/ from a fit with a simple Gaussian, shown in Fig. 3 a. This value has now to be distributed among the inequivalent interfaces. For equivalent interfaces each of them would show a roughness of σ = 0.23 Å.

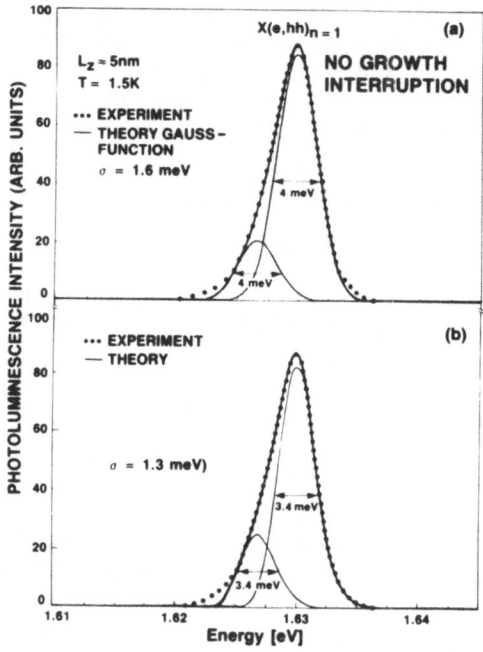

Fig. 3 T = 1.5 K luminescence spectrum of a 5 nm QW and theoretical fit according Eq. (1) neglecting Lorentzian broadening (b). A perfect agreement is achieved. Part (a) shows a fit with a simple Gaussian for comparison. The origine of the doublet structure is discussed in Ref.[14].

Details are given elsewhere /10/. The origin of the doublet character of the luminescence line in Fig. 4 is not related to structural effects and is discussed elsewhere /14/.

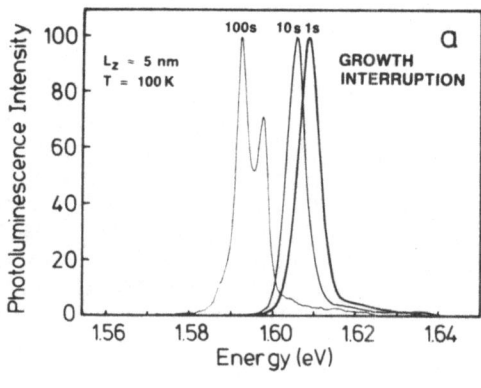

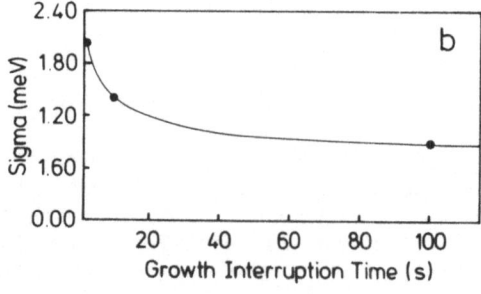

Fig. 4 (a) Comparison of T = 100 K photoluminescence spectra of 5 nm wide QW's grown with growth interruptions at both interfaces of 1 s, 10 s, 100 s, respectively. Part (b) shows the variance of the energy broadening function versus growth interruption time.

4. Photoluminescence Spectra of Growth Interrupted Quantum Wells: Evidence for Suppression of Carrier Thermalization

Fig. 4 a shows T = 100 K spectra of $\approx$ 5 nm wide QW's grown at 620° C at a rate of 2.8 Å/s with interruption of the growth Δt at both interfaces of 1 s, 10 s, 100 s, respectively (see Sect. 2). Obviously the statistical contribution to the width shrinks appreciably with increasing growth interruption time. A theoretical analysis according to Eq. (1) yields variances of the energy broadening function of 2 meV, 1.4 meV and 0.9 meV, respectively, yielding variances of the L_z distribution function of 0.62 Å, 0.44 Å and 0.28 Å, respectively, using Fig. 2. The results are shown in Fig. 4 b. The roughness of one interface would be 0.7 of these values for equivalent interfaces, which is usually not the case.

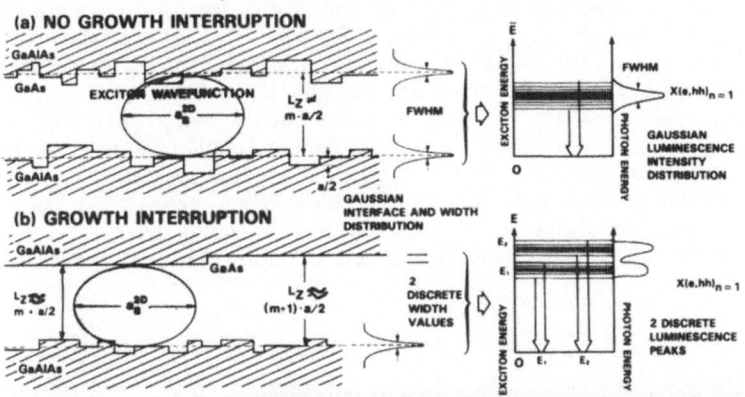

Fig. 5 Model for the interface disorder of a AlGaAs/GaAs QW and for the energy state(s) of the X(e, hh)-exciton in such a well grown (a) without and (b) with interruption of the growth. The GaAs surface smoothes more rapidly than the AlGaAs one.

The Δt = 100 s spectrum shows a doublet. A similar spectrum taken at lower temperature shows a single line (we ignore here for simplicity the "intrinsic" doublet splitting of the heavy hole exciton reported elsewhere /14/). At higher temperatures no third line on the high energy side emerges. Apparently the islands at the GaAs surface (top interface) are larger than the exciton diameter /1,10/ whereas the islands at the AlGaAs surface (bottom interface) are smaller. Figure 5 shows this situation schematically. The disappearance with decreasing temperature of the second line is indicative of thermalization in a two-level-system and is not at all surprising for given lifetimes of the order of 0.5 ns /12/.

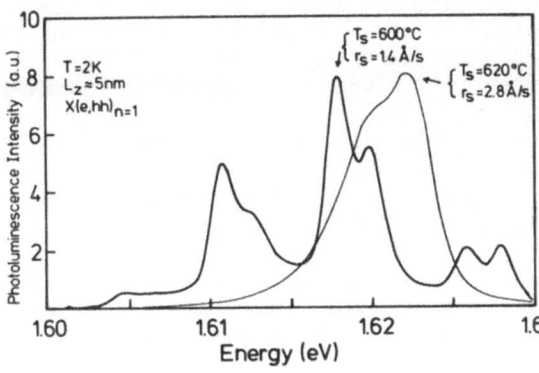

Fig. 6 Comparison of T = 2 K photoluminescence spectra of two samples grown with similar growth interruption times but at different growth rates and temperatures: a) T_S = 600° C, r_S = 1.4 Å/s, b) T_S = 620° C, r_S = 2.8 Å/s

Fig. 6, in contrast, presents a very surprising result. The 2 K photoluminescence spectra of two $L_z \approx 5 - 6$ nm QW's grown with almost the same interruption time of 100 s (# 792) and 120 s (# NAK1) at the interfaces are compared (see Sect. 2). We recognize one component (with doublet finestructure /14/) and a triplet (with doublet finestructure /14/). Apparently the excitons do not thermalize any more in the quantum wells of sample NAK1. Similar results are obtained for the other three SQW's of that sample (see Fig. 1 of Ref. 14). Apparently the lateral extension of islands at both interfaces of that sample is equal to or larger than the diffusion length L_D:

$$L_D = (\mu kT \tau /e)^{1/2} \hspace{4cm} (5)$$

Here T is the temperature, τ is the lifetime and μ is the mobility. For T = 2 K, τ = 0.5 ns /12/ and $\mu \approx 1 - 2$ x 10^5 cm²/Vs, as limiting value of the ambipolar mobility, which is taken to be equal to some of the best known hole mobilities /15/, we get $L_D \lesssim 0.9 -1.3$ µm. The extension of these very smooth islands must be thus larger than ≈ 1 µm. Spot CL spectra showing single peaks at different photon energies for different spots corroborate this conclusion. Consequently, the direct observation of such interface islands by CLI should be possible. The next section proves that we indeed are able to obtain such images. In that section we will also discuss the influence of the various growth parameters on island size.

5. Images of Growth Islands at Heterointerfaces

Fig. 7 shows CLI images of sample NAK 1 grown at 600° C at r_s = 0.14 nm/s. The spectrometer is set at a fixed wavelength given by the wavelength of maximum intensity of one of the PL triplet components shown in Fig. 6. Then the variation of CL-luminescence intensity as a function of position of excitation is recorded. This procedure is repeated alltogether three times. The black/white scale of the three pictures is adjusted such that the sum of all white areas covers or sligthly overcovers the total surface. Then the bright areas in Fig. 7 show the extension of quantum well columns of thickness close to 18, 19 and 20 monolayers, respectively. At the edge of each bright area a step of 2.8 Å height occurs at one of the interfaces. From a larger number of such experiments we find that the mean size of the islands is 6 - 8 µm. Diffusion effects influence the size of the islands reported here only in a minor way, as shown by time resolved experiments. Details of these experiments are given elsewhere.
Fig. 8 shows CLI images of sample 512 grown at 660° C at r_s = 0.17 nm/s. Obviously the islands are much smaller. Their mean size is determined to ≈ 2 µm. An increase of the growth rate and an increase of the temperature lead to a reduction of the interface island size and to an increase of interface roughness indicating the transition from planar to three-dimensional growth. The influence of growth rate is particularly important at low substrate temperatures of the order of 600° C - 620° C as illustrated by Fig. 7. A detailed discussion of the separate influence of both growth parameters in terms of surface kinetics will be given elsewhere.
Time resolved cathodoluminescence images directly visualize the lateral diffusion of excitons. Their lateral diffusion velocity was determined in a separate study /16/ for the samples of Fig. 7 as 10^5 cm/s.

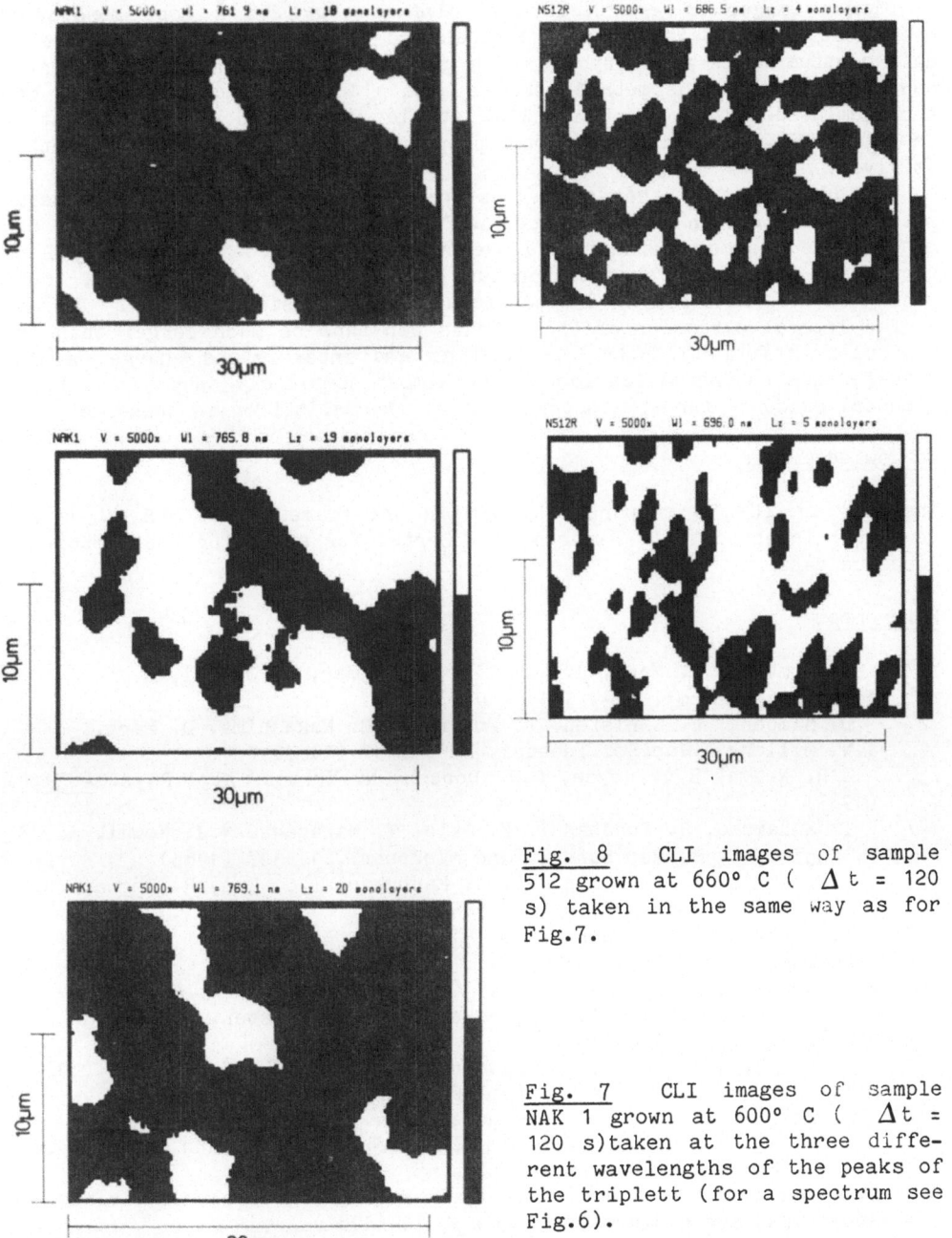

Fig. 8 CLI images of sample 512 grown at 660° C (Δt = 120 s) taken in the same way as for Fig.7.

Fig. 7 CLI images of sample NAK 1 grown at 600° C (Δt = 120 s)taken at the three different wavelengths of the peaks of the triplett (for a spectrum see Fig.6).

6. Conclusions

Quantitative information on the reduction of roughness of the quantum well interfaces with increasing interruption time of the growth process is obtained from detailed analysis of luminescence lineshapes. A full theory of the shape of luminescence from quantum wells and its dependence on structural properties of the interfaces is given. For the specific example of GaAs/AlGaAs heterostructures grown at 620° C at a rate of 2.8 Å/s a reduction of interface roughness from 0.31 Å to 0.14 Å is observed for an increase of interruption time from 1 s to 100 s. Earlier reports by us on the inequivalence of the top/bottom interfaces are corroborated.

Direct images of growth islands differing by 2.8 Å (one monolayer) height at GaAs/AlGaAs heterointerfaces and of the columnar structure of GaAs quantum wells grown by MBE with growth interruption at the interfaces are reported. The method used to obtain these images is scanning cathodoluminescence of quantum wells. Atomic scale structural information is expressed in terms of spectral properties. An outline of this method is given.

The dependence of the lateral extension of such islands on the parameters of MBE growth is investigated. For fixed GaAs growth rate $r_s = 0.5$ monolayer/s the mean island size decreases from 6 μm - 8 μm to $\approx$ 2 μm upon an increase of growth temperature from $T_g = 600°$ C to 660° C, indicating the transition from 2-dimensional to 3-dimensional growth.

The lateral extension of the islands can thus be much larger than the ambipolar diffusion length $L_D \lesssim$ 1 μm and inter-island charge carrier transfer due to thermalization can be almost completely suppressed. Experimental evidence for this suppression of thermalization is presented.

Acknowledgements

The work at TUB is supported by DFG in the framework of SFB 6. We are very much indebted to R. Bauer and D. Oertel for supplying the photoluminescence data.

References

/1/ D. Bimberg, D. Mars, J.N. Miller, R. Bauer, D. Oertel, J.Vac.Sci.Technol. B 4, 1014 (1986)

/2/ D. Bimberg, J. Christen, T. Fukunaga, H. Nakashima, D. Mars, J.N. Miller, J.Vac.Sci.Technol. B 5, 1191 (1987)

/3/ J.H. Neave, B.A. Joyce, P.J. Dobson, N. Norton, Appl.Phys. A 31, 1 (1983)

/4/ T. Sakamato, H. Funabashi, K. Ohta, T. Nakagawa, N.J. Kowai, T. Kojima, Y. Bando, Superl. and Microstr., 1, 347 (1985)

/5/ B.F. Lewis, F.J. Grunthamer, A. Madhukar, T.C. Lee, R. Fernandez, J.Vac.Sci.Technol. B 3, 1317 (1985)

/6/ A. Steckenborn, H. Münzel, D. Bimberg, J.Luminescence 24/25, 351 (1981) and Inst.Phys.Conf., Ser.60, 185 (1981)

/7/ D. Bimberg, H. Münzel, A. Steckenborn, J. Christen, Phys.Rev. B 31, 7788 (1985)

/8/ J. Singh, K.K. Bajaj, S. Chaudhuri, Appl.Phys.Lett. 44, 805 (1984) and J. Singh, K.K. Bajaj, J.Appl.Phys. 57, 5433 (1985)

/9/ R. Hull, K.W. Carey, J.E. Fouquet, G.A. Reid, S.J. Rosner, D. Bimberg, D. Oertel, Proc.Int.Symposium on GaAs and Related Compounds, Las Vegas 1986, in print

/10/ D. Bimberg, D. Mars, J.N. Miller, R. Bauer, D. Oertel, J. Christen, Superl.and Synth.Microstructures 3, 79 (1987) and Proc. MSS III, J.de Physique, in print

/11/ P.T. Landsberg, Proc.Phys.Soc., A 62, 806 (1949) and Phys.Stat.Sol. 15, 623 (1966)

/12/ J. Christen, D. Bimberg, A. Steckenborn, G. Weimann, Appl.Phys.Lett. 44, 84 (1984) and D. Bimberg, J. Chrsiten, A. Werner, M. Kunst, G. Weimann, W. Schlapp, Appl.Phys.Lett. 49, 76 (1986)

/13/ G. Bastard, Phys.Rev. B 24, 5693 (1981)

/14/ R. Bauer, D. Bimberg, J. Christen, D. Oertel, D. Mars, J.N. Miller, T. Fukinaga, H. Nakashima, Proc. 18th Conf. Phys. Semic., Stockholm 1986, in print

/15/ G. Weimann, private communication

/16/ D. Bimberg, J. Christen, T. Fukunaga, H. Nakashima, D.E. Mars and J.N. Miller, Superlatt. and Microstructures 4 (1988), in print

CHARACTERIZATION AND DESIGN OF SEMICONDUCTOR LASERS USING STRAIN

A. R. Adams, K. C. Heasman and E. P. O'Reilly

Physics Department, University of Surrey, Guildford, Surrey
GU2 5XH, UK

INTRODUCTION

Hydrostatic pressure increases the direct band gap of III-V semiconductors and decreases the sub-band gaps to the conduction band satellite minima. By contrast uniaxial stress removes the cubic symmetry of the lattice and can significantly alter the properties of the valence band which is relatively insensitive to hydrostatic pressure. In this paper we consider how these effects can be applied to a variety of semiconductor lasers. Hydrostatic pressure tunes the operating wavelength of Fabry-Perot lasers. On the other hand, in distributed feedback (DFB) lasers, the wavelength is held almost constant against changes in pressure by the period of grating. Here we show how measurements of the threshold current with pressure in these different structures give information about the internal loss mechanisms occuring in long wavelength devices (1.3-1.65μm) which are responsible for the T_0 problem. Having characterised these mechanisms we suggest laser structures with inbuilt strain in quantum well structures which should have greatly improved performance. Pressure measurements on GaAs lasers indicate that there are no significant optoelectronic loss mechanisms in standard heterostructure devices or quantum well lasers with wide wells. This is not the case for narrow well devices where considerable losses can occur particularly at high pressure. A model for this effect is tentatively suggested.

Most of the work described in this paper was performed at Surrey and the paper reviews past work to put in context the new results which are also presented here.

LONG WAVELENGTH LASERS AND THE T_0 PROBLEM

Although a large variety of simple and complex GaInAsP/InP laser structures have been made in many laboratories throughout the world using a whole range of growth techniques, they all suffer from a threshold current that is very temperature sensitive. The threshold current is often expressed in the form

$$I_{thr}(T) = I_{thr}(T') \exp \left\{ (T-T')/T_0 \right\} \qquad (1)$$

where T_0 is the characteristic temperature which is used to express the temperature sensitivity of the threshold current. Although extrapolation of low temperature data suggests that T_0 should be approximately 150K at room temperature for a loss free GaInAsP laser[1], measured T_0 values lie normally in the range $T_0 \approx 55 \pm 20^{\circ}$C. It is important not only for the thermal instability it creates which can be countered by suitable if complicated heat sinking but also because it arises from an intrinsic inefficiency. Several models[2-8] have been proposed to explain the effect which appears to be independent of the growth technique, device doping or structure, suggesting that the main loss mechanisms within the laser are intrinsic. Loss due to deep levels and surface recombination would therefore appear to be unlikely in determining the low T_0.

Traditionally the T_0 problem has been studied as a function of temperature. In the first half of this review we will concentrate on how hydrostatic pressure has been used as an additional technique to study the loss processes occuring in semiconductor lasers, concentrating in particular on the relative strengths of the various mechanisms responsible for the low T_0 in the quaternary lasers.

Continual improvements in high pressure apparatus design[9,10] and an increasing awareness of its usefulness for material and device characterisation has created a lot of interest in this area. Typically, hydrostatic pressure increases the direct band gap, E_g, of III-V semiconductors and their alloys at about 10meV/kbar. Hydrostatic pressure can therefore provide a continuously variable and reversible method of approximating to a composition change in a semiconductor alloy. Thus the operating wavelength of semiconductor lasers can be shortened whilst keeping other important device parameters constant, a feature which cannot be achieved in a series of lasers grown from wafers of slightly different active layer composition. In GaInAsP/InP lasers all the loss mechanisms proposed to explain the high temperature sensitivity of the threshold current are very sensitive to changes in the direct band gap, and therefore pressure is a very convenient tool for analysing these loss mechanisms.

A piston and cylinder apparatus was used at Surrey with the devices mounted on the top piston butt coupled to a large-core optical fibre fed through a high pressure seal[11]. The lasers are pulsed usually with duty cycles <0.02%. The pressure within the high pressure system was monitored by a manganin coil attached to the bottom piston. Results from this system are reversible and repeatable showing that, within the pressure range used, no significant pressure induced damage occurs. Castor oil has a refractive index at 300K of 1.48. Immersion into the caster oil reduces the end facet reflectivity and thereby raises the threshold current. To reduce this effect, slightly longer length lasers of length 300μm to 400μm were generally measured. For GaInAsP lasers of this length the threshold current increased by around 10 to 15%. With quantum well lasers with low optical confinement factors the increase can be significantly higher.

We will now describe the three loss mechanisms that are most often referred to in the literature as the cause of the T_0 problem and consider the evidence about their relative importance.

Intervalence Band Absorption

Adams et al[3] proposed that intervalence band absorption (IVBA) could account for the increase in the threshold current and the decrease in quantum differential efficiency with increasing temperature in the long wavelength GaInAsP lasers. Intervalence band absorption involves the reabsorption of the laser radiation by electronic transitions from the split-off band into holes which have been injected into the heavy hole band under forward bias.

The absorption α_{ac} is directly dependent on the hole density available at energy E_1, with B_1 a constant of proportionality

$$\alpha_{ac} = \frac{B_1}{1+\exp\dfrac{(E_1-E_F)}{k_B T}} \qquad (2)$$

Henry et al[12] reported that direct measurement of IVBA in the quaternary alloy had showed that IVBA neither had the magnitude nor the temperature dependence to explain the low T_0 values. However, more recent measurements would suggest that both the magnitude and temperature dependence is a lot larger than Henry concluded. Mozer et al[13] used two techniques to observe IVBA in quaternary material. Firstly they observed $1\mu m$ radiation from $1.3\mu m$ lasers which is due to electron transitions from the conduction band to the spin split-off band caused by the pumping of holes into the split-off band by either IVBA or CHHS Auger processes. Above threshold the $1\mu m$ emission was seen to increase by 60% between I_{thr} and $2I_{thr}$. As the Auger process saturates above threshold, because the carrier density is pinned by the stimulated emission, they determined that it was IVBA that was populating the spin split off band above threshold and α_{ac} was estimated to be $100cm^{-1}$ for a hole concentration as low as $8\times10^{17}cm^{-3}$. The second experiment reported by Mozer et al[13] involved direct measurements at $1.3\mu m$ of the absorption caused by free carriers induced in a single epitaxial layer by radiation from a Nd-YAG laser. The laser intensity was modulated giving a modulated absorption signal which was therefore free from reflectivity corrections. They observed very strong absorption ($400cm^{-1}$) which increased swiftly with temperature.

Recently[14] a careful analysis of absorption as a function of wavelength, carrier density and temperature in $1.3\mu m$ quaternary material showed an absorption of $130-170cm^{-1}$ at room temperature at threshold, of which $50-80cm^{-1}$ was attributed to IVBA; the threshold carrier concentration unfortunately was not stated. Hauser et al[(15)] obtained evidence that suggested that IVBA is more important in the longer wavelength devices. Their temperature measurements of the threshold carrier density and differential quantum efficiency revealed a strong influence of intervalence band absorption for the case of $1.65\mu m$ lasers, suggesting also that its magnitude was $120cm^{-1}$ at room temperature for a carrier density of approximately $1.2\times10^{18}cm^{-3}$.

Childs et al[16] using a pseudopotential band structure and matrix elements evaluated along 21 k-space directions through the Brillouin zone, calculated that at $1.6\mu m$ the IVBA absorption coefficient is about $39cm^{-1}$ for 1×10^{18} holes and has a temperature coefficient at room temperature of $0.45\%K^{-1}$ which they suggested was insufficient to explain the observed temperature sensitivity of lasers. Our recent assessment of IVBA[17] absorption showed however that, when the rate of change of threshold carrier density with temperature is taken into account, the values calculated by Childs et al can account almost entirely for the low T_0's observed experimentally.

Auger recombination

Since the pioneering work of Beattie and Landsberg[18], it is generally believed that Auger recombination can be an important nonradiative loss mechanism in narrow gap semiconductors. Although there are many different forms of Auger recombination, the discussion here will be limited to the two types of band to band Auger recombination which have been cited as the most likely causes of the anomalous temperature dependence of light sources made from the quaternary alloy. The phonon assisted Auger recombination process which was recently theoretically predicted by Haug[19] to be nearly as important as the direct band to band processes has not been considered here because of its weaker temperature dependence. Theoretical and experimental values for the Auger coefficient in the quaternary alloy cover a wide range from less than $1\times10^{-29}cm^6/sec$ to greater than $1\times10^{-28}cm^6/sec$. The larger values are found either in work performed early on or else on the smaller band gap materials. The Auger coefficient for the larger band gap material GaAs is approximately $10^{-31} \mathrm{cm}^6/\mathrm{sec}$, consequently the effect of Auger recombination on the threshold current of GaAs lasers is considered negligible.

Initial theory using a parabolic band structure, suggested that the CHCC Auger process would be the dominant Auger process above room temperature, and was proposed by Dutta and Nelson[4] to explain the temperature dependence of the threshold current in 1.3 and 1.5μm quaternary lasers. In the CHCC process, a Conduction electron and valence Hole recombine across the band gap, exciting a Conduction electron to an empty Conduction state. Since these initial predictions, calculations assuming a more realistic non-parabolic band structure[20] indicate that the activation energy of the CHHS process is the most important at room temperature. For CHHS, recombination across the band gap excites a Hole into the Spin-split-off band.

According to Burt et al[21], the effective mass sum rule previously used to estimate Auger recombination overestimates the square of the modulus of the overlap integral by at least an order of magnitude. Thus Dutta and Nelson's[4] and Sugimura's[22] initial estimates of Auger recombination are at least one order of magnitude too large[21]. Very recently, Bardyszewski and Yevick[23] using refined expressions for the overlap integrals obtained Auger coefficients in agreement with recent experimental values, and at least a factor of five smaller than those quoted by Dutta and Nelson. All the previously mentioned calculations of the Auger theory used the k.p perturbation theory to calculate the band structure. From a simultaneous study of the electron effective mass and the band gap, Shantharama et al[24] observed that the rate of increase of the conduction band effective mass at the band minimum with pressure was at least 50% larger in the quaternary than 3 band or multiband formulations of the k.p theory would allow. Sarkar et al[25] have measured the conduction band effective mass as a function of temperature and magnetic field and concluded that the non-parabolicity of the conduction band is approximately double that expected on the basis of 3 band k.p theory and that it could not be explained by consideration of higher bands. Consequently it was tentatively concluded that alloy disorder profoundly affects the k.p interaction in semiconductors. Clearly then it is not easy to obtain a totally reliable theoretical evaluation of either Auger or IVBA until there is a better understanding of the effects of alloy disorder and non-parabolicity.

Carrier leakage to the confinement layers

Yano[5] proposed that carrier leakage to the confinement layers over the heterojunction barrier was significant in reducing the T_0 of 1.3μm quaternary lasers above room temperature. Although numerous papers have been published that show that both electron and hole leakage exist, and exist in different degrees dependent on such factors as the doping levels in the active and p-InP confinement layers, no experimental evidence is available that confirms Yano's specific T_0 predictions ($T_0 \approx 70 \pm 10 \mathrm{K}$ at T=298K). Chen[26] has reported that lasers with T_0's as high as 90K are possible in a laser free from leakage effects. Yano's calculations assumed a parabolic conduction band and various quaternary parameters that overestimated the effect of thermal leakage. Carrier leakage may be enhanced due to the presence of hot carriers in the active region as a consequence of either Auger or IVBA processes. We have undertaken Monte Carlo simulations of the carrier leakage in 1.3μm and 1.5μm lasers at room temperature which show carrier leakage to be very small, and more importantly, primarily resulting from Auger or IVBA produced hot carriers, rather than carriers thermally excited over the barrier[27]. Experimental evidence by Chen et al[28], Zhuang et al[29] and Yamakoshi et al[30] would support this conclusion.

There is other evidence that would suggest that carrier loss is of secondary importance as far as the T_0 is concerned. Lasers with very thick active regions also have low T_0's. Nahory et al[31] showed that varying the confinement layer material of similar wavelength lasers had no noticeable effect on the low temperature dependence of the characteristics.

INVESTIGATIONS USING HYDROSTATIC PRESSURE

a) Fabry Perot Devices

The threshold current in $GaAs/Ga_{0.7}Al_{0.3}As$ oxide stripe lasers has been measured recently at room temperature (figure 1) and shows an increase with increasing pressure, in agreement with our earlier results[32]. In a loss free laser, the threshold current is expected from radiative recombination theory to have the following pressure variation[32].

$$\frac{1}{I_{thr}} \frac{dI_{thr}}{dP} = \frac{2}{E_g} \frac{dE_g}{dP} \qquad (3)$$

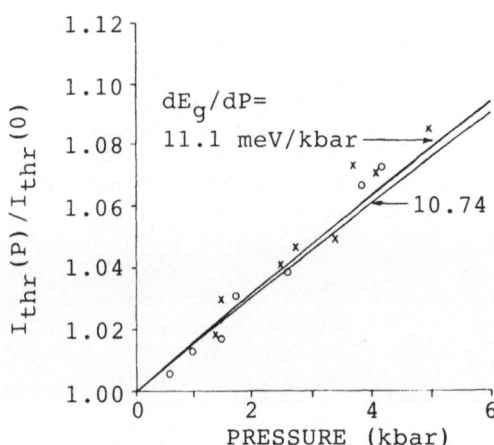

Fig. 1. The pressure dependence of the threshold current in GaAs/AlGaAs oxide stripe lasers at room temperature.

Figure 1 shows that equation 5 is in reasonable agreement with the experimental data assuming the GaAs pressure coefficients of 11.1meV/kbar quoted by Shantharama[33] or 10.74meV/kbar quoted by Wolford[34]. The agreement with experiment suggests that pressure dependent loss mechanisms are unimportant in these GaAs/GaAlAs lasers.

In contrast to the GaAs lasers, the threshold current in $1.3\mu m$ quaternary lasers decreased markedly with increasing pressure as shown in figure 2. The threshold current decrease with pressure was attributed to a loss mechanism in the laser which was becoming weaker as the pressure was increased. Measurement made at room temperature and at higher temperatures, (shown in figure 3) indicate that the threshold current decreased faster with pressure at the higher temperature, showing that the T_0 of the $1.3\mu m$ quaternary lasers increases with pressure. The effect of pressure on the $1.3\mu m$ GaInAsP lasers measured is to eliminate the loss mechanism or mechanisms responsible for the low T_0 in these lasers.

We also showed that for $1.3\mu m$ GaInAsP oxide stripe lasers with n+ doped active regions the threshold current again reduced with hydrostatic pressure, although the effect was less marked for n+ active layers than for nominally

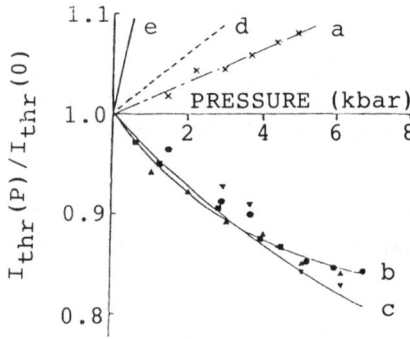

Fig. 2. A comparison between the room
temperature pressure dependence
of the threshold current in $1.3\mu m$
lasers and a GaAs laser. Lines
are calculated variation assuming
a) radiative recombination
b) IVBA c) Auger d) impurity
scattering e) barrier leakage

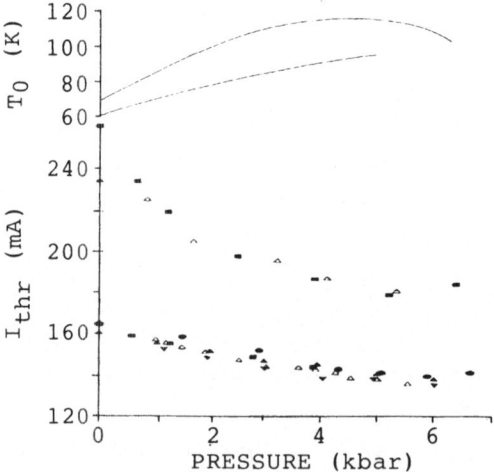

Fig. 3. Comparison between the threshold
current pressure dependence at room
temperature and at 315K and 322K
shows T_0 increases with pressure in
$1.3\mu m$ undoped oxide stripe lasers

284

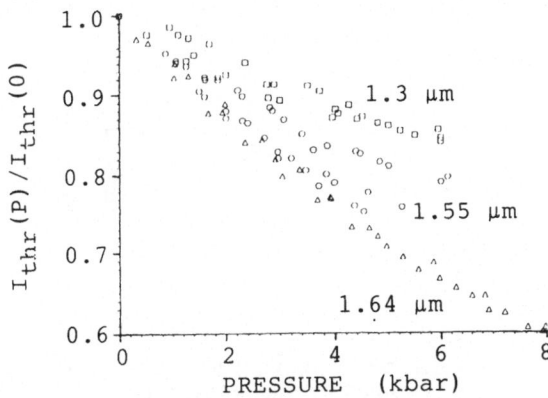

Fig. 4. Variation of the normalised
threshold current versus pressure
for two (□) 1.3μm, three (o) 1.55
μm and three (▲) 1.64μm undoped
oxide stripe and IRW lasers at
room temperature.

undoped ones[36]. Again the temperature sensitivity was also reduced with
measurements on GaAs and 1.3μm GaInAsP were repeated on other oxide stripe
and inverted rib waveguide laser structures and the results are shown in figure 4.
The agreement with the earlier data is very good. The data is also compared with
recent measurements on 1.64μm GaInAs/InP oxide stripe lasers and IRW and oxide
stripe 1.55μm GaInAsP/InP lasers. All the lasers shown in figure 4 have nominally
undoped active regions.

The decrease in threshold current is inconsistent with Horikoshi and
Furukawa's model[8] which attributed the temperature sensitivity of the threshold
current to nonradiative recombination centres 0.3eV above the conduction band
edge in 1.3μm material. Such levels can exist but would usually exhibit 'X' or 'L'
like pressure coefficients, leading to an extremely steep increase in threshold
current with pressure for the 1.3μm lasers.

Photoluminescence measurements of the direct band gap of GaInAsP material
as a function of pressure (figure 5), indicate that its pressure coefficient is larger
than that of InP If carrier leakage to the confinement layers is the dominant
loss mechanism in the lasers, as suggested by Yano[5], again the threshold current
would be expected to increase with pressure, as the barrier decreases but clearly
this does not seem to be the case.

The theoretical model of the effect of IVBA on the threshold current has
been given in some detail for 1.6μm GaInAsP lasers[2]. The threshold current is
expected to decrease with pressure. This mainly arises because increasing the
photon energy with increasing pressure also increases the k value at which vertical
transitions between the split-off and heavy hole bands can occur. As the vertical
transitions move away from the zone centre there are fewer holes into which
electrons from the split-off band can be promoted. Consequently the absorption
coefficient decreases with increasing pressure.

The Auger rate A can be expressed as[4]

$$A \propto \exp(-\Delta E/k_B T) \qquad (4)$$

where for the CHCC process

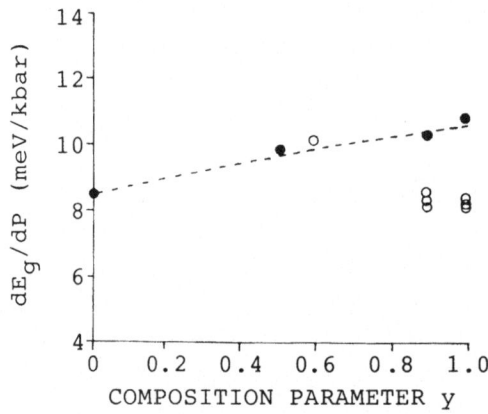

Fig. 5. The pressure coefficient of the direct band gap of (GaIn)(AsP) measured by photoluminescence compared with that derived by the shift in lasing wavelength with pressure.

$$\Delta E_{(CHCC)} = E_T - E_g \qquad (5)$$

where E_T is a threshold energy. It arises from the simultaneous conservation of energy and momentum for the four particle states involved. E_T for the CHCC process is given by

$$E_T = \frac{2m_C + m_{HH}}{2m_C + m_{HH} - m_C(E_T)} E_g \qquad (6)$$

where m_C, m_{HH} are the conduction and heavy hole zone centre effective masses and $m_C(E_T)$ the effective mass of the excited carrier. Combining Eq.(5) and (6) we get

$$\Delta E_{(CHCC)} = \frac{m_C E_g}{2m_C + m_{HH} - m_C(E_T)} \qquad (7)$$

For the CHHS process

$$\Delta E_{(CHHS)} = E_T - (E_g - \Delta) \qquad (8)$$

and

$$E_T = \frac{2m_{HH} + m_C}{2m_{HH} + m_C - m_S(E_T)} (E_g - \Delta) \qquad (9)$$

so that

$$\Delta E_{(CHHS)} = \frac{m_S(E_T)}{2m_{HH} + m_C - m_S(E_T)} (E_g - \Delta) \qquad (10)$$

These expressions show that the Auger rates are very dependent on the band gap and band structure of the material. Sugimura[20] showed that the Auger rate for the CHCC process especially is very dependent on the non-parabolicity of the conduction band. The equations show that the Auger recombination rates decrease as the band gap increases with increasing pressure.

Our initial calculations for 1.3µm quaternary lasers showed that the room temperature threshold current data could be explained by the reduction of either Auger recombination or IVBA[32], although when the pressure variation of the threshold current was also considered at temperatures above room temperature IVBA seemed to describe the data better[35]. In the earlier analysis of the threshold current variation with pressure, each loss mechanism was considered separately. In practice, as we will show, probably both IVBA and Auger recombination are influencing the threshold current and influencing each other.

The threshold current decreases relatively linearly with pressure to a value of 0.6±0.02 at 8kbar for the three oxide stripe 1.64µm GaInAs lasers. This decrease is considerably larger than that observed for 1.55µm, which in turn is larger than that observed for 1.3µm quaternary devices. Both IVBA and Auger recombination would predict this faster decrease in the longer wavelength lasers as both mechanisms are expected to be stronger because both the band gap decreases and the spin-orbit splitting increases. On the other hand, the heterojunction barrier is larger in the longer wavelength lasers. Therefore, the fact that the threshold current decreased with pressure and decreased faster in the longer wavelength lasers is further evidence that carrier leakage due to thermal excitation over the heterojunction barrier has negligible effect on the threshold current at room temperature in these lasers.

The dotted lines in figure 4 are the pressure variations expected assuming the radiative and Auger coefficients given by Haug[19]. As can be seen, the decrease with pressure is significantly less than the experimentally measured variations. Much better agreement should be obtained by including the influence of IVBA, particularly for the longer wavelength materials.

A more direct method for distinguishing between the influence of IVBA and the other loss mechanisms is to simultaneously measure the pressure dependence of the threshold current and carrier lifetime at threshold. The threshold current I_{th} is related to the lifetime, τ, by the expression,

$$I_{thr} = n_{thr} e V_a / \tau \qquad (11)$$

where V_a is the volume of the active region, e the electronic charge and n_{th} the injected carrier density at threshold. If Auger recombination, loss over the barrier or recombination through defects are, either singly or together, the dominant loss mechanisms influencing the threshold current, the pressure dependence of the lifetime should be inversely proportional to the change in threshold current. IVBA on the other hand, is an optical absorption process which alters the threshold current by changing the carrier density, n_{thr} required to reach threshold.

These measurements have been made on oxide stripe and inverted rib waveguide, 1.55µm lasers, with nominally undoped active regions. The carrier lifetime was measured by a double pulse time delay technique[37,38,39]. The measured lifetimes shown in figure 6 have been normalised to the 6kbar point because τ could be more accurately determined at high pressure. The dashed line in figure 6 shows the expected variation of τ if the measured decrease in I_{thr} is due to Auger or any other nonradiative recombination mechanisms causing τ to increase with pressure. The solid line is the variation of carrier lifetime expected if the only loss mechanism is IVBA. In this case τ increases because the radiative recombination lifetime increases with pressure. For a nominally undoped active region, the lifetime is inversely proportional to the injected carrier concentration, so that

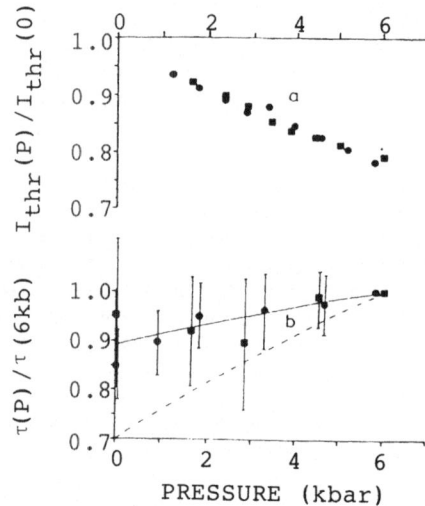

Fig. 6. Variation of (a) normalised
threshold current and (b)
recombination lifetime versus
pressure for 1.55μm GaInAsP
lasers: —— IVBA; - - - Auger,
recombination at defects, etc

$$\frac{1}{\tau} \frac{d\tau}{dP} = \frac{1}{2I_{thr}} \frac{dI_{thr}}{dP} \qquad (12)$$

Although the experimental uncertainty in the measurements is large, the results clearly indicate that the decrease in threshold current with pressure, for these undoped lasers, is closer to that predicted assuming IVBA as the most important of the laser loss mechanisms.

If IVBA is the dominant loss mechanism in 1.55μm lasers as indicated by the lifetime measurements, the threshold carrier density is expected to decrease with pressure due to the reduction in gain necessary to overcome the losses. Figure 5 compared the pressure coefficient of the direct band gap as measured by Lambkin[40] and Prins[41] with measurements of the pressure coefficient of the lasing wavelength. From the relationship

$$\frac{d\lambda}{dP} = \frac{-\lambda^2}{hc} \frac{dE_g}{dP} \qquad (13)$$

the lasing wavelength variation can be converted to a band gap variation. The values of 8.4±0.3meV/kbar and 8.3±0.3meV/kbar determined for 1.55μm GaInAsP and 1.64μm GaInAs lasers respectively are significantly less than the direct band gap pressure coefficient, indicating that a reduction in the threshold carrier density with pressure may have occured. In order to quantify this, calculations of the peak gain photon energy have been undertaken following the approach of Osinski and Adams[42] and the maximum gain was also calculated at various pressures using the gain theory of Asada and Suematsu[43,44]. The dominant effect appears to be due to the Fermi levels decreasing in the bands with pressure, in response to the decrease in gain required to reach lasing conditions as IVBA is decreased, but the large increase in conduction band effective mass with pressure[33] may also contribute to the effect.

The pressure coefficient of the lasing wavelength in 1.3μm quaternary lasers is 10±0.5meV/kbar[45], which is in good agreement with the pressure coefficient of the direct band gap measured by photoconductivity measurements on LPE grown epitaxial layers and on a 1.3μm surface emitting LED (10±0.5meV/kbar[27]). The equivalence of the results indicates that the threshold carrier density did not decrease significantly with pressure and that IVBA may not be as important in the shorter wavelength quaternary devices.

Another consequence of eliminating IVBA with increasing pressure is that the quantum differential efficiency should increase[2]. Because not all of the light emitted from the laser is coupled into the fibre, and because the proportion of light coupled may not be constant with pressure, quantum differential efficiency measurements at present are not particularly reliable. However in the quaternary GaInAsP lasers the quantum differential efficiency does increase with pressure and increases substantially more in the longer wavelength devices[46], in contrast with GaAs lasers where the quantum differential efficiency does not increase with pressure. Mozer et al[14] recently showed that in a 1.3μm GaInAsP laser the quantum differential efficiency increased **substantially** with increasing pressure whilst the threshold current decreased.

The CHHS and IVBA processes are similar in that both involve transitions from the heavy hole band to the split-off band. Recently, Mozer et al[47,14] and Nanomura et al[48] have provided significant evidence to conclude that the valence band plays a vital role in quaternary lasers. Significant population of holes in the split-off band, has been observed in lasers operating both below and above threshold. Are these holes created by IVBA or CHHS Auger processes? Through observation of the 1μm emission from the recombination of holes in the split-off band and electrons in the conduction band in 1.3μm lasers, Mozer et al[4] concluded that below threshold the CHHS Auger mechanism was mainly responsible for the population of holes, while far above threshold when the Auger process had saturated due to the pinning of the carrier density by the stimulated emission, significant population of the split-off band was observed as the injected current increased. This was interpreted as direct evidence of the existence of IVBA. From estimates of the integrated intensity of the $E_0 + \Delta_0$ luminescence below and above threshold, Mozer et al concluded that the CHHS process was the dominant process in the lasers that they measured. More recently[14], based on additional information derived from differential quantum efficiency measurements as a function of pressure they concluded that 50% of the injected carriers are lost by non-radiative Auger recombination compared to 23% due to IVBA.

Nanomura et al[48] also observed evidence of hole excitation into the split-off band by observing long wavelength 4.3-5.0μm emission associated with recombination of holes in the split off band with electrons in the light hole valence band from both 1.3 and 1.5μm GaInAsP lasers. They also observed acoustic signals originating from the LO phonons emitted by the excited holes as they relax to the band minima. Simultaneous observation of the acoustic signal and the 4-5μm emission showed that both phenomena were closely related to the split-off band. The main mechanism responsible for these phenomena was attributed to the CHHS Auger recombination in the low current region $I \approx 0.5 \, I_{thr}$, however IVBA was believed to be the main phenomenon occuring at and above threshold, which is contrary to Mozer's conclusions.

Chen[28] has measured spatially resolved emission from the n and p InP confinement layers (resulting from hot carrier leakage). The emission intensity from n-InP was observed to continue to increase above threshold, although the slope of the intensity versus injection current reduced appreciably at threshold which they interpreted to mean that the CHHS process was probably more dominant than IVBA.

Zhuang et al[29] showed that the hot carrier generation of holes in the n-InP confinement layer varied proportionally to n^3, and thus concluded that Auger recombination was chiefly responsible for this leakage loss in 1.3μm lasers.

b) Distributed feedback lasers

Here we present pressure measurements on GaInAsP distributed feedback (DFB) lasers, where the operating wavelength is determined not by the band gap of the active layer material but by continuous reflections from periodic changes in refractive index along the optical waveguide. The DFB lasers measured were ridge waveguide devices grown by MOCVD at STC Technology (Harlow) to operate at $1.52\mu m$ wavelength[49]. They have quarter wave shifts incorporated into the gratings and antireflection coatings on both facets. The grating is within the anti-melt back layer.

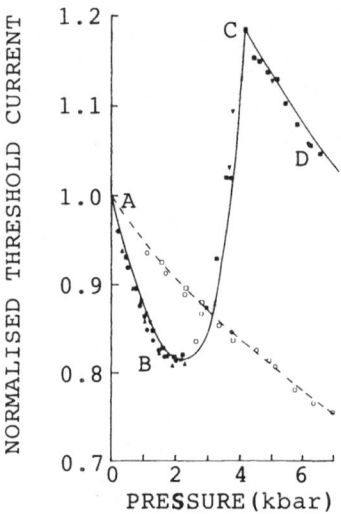

Fig. 7. Pressure dependence of I_{thr} in a DFB laser compared to a Fabry-Perot. Solid line: theory fit to DFB data. Closed and open symbols are DFB and Fabry-Perot data respectively.

In figure 7, the pressure variation of the normalised threshold current for a DFB laser is compared with a conventional Fabry-Perot $1.55\mu m$ wavelength oxide stripe laser. When pressure is applied, the gain spectrum moves to shorter wavelength as the band gap increases. Section A to B in figure 7 occurs as the peak gain moves toward the operating wavelength. This leads to a reduction in the threshold carrier density n_{thr} which is further enhanced by the reduction in IVBA which is proportional to n_{thr}. The threshold current also decreases due to the decrease of Auger recombination with increasing band gap and decreasing n_{thr}. In section B to C the peak gain is moving to even higher photon energy so the operating wavelength moves into the long wavelength tail of the gain spectrum requiring higher carrier densities to reach lasing threshold. The sharp cusp at C occurs when the laser jumps to a Fabry-Perot mode between the end facets when the round trip gain at this wavelength equals that in the DFB mode. In the region C to D the laser acts as a normal Fabry-Perot device with low reflectivity facets. This interpretation has been confirmed by careful observation of the lasing mode relative to the spontaneous emission out of the end facet.

Three other DFB ridge waveguide lasers fabricated from the same slice of material but with gratings of slightly different period were also studied[27]. All three showed a similar change over from DFB to Fabry-Perot operation with pressure, and the region from A to B increased as the mismatch between the DFB wavelength and the peak of the spontaneous emission spectrum increases supporting our model.

The lasing wavelength variation with pressure is shown in figure 8. It shows the slow shift with pressure when the laser is acting as a DFB, and the jump to shorter wavelength when the device changes over to operate as a Fabry-Perot laser. Once the laser is operating in a Fabry-Perot mode the wavelength decreases much faster with pressure in good agreement with other Fabry-Perot 1.55μm GaInAsP lasers. Again the wavelength shift is slightly less than that expected from the band gap pressure coefficient.

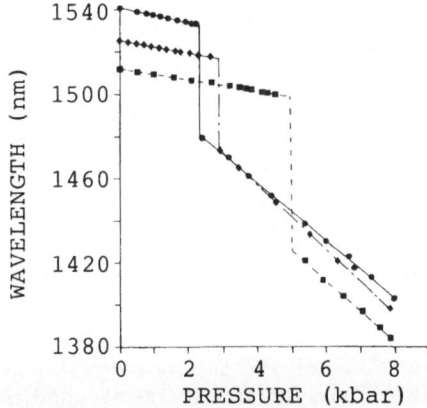

Fig. 8. Pressure dependence of lasing wavelength for three DFB lasers with gratings of slightly different period. The switch from DFB to Fabry-Perot operation can be clearly seen.

A theoretical fit to the above data shown by the solid line in figure 7 was obtained with the following parameters: an intervalence band absorption coefficient, α, of 40cm^{-1} per 1×10^{18} holes[16], a CHHS Auger coefficient of 5×10^{29}cm^6/sec, a conduction band non-parabolicity 2.5 times larger than the value suggested by k.p theory[25], a radiative recombination coefficient of 5×10^{-11}cm^3/sec and a total optical loss including the end facets of 103cm^{-1}. The fit clearly shows that a model including both IVBA and Auger recombination (CHHS) as the loss mechanisms is required to understand the results, however because of the number of interdependent parameters involved the possible ranges they may take have not yet been determined. The results also give an unambiguous measure to the crystal grower of the mismatch between the DFB period and the peak of the gain spectrum.

So far only pressure measurements on lasers have been mentioned. To the authors' knowledge there are no published measurements of the effect of pressure on double heterostructure light emitting diodes. The saturation of the output intensity of GaInAsP/InP and GaAs/GaAlAs surface emitting LEDs as the drive current increases is an interesting problem that is not totally understood. The output power saturation of surface emitting LEDs due to a loss mechanism proportional to n^{2-4} was immediately attributed to the 3 body Auger process[53].

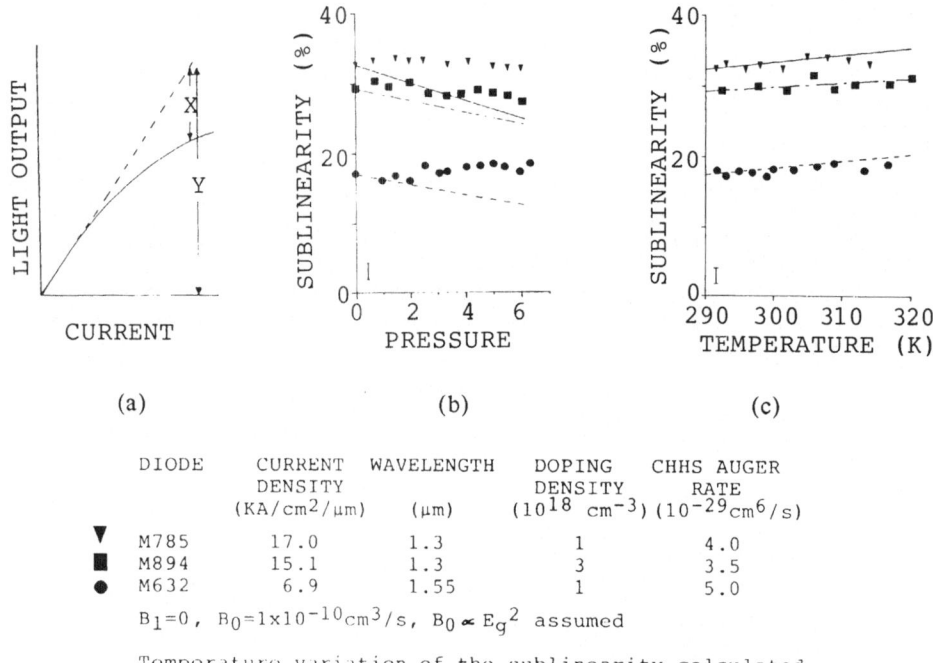

DIODE	CURRENT DENSITY ($KA/cm^2/\mu m$)	WAVELENGTH (μm)	DOPING DENSITY (10^{18} cm^{-3})	CHHS AUGER RATE ($10^{-29}cm^6/s$)
▼ M785	17.0	1.3	1	4.0
■ M894	15.1	1.3	3	3.5
● M632	6.9	1.55	1	5.0

$B_1=0$, $B_0=1\times10^{-10}cm^3/s$, $B_0 \propto E_g^2$ assumed

Temperature variation of the sublinearity calculated according to the expression given by Temkin et al.

Fig. 9. (a) Sublinearity of a diode is defined as X/Y. (b) Pressure and (c) temperature variation of the sublinearity of (GaIn)(AsP)/InP surface emitting diodes for the operating conditions given above.

The investigation of the saturation of surface emitting LEDs by Temkin et al[54] showed however that the saturation is independent of temperature. Figure 9 shows for the first time results we have obtained on 1.3μm and 1.55μm LEDs and it can be seen that the saturation is also pressure independent. These measurements are difficult to explain if an Auger process alone is the cause of the saturation as first proposed. Recently Olshansky[55] has indicated that the n^3 dependent loss mechanism responsible for the saturation could be due partly to a carrier concentration dependent radiative recombination coefficient rather than just to Auger recombination.

Alternatively Goodfellow et al[56] and Dutta and Nelson[57] developed models that involved Auger and superluminescence along the plane of the active layer. In surface emitters where there is significant in plane superluminescence, IVBA could become important. For edge emitting and superluminescent LEDs IVBA could be decreasing the light output significantly, but more importantly it may be causing the light output to be more temperature sensitive than the surface emitter as indicated by the results of Dutta et al[58].

STRAIN ENGINEERING OF VALENCE BANDS

We now describe a laser structure in which the loss mechanisms discussed above are effectively eliminated.

An ideal semiconductor laser should have a band structure in which both the conduction and the valence band has a low effective mass[59,60]. In practice, conventional semiconductor lasers, as described above, satisfy only one of these requirements. The conduction band has a low effective mass m_c but the highest valence band is always heavy-hole-like, with a large effective mass, m_h. The high density of valence states increases the carrier density required for population inversion, and is primarily responsible for the two major loss mechanisms in long-wavelength lasers, namely Auger recombination[4] and intervalence band absorption[2]. A laser in which the highest valence band has a low effective mass should therefore have considerable advantages, with reduced carrier concentration for population inversion, and the virtual elimination of the major loss mechanisms at 1.3μm and at longer wavelengths. The combination of these factors would imply a dramatic reduction in the threshold current density and would reduce the temperature sensitivity described above. Also, the temperature sensitivity of the quantum efficiency would be reduced; a factor of considerable importance for lasers operating at several times their threshold current.

Light-hole behaviour can be partially achieved in a strained bulk semiconductor. Axial strain splits the degeneracy of the valence band maximum, and gives rise to an anisotropic valence band structure[61]. For a biaxial compression, the highest band is heavy-hole-like along the strain direction, and comparatively light in the other two dimensions. The valence band density of states of a bulk strained laser would be dominated by the heavy-hole mass along the strain axis. If we consider growth of a strained quantum well structure, where the quantum well is under biaxial compression, the confinement energy of the highest valence state is determined by its (heavy) mass along the growth (and strain) direction, while the band dispersion is determined by the low effective mass in the well plane. It can thus be predicted theoretically[62] and has been shown experimentally[63] that the highest valence band of a strained-layer structure can have a relatively low in-plane effective mass. This implies that strained-layer structures are ideal candidates for low threshold current, high T_0, semiconductor lasers. In the remainder of this section we first outline the design criteria for a strained-layer laser and then describe some specific structures which are possible candidates for strained lasers.

Optimisation of Laser Structure

In choosing a model laser structure, we seek light-hole behaviour at the valence band maximum over several kT at room temperature. This requires as large a splitting as possible between the two highest hole subbands. The largest subband splittings are found in thin, strained wells. However, for a given in-plane effective mass, m_v, the carrier density for population inversion, n_0, is inversely proportional to the well width, L_z. This follows because population inversion occurs at an approximately constant carrier density per unit area, N_{2D}^o, and the volume carrier density, n_0, depends on N_{2D}^o as N_{2D}^o/L_z. Further, the optical confinement factor, Γ, decreases with decreasing well width, being proportional to L_z for SCH lasers, and L_z^2 for quantum well lasers. Both these factors work against very thin wells. Finally we are also limited by the degree of strain which can be incorporated in a given well. The ideal laser structure is therefore found by balancing a number of competing effects.

We consider growth along the (001) direction. Under biaxial compression in the x-y plane, the two horizontal axes experience equal strains, $\epsilon_{xx} = \epsilon_{yy}$, while the strain along the growth direction, ϵ_{zz} , is of opposite sign and is given by

$$\epsilon_{zz} = - 2(C_{12}/C_{11}) \; \epsilon_{xx} \qquad (14)$$

where C_{11} and C_{12} are the the elastic stiffness constants. The axial component of this strain lifts the degeneracy of the valence band maximum. The heavy-hole band energy S is shifted upwards as

$$S = b \; (\epsilon_{xx} - \epsilon_{zz}) \qquad (15)$$

For small strain, the light-hole band edge shifts down equally in energy; larger strain induces mixing between the light-hole and spin-split-off bands, as described by the Luttinger-Kohn Hamiltonian[64]. b is the (001) axial strain deformation potential, with values typically ~ -1.5 in III-V semiconductors[65]. A 1% compressive strain in the x-y plane lifts the heavy-hole band edge of order 60 - 80meV above the light-hole band edge; this is approximately the minimum strain required for significant light-hole behaviour in the well plane.

The splitting between the highest heavy-hole state, HH1, and light-hole state, LH1, is determined predominantly by the axial strain in the structures we are considering, and for sufficiently large strain the second highest valence subband will be the second heavy-hole state, HH2. The HH1-HH2 splitting depends strongly on the well width, varying roughly as $1/L_z^2$. We calculate that in general this splitting will be less than 50meV for $L_z > 50$Å and therefore estimate ~ 50Å as the upper useful strained quantum well width. We note that monolayer fluctuations in well width have an increasingly deleterious effect with decreasing L_z, causing significant broadening of the band edges. In GaAs lasers, such effects are important for L_z ~ 25Å[66]; we therefore presume 25 - 50Å as the useful range of well widths for strained layer lasers.

Most experimental work on III-V strained-layer structures has to date focussed on the growth of $Ga_{1-x}In_xAs$ on GaAs. Fritz et al.[67] showed that good quality growth can be achieved in this system as long as the product of the percentage lattice mismatch, ϵ, times L_z is less than 100Å%. More recently, Andersson et al.[68] achieved ϵL_z of order 200Å%. For the extreme case of InAs on GaAs, Grunthaner and coworkers[69] have grown GaAs-InAs superlattices with InAs layers up to 15 monolayers thick. Based on these data, we believe that the specific structures which we now consider should all be within the limits of good quality growth.

Specific Laser Structures

Based on the above considerations, we have suggested a number of specific structures for strained-layer lasers operating at 1.55μm. To demonstrate the idea of a strained laser, we considered growth of 40Å quantum wells of strained $Ga_{.33}In_{.67}As$ between unstrained $Ga_{.7}In_{.3}As$ barriers[70]. This structure was chosen to have an optical gap of 0.8eV and a lattice mismatch of 2.5% between well and barrier, so that the ϵL_z product equalled 100Å%. The highest valence band was calculated to be light-hole-like over more than 2kT at room temperature, and the threshold current density could be less than 200Acm^{-2}. Such a laser would be difficult to grow in practice, as the $Ga_{.7}In_{.3}As$ barriers are not lattice-matched to either of the common III-V substrates, GaAs or InP. We estimate that a low-threshold-current 1.55μm GaAs-based laser could be achieved with 35Å strained InAs wells between unstrained GaAs barriers[71]. The lattice-mismatch between well and barrier, ϵ, in this system is substantially greater than required, and as the stored strain energy increases as ϵ^2, this laser may be close to or beyond the limits of good growth. The most suitable 1.55μm strained-layer lasers are probably those grown on InP substrates[72,73].

InP-Based 1.55μm Strained-Layer Laser

InP is the predominant substrate for growth of 1.55μm lasers. We have proposed a specific strained-layer, separate confinement heterostructure (SCH) laser for operation at this wavelength[72]. The structure is closely related to a conventional SCH laser, with the well material replaced by strained layers. The wells consist of $Ga_{.17}In_{.83}As$ layers, each 35Å wide. Quantum confinement is achieved by sandwiching the wells between quaternary $Ga_xIn_{1-x}As_yP_{1-y}$ barriers, lattice-matched to InP. Separate optical confinement is achieved with an InP cladding above and below the quaternary region. We did not attempt to determine the optimum quaternary composition for the barrier but found that with x = 0.14 and y = 0.3, neither the optical nor quantum confinement is significantly degraded.

The quantum well valence subband structure was calculated using the

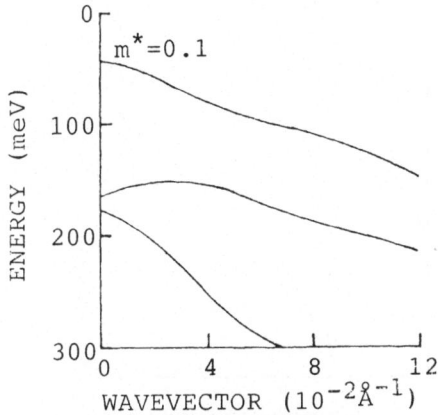

Fig. 10. Valence subband structure
in the well plane for 35Å
$Ga_{.17}In_{.83}As$ well between

$Ga_{.17}In_{.83}As_{.3}P_{.7}$ barriers.

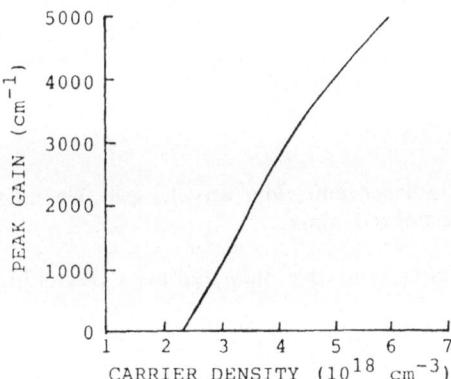

Fig. 11. Variation of peak gain with
carrier concentration for the
band structure in Fig. 10

Luttinger-Kohn 6x6 Hamiltonian[64] in the axial approximation[74], and is presented in figure 10. The splitting between the two highest valence band states at the zone centre (HH1 and HH2) is 123meV, and the highest band is light-hole-like ($m_v^* = .10$) over about 50meV (~ 2kT at room temperature).

We calculated the gain spectrum for this band structure, following Asada[75]. The maximum gain as a function of the volume carrier density, n, at room temperature is shown in figure 11. Transparency is achieved at a carrier concentration, n_{o}, ~ $2 \times 10^{18}cm^{-3}$. The differential gain above transparency, $\beta = \frac{dg}{dn}$, is large and the gain exceeds $1000cm^{-1}$ by a carrier density of $3 \times 10^{18}cm^{-3}$. Lasing occurs when the maximum gain equals the threshold gain, g_{thr}, given by

$$\Gamma\ g_{thr} = \Gamma\ \alpha_{in} + (1 - \Gamma)\ \alpha_{ex} + 1/L\ \ln(1/R) \qquad (16)$$

We assume the cavity length L = 300μm, and that the plane wave reflectivity R = 0.31. The external losses are set at $\alpha_{ex} = 10cm^{-1}$. The internal losses α_{in} are expected to be considerably reduced in comparison to unstrained quantum wells, due to the virtual elimination of intervalence band absorption. We thus take $\alpha_{in} =$

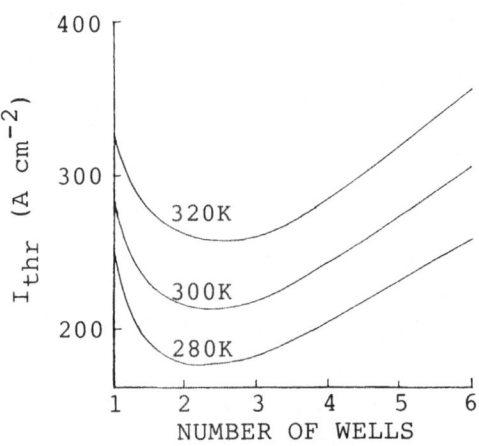

Fig. 12. The threshold current density
for varying number of 35Å
wells, assuming negligible
Auger recombination.

10cm^{-1}. Γ is the optical confinement factor, and for a SCH laser is proportional to the product of the well thickness L_z and the number of wells, N_w, i.e. $\Gamma = \gamma N_w L_z$[76], with γ dependent on the difference in refractive index of the barrier, n_b, and of the cladding, n_c. We estimate γ as

$$\gamma = 4(n_b{}^2 - n_c{}^2)^{\frac{1}{2}}/\lambda_o \qquad (17)$$

where $\lambda_o = 1.55\mu m$ is the laser emission wavelength. We calculated $\gamma = 2.09 \times 10^{-4}$ Å^{-1} for the structure considered above.

The threshold current, assuming only radiative recombination, is

$$J_{thr} = e \, B \, N_w L_z \, n_{thr}{}^2 \qquad (18)$$

where B is the radiative recombination coefficient, set equal to $10^{-10} cm^3 s^{-1}$. For a small number of wells, N_w, the low value of optical confinement factor, Γ, implies a high value of the threshold carrier density, n_{thr}, and gain saturation effects can become important. For larger N_w values, n_{thr} approaches n_o, the inversion carrier density, and, from equation (18), the threshold current density increases linearly with N_w.

The radiative current density as a function of well number N_w is shown in figure 12, at 280K, 300K and 320K. We find the lowest threshold current density for two wells. The radiative contribution to the total current density is then less than $220 Acm^{-2}$. Assuming negligible Auger recombination, the threshold current density is down by nearly an order of magnitude over conventional long wavelength lasers and is comparable to the best short wavelength lasers. In addition, the calculated T_o values are over 100K, implying much improved temperature stability over conventional lasers, where $T_o \sim 60K$.

We have not attempted a calculation of the Auger recombination rate at threshold, as calculated Auger rates are very sensitive to fine detail of the band structure at large wavevector k, where the Luttinger-Kohn Hamiltonian is no longer expected to be a good approximation. We note however that in our calculations less than 5% of the holes are in the second subband at threshold. Auger recombination will increase the threshold current density above the calculated value, but we expect that with the light-hole-like cap at the valence band maximum the threshold current density will still be substantially lower than that in conventional lasers.

296

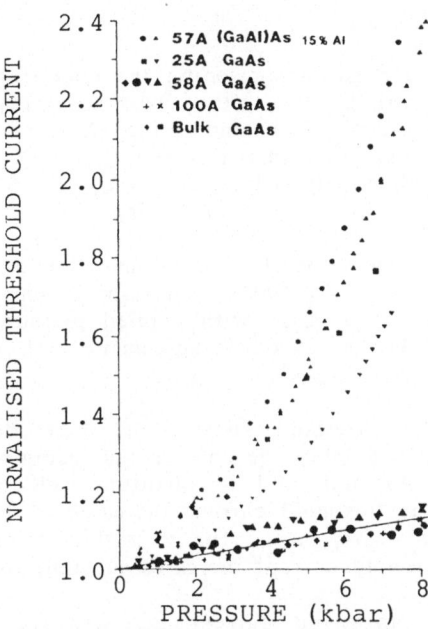

Fig. 13. Pressure dependence of the normalised threshold current in (GaAl)As quantum well and bulk lasers. Solid line from equation (3).

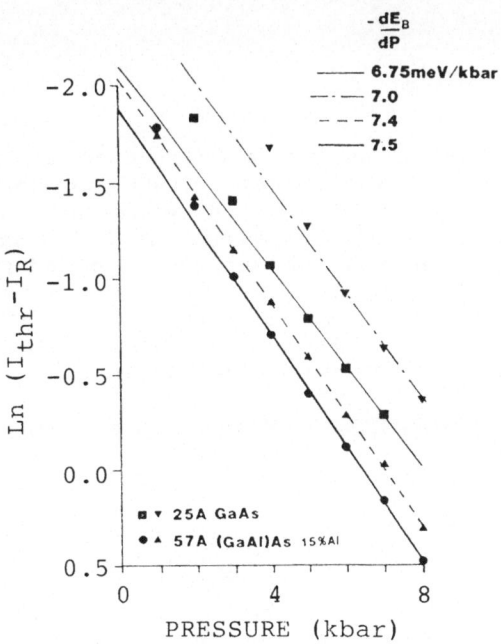

Fig. 14. Pressure dependence of the activation energy derived from the variation of the nonradiative current with pressure, at high pressures.

There is widespread interest in shortening the operating wavelength of GaAs semiconductor lasers. This can be done either by incorporating aluminium into the wells or by reducing the well width and using the quantum size effect, but unfortunately it is found that this increases the threshold current density. The cause was not well understood, interpretation being complicated by data from devices of different structure, growth and processing.

Pressure measurements have therefore been made on a range of devices with wells of different width ($25\text{Å}<L_z<100\text{Å}$) and aluminium content. The lasing wavelength for all the lasers decreased with applied pressure at a rate equivalent to $dE_o/dP = 10.8\pm0.3\text{meV/kbar}$ in excellent agreement with the band gap pressure coefficient of GaAs[33,34].

Figure 13 shows the threshold current of these devices increased with pressure, and increased more quickly the shorter the lasing wavelength. The solid symbols show that, in GaAs bulk and in quantum well lasers with GaAs well widths greater than 60Å, the threshold current increased with pressure by less than 2% per kbar. This agrees with theory for a lossless laser but for the shorter wavelength devices the threshold current increases substantially more.

We assume that the threshold current I_{thr}, consists of two currents 1) a radiative recombination current I_R, 2) a non-radiative current I_B proportional to the number of carriers thermally activated out of the well over an energy barrier E_B so that we may write

$$I_{thr} = I_R + C \exp (-E_B/k_BT) \tag{19}$$

where C is a constant, k_B is Boltzmanns constant and T is the temperature. With the condition that C is pressure independent the threshold current pressure variation can be related to the change in barrier energy by

$$\frac{dE_B}{dP} = -k_BT \frac{d}{dP} (ln_e(I_{thr} -I_R)). \tag{20}$$

Figure 14 shows that the increase in threshold current can be characterised by an activation energy which decreases with pressure by approximately 7meV/kbar. This indicates that the level to which the electrons are activated moves away from the top of the valence band by 4meV/kbar which is charactersitic of the L minima in all zincblende semiconductors[77]. Very recent results on other quantum well lasers show that the barrier energy is decreasing slightly faster than 7meV/kbar, this would indicate that electrons are also being promoted into the X minima. Our pressure measurements confirm Sugimura's proposal[78] that nonradiative recombination from higher lying minima is a problem in short wavelength (GaAl)As lasers.

CONCLUSIONS

We have shown that hydrostatic pressure is a very useful tool for determining the major loss mechanisms in optoelectronic devices, concentrating in particular on its application to the study of III-V semiconductor lasers. For short wavelength lasers, the dominant loss mechanism is shown to be transfer of electrons to the higher conduction band minima at L and X. The dominant loss mechanisms in long wavelength lasers, Auger recombination and intervalence band absorption, are both due to the large effective mass associated with the highest valence band. This effective mass can be significantly reduced in a strained layer structure, suggesting long-wavelength strained-layer lasers could confer significant benefits for long-distance optical communications. Strain can thus play a major role both in characterisation and design of III-V semiconductor devices for optoelectronic applications.

ACKNOWLEDGEMENTS

We are grateful for valuable discussions and the supply of devices from P D Greene, J Whiteaway, C Armistead, B Garrett and R Glew of STC Technology (Harlow), P Blood and E D Fletcher of Philips Research Laboratories (Redhill) and P Williams of Plessey (Caswell). We are indebted to J D Lambkin, A D Prins and D J Dunstan for photoluminescence data prior to publication and to W Batty, U Ekenberg, A Ghiti and P G Willis for their contributions to the theoretical and experimental work. One of us (KCH) is grateful to Philips for financial support.

REFERENCES

1. Horikoshi Y, p.379ff in "GaInAsP alloy semiconductors", T P Pearsall, ed., J Wiley and Sons, New York/London (1982).
2. Asada M, Adams A R, Stubkjaer K E, Suematsu Y, Itaya Y and Arai S, IEEE JQE 17(5):611 (1981).
3. Adams A R, Asada M, Suematsu Y and Arai S, Jpn J Appl Phys 19:L621 (1980).
4. Dutta N K and Nelson R J, J Appl Phys 53:74 (1982).
5. Yano M, Imai H and Takusagawa M, IEEE JQE 17:1954 (1981).
6. Ridley B K, Solid State Electron 21:1313 (1978).
7. Wang De-ning and Shui Hai-long, 8th IEEE Int Conf Semi Lasers, Ottawa 64 (1982).
8. Horikoshi Y and Furukawa Y, Jpn J Appl Phys 18:4 (1979).
9. Dunstan D J and Scherrer W, Rev Sci Inst., in press.
10. Dunstan D J and Spain I L, to be submitted to Rev Sci Inst.
11. Lambkin J, Gunney B J, Lancefield D, Bristow F G and Dunstan D J, to be published in J Phys E.
12. Henry C H, Logan R A, Merritt R F and Laongo J P, IEEE 19:947 (1983).
13. Mozer A, private communication.
14. Mozer A, Hauser S and Pilkuhn M H, IEEE JQE 21:719 (1985).
15. Hauser S, Zielinski E, Asada M, Schweizer H, Burkhard H and Kuphal E, ESSDERC (1987).
16. Childs G N, Brand S and Abram R A, Semicond Sci Technol 1:pp.116-120 (1986).
17. Adams A R, Heasman K C and Hilton J, Semicond Sci Technol 2:7 (1987).
18. Beattie A R and Landsberg P T, Proc Royal Soc London 249:16 (1959).
19. Haug A, Appl Phys Lett 42:512 (1983).
20. Sugimura A, IEEE JQE 19:930 (1983).
21. Burt M G, Brand S, Smith C and Abram R A, J Phys C 17:6385 (1984).
22. Sugimura A, IEEE JQE 17:441 (1981).
23. Bardyszewski W and Yevick D, IEEE JQE 21:1131 (1985).
24. Shantharama L G, Nicholas R J, Adams A R and Sarkar C K, J Phys C 18:L443 (1985).
25. Sarkar C K, Nicholas R J, Portal J C, Razeghi M, Chevrier J and Massies J, J Phys C 18:2667 (1985).
26. Chen T R, Chang B, Chiu L C, Yu K L, Malgalits and Yariv A, Appl Phys Lett 43:217 (1983).
27. Heasman K C, PhD thesis, University of Surrey, England, 1985.
28. Chen L H, Mattos J C V, Prince F L and Patel N B, Appl Phys Lett 44:520 (1984).
29. Zhuang W, Zheng B, Xu J, Li Y, Xu J and Chen P, IEEE JQE 21:712 (1985).
30. Yamakoshi S, Wada O, Umeba I and Sakurai T, Jap Soc Appl Phys 3-H-12 (1982).
31. Nahory R E, Pollack M A and de Winter J C, Electron Lett 15:695 (1979).

32. Patel D, Adams A R, Greene P D and Henshall G D, Elect Lett 18:12 (1982).

33. Shantharama L G, Adams A R, Ahmad C N and Nicholas R J, J Phys C 17:4429 (1984).

34. Wolford D J, Mariette H and Bradley J A, Inst Phys Ser GaAs and related compounds, 74:2754 (1984).

35. Patel D, Adams A R, Greene P D and Henshall G D, Elect Lett 18:527 (1982).

36. Greene P D, Patel D, Adams A R, Allen E M and Henshall G D, Inst Phys Conf Ser GaAs and Rel Compounds 65:265 (1982).

37. Ripper J E, J Appl Phys 43:pp.1762 (1972).

38. Dixon R W and Joyce W B, J Appl Phys 59:pp.4591 (1979).

39. Stubkjaar K, Asada M, Arai S and Suematsu Y, Jap J Appl Phys 20:pp.1499 (1981).

40. Lambkin J and Dunstan D J, submitted to Solid State Commun.

41. Prins A D and Dunstan D J, submitted to Phil Mag Lett.

42. Osinski M and Adams M J, IEE Proc 129:229 (1982).

43. Yamada M and Suematsu M, J Appl Phys 52:2653 (1981).

44. Asada M and Suematsu Y, International Quantum Electronics Conference MII-3, Anaheim, CA, June 1984.

45. Hayes J R, Patel D, Adams A R and Greene P D, J Elect Mater 11:1 (1982).

46. Heasman K C, Adams A R and Willis P G, to be published.

47. Mozer A, Romanek K M, Schmid W, Pilkuhn M H and Schlosser E, Appl Phys Lett 41:964 (1982).

48. Nanomura K, Suemune I, Yamanishi M and Mikoshiba N, Jap J Appl Phys 22:L556 (1983).

49. Armistead C J, Wheeler S A and Plumb R G, J Appl Phys 22:1145 (1986).

50. Hayes S R, Hatton P and Adams A R, to be submitted to Electron Lett.

51. De Meis W M, Harvard Univ Tech Report HP-15 Chapter 3, p.10 (1965).

52. Heasman K C, Adams A R, Armistead C J and Whiteaway J, submitted to IEEE Quantum Electronics.

53. Uji T, Iwarnoto K and Lang R, Appl Phys Lett 38:193 (1981).

54. Temkin H, Chin A K, Di Guiseppe M A and Keramida V G, Appl Phys Lett 39:405 (1981).

55. Olshansky R, Su C B, Manning J and Powazanik W, IEEE JQE 20:838 (1984).

56. Goodfellow R C, Carter A C, Rees G J and Davis R, IEEE Trans Electron Devices, ED 28:365 (1981).

57. Dutta N K and Nelson R J, IEEE QE 18:375 (1982).

58. Dutta N K, Ndoun R J, Wright P D, Besomi P and Wilson R B, IEEE Trans Electron ED30 360 (1983).

59. Adams A R, Electronics Letters 22:249 (1986).

60. Yablonovitch E and Kane E O, IEEE J. Lightwave Technology, LT-4:504 (1986).

61. Pikus G E and Bir G L, Sov Phys - Solid State 1: 1502 (1959).

62. Osbourn G C, Superlattices Microst 1:223 (1985).

63. Schirber J R, Fritz J J and Dawson L R, Appl.Phys.Lett. 46:187 (1985).

64. Luttinger J M and Kohn W, Phys. Rev. 97:869 (1955).

65. Madelung O, ed. "Numerical Data and Functional Relationships in Science and Technology", Group III, Vol. 17, Springer-Verlag, Berlin (1982).

66. Blood P, Colak S and Kucharska A I, Appl Phys Lett 52:599 (1988)

67. Fritz I J, Picraux S T, Dawson L R, Drummond T J, Laidig W G and Anderson N G, Appl.Phys. Lett. 46:967, (1985).

68. Andersson T G, Chen Z G, Kulalovskii V D, Uddin A and Vallin T J, Appl. Phys. Lett. 51:752, 1987

69. Grunthaner F et al., presented at Superlattices, Microstructures and Microdevices, Gothenburg 1986

70. O'Reilly E P, Heasman K C, Adams A R and Witchlow G P, Superlattices and Microstructures 3:99 (1987).

71. Heasman K C, O'Reilly E P, Witchlow G P, Batty W and Adams A R, SPIE

Vol. 800, "Novel Optoelectronic Devices", p.50 (1987).

72. Ghiti A , Batty W, Ekenberg U and O'Reilly E P, SPIE Vol. 861, to be published

73. Yablonovitch E and Kane E O, IEEE J. Lightwave Technology, to be published.

74. Altarelli M, Ekenberg U and Fasolino A, Phys. Rev. B 32:5138 (1985).

75. Asada M, Kameyama A and Suematsu Y, IEEE J. Quantum Electronics QE-20:745, (1984).

76. Eliseev P G and Drakin A E, Sov. J. Quantum Electronics 14:119 (1984).

77. Camphausen D L, Connell G A N and Paul W, Phys Rev Lett 26:184 (1971).

78. Sugimura A, IEEE J. Quantum Electronics QE-20:336 (1984)

PHOTOREFLECTANCE AND PHOTOLUMINESCENCE OF STRAINED In$_x$Ga$_{1-x}$As/GaAs SINGLE QUANTUM WELLS

D.J. Arent, K. Deneffe, C. Van Hoof, J. De Boeck, R. Mertens, G. Borghs

Interuniversity Microelectronics Center, v.z.w.

Kapeldreef 75, B-3030, Leuven Belgium

ABSTRACT

Molecular Beam Epitaxy (MBE) grown single strained layer quantum wells composed of GaAs/In$_x$Ga$_{1-x}$As/GaAs have been characterized at room temperature by photoreflectance and at 6K and 77K by photoluminescence. Excellent agreement between experimentally determined quantum transitions and theory is achieved utilizing a band offset ratio of 85:15 (conduction band:valence band) and a contribution of the hydrostatic compression to the valence band movement corresponding to the pressure sensitivity of the spin orbit band. Analysis of low temperature data indicate that strain induced band changes are not temperature dependent and data obtained at 77K leads to an empirical equation describing the non-strained band gap energy as a function of In fraction, and which differs slightly from that for bulk InGaAs crystals.

I. INTRODUCTION

Semiconductor structures composed of lattice mismatched epitaxial layers represent an important category of materials for use in micro and optoelectronic devices. These structures allow for wide variation in mechanical and electronic properties while incorporating strain arising from large lattice mismatch. The growth of high quality dislocation free material is only possible for layers which do not exceed a critical layer thickness[1]. Layers thicker than the critical value will accommodate lattice mismatch in misfit dislocations rather than elastic strain and therefore exhibit degraded device performance. For thin layers grown pseudomorphically, the strain induces a significant change in the electronic band structure, thus allowing for control of the band gap and associated quantum transitions by altering the ternary compositions (strain) and layer thickness (quantum confinement). Single strained quantum well (SSQW) consisting of a thin strained layer grown between two layers of the same material represent the most simple strained system. Here, growth conditions and the limit of only one layer provide little ambiguity in accurately knowing the structural characteristics and is thus an excellent system for fundamental studies.

Of the III-V semiconductor systems under current investigation, the InGaAs/GaAs system is of both fundamental importance for the study of strained systems and for use in high quality optical[2] and electronic devices[3]. Initial studies showed that two types of SSQW could exist, one with the light hole band confined in the energy well defined by the GaAs valence band, and another where the light hole band is virtual within the GaAs valence band continuum[4]. More recently, however, some discrepancies in the calculation of allowed transitions, especially in the determination of the band offset parameter Q_e have been noted[5-9]. Calculations and experimental results have also shown the critical layer thickness in this system to be approximately 20nm at room temperature[10-14].

We present here the results of investigations of InGaAs/GaAs SSQWs as a function of composition, layer thickness and temperature which help clarify the discrepancies in calculating the expected transitions in InGaAs/GaAs SSQWs. After a brief presentation of experimental methods in Section II, Section III reviews the current theory and presents calculations used for this study. In Section IV, results for photoreflectance experiments performed at room temperature and low temperature photoluminescence are discussed.

II. EXPERIMENTAL METHOD

All samples studied were grown by molecular beam epitaxy (MBE) on undoped GaAs substrates. Following standard cleaning procedures, a 1.0μm GaAs buffer layer was grown followed by the InGaAs layer and finally a 50nm capping GaAs layer. The InGaAs layers were grown at 545°C and the GaAs buffer layers at 585°C. The In mole fraction varied from 0.05-0.30 and layer thicknesses from 2.5nm to 20nm. The In fraction and thicknesses were determined from flux measurements and confirmed by Auger and microprobe analysis. Photoreflectance (PR) measurements at 300K were carried out using standard lock-in detection techniques and electronically divided responses. All low temperature optical measurements were performed in a helium cooled cyrostat. Photoluminescence excitation was provided by the unexpanded 514nm line of an Ar+ laser operating at 400mW.

III. THEORY

To determine the band positions of the SSQW, one must consider both strain and confinement contributions. The details of the strain induced changes to the band positions of layered semiconductor structures have been presented elsewhere[8,15,16] and require only brief review. For InGaAs pseudomorphically grown on GaAs the In concentration must be kept low (<40%) and we assume since the GaAs substrate is very thick compared to the InGaAs layer that all strain is accommodated in the InGaAs. In this case, the energy gap of the InGaAs is always smaller than that for GaAs and the lattice constant always larger. The strain, therefore is represented by a hydrostatic compression term plus a uniaxial tension term and always decreases the effective difference in band gaps. The energy shifts calculated using a strain Hamiltonian which describes strain in the [100] and [010] directions yield[15]

$$\delta E_0(1) = E_H + E_S \tag{1}$$

$$\delta E_0(2) = E_H + \Delta_0/2 - E_S/2 - 1/2\left[\Delta_0^2 + 2\Delta_0 E_S + 9E_S^2\right]^{1/2} \tag{2}$$

$$\delta E_0 + \Delta_0 = E_H + \Delta_0/2 - E_S/2 + 1/2\left[\Delta_0^2 + 2\Delta_0 E_S + 9E_S^2\right]^{1/2} . \tag{3}$$

where $E_H = a(2-K)\varepsilon$ and $E_S = b(1+K)\varepsilon$ are the hydrostatic energy shift and the spin orbit energy shift, respectively, a and b are deformation potentials, ε is the strain value, Δ_0 is spin-orbit splitting energy, and K relates the elastic moduli as given by Pollak[15]. All strain tensor components are found by linearly interpolating between the values for InAs and GaAs using values published in the Landolt-Bornstein tables and are summarized in Table I as a function of In mole fraction. The hydrostatic strain contributes to both conduction and valence bands. As indicated by eqs (1-3), the valence band and the spin orbit band undergo the same hydrostatic shift under pressure. The relative contribution of the hydrostatic term to these bands is determined by, ΔQ_H, the ratio between the total change in the band gap under pressure to the change in the spin orbit band, or

$$\Delta Q_H = [\delta E_g/\delta P / \delta(E_g+\Delta_0)/\delta P]-1 \tag{4}$$

and is found to be 11%. E_S contributes only to splitting of the valence band states,|3/2,1,2> and |3/2,3/2> and induces a mixing of the |3/2,1/2> and |1/2,1/2> bands.[15]

The quantum levels in the respective electron and (heavy) hole wells are determined by standard calculations. As shown in Fig.1, type II SSQWs are formed for thin InGaAs

Table I. Material parameters for $In_xGa_{1-x}As$ as a function of In Fraction.

a_0 (Å)	dE_g^0/dp (10^{-6}eV/kg cm²)	b (eV)
$5.6536 + 0.4054x$	$-9.77(1-.386x)$	$-2.00(1-.100x)$

Δ (eV)	C_{11} (10^{11}dyn/cm²)	C_{12} (10^{11}dyn/cm²)
$0.341(1+.1144x)$	$12.0(1-.308x)$	$5.38(1-.167x)$

Me (Mo)	Mhh (Mo)	Mlh (Mo)
$0.067(1-.642x)$	$0.450(1-.089x)$	$0.088(1-.716x)$

K
$0.89(1+.212x)$

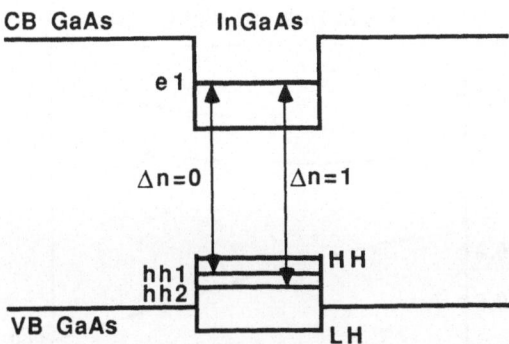

Fig. 1   Energy band representation of an GaAs/InGaAs/GaAs strained single quantum well. CB and VB are the Conduction Band and Valence Band, respectively. e1, hh1, and hh2 indicate confined electron and hole quantum levels, respectively. The allowed ($\Delta n=0$) transition is between the first conduction band level and first heavy hole band. A forbidden transition ($\Delta n \neq 0$) is indicated between the first electron level and second hole level. The light hole (LH) valence band resides in the GaAs valence band continuum and is not confined in a single well structure.

compositions lattice matched to GaAs (In concentrations < 0.4), and the light hole band is unconfined in the GaAs valence band states[3], but may be confined in a strained layer superlattice structure in the potential defined by the InGaAs heavy hole and the GaAs valence band maxima. The composition dependent band gap E_g^0 of the unstrained $In_xGa_{1-x}As$ material at 300K is given by[17]

$$E_g^0 = 1.425 - 1.501x + 0.436x^2 \text{ (eV)}. \tag{5}$$

Due to very small differences in the linear expansion coefficients for InAs and GaAs (and even smaller differences for the low In percent alloys), the additional thermal strain is found to be negligible compared to the large lattice mismatch strain and is therefore not included in the calculations. Further discussion of the low temperature results are presented in Sec. IV.

The results of the strain induced band shift calculations are presented in Fig 2 where the effect of the hydrostatic strain is clearly seen in the movement of the heavy hole band edge up in energy and the degenerate valence band states down. Strong mixing of the $|3/2,1/2\rangle$ and $|1/2,1/2\rangle$ bands, as mentioned above, induces further energy changes in the $\delta E_0(2)$ and $\delta E_0+\Delta_0$ valence band states.

IV. RESULTS AND DISCUSSION

Fig. 3 indicates a typical photoreflectance spectrum and associated fit at 300K for an InGaAs SSQW. The spectra are fitted by non linear least squares analysis to the Aspnes third derivative functional form (TDFF) which has been shown to adequately model the observed phenomena[18]. Recently it has been shown that the application of TDFF to PR spectra of multiple quantum wells is physically inappropriate.[19,20] In contrast with bulk semiconductors where the Franz-Keldysh effect is responsible for the change in the dielectric function, in quantum wells (single or mulitiple) this change is due to excitonic effects. On the other hand, the lineshapes produced using the TDFF sufficiently mimic the excitonic description and are essentially indistinguishable from the first principle derived excitonic formulations at temperatures higher than about 110K[19,20]. This implies that the values for the energy levels derived with the TDFF are correctly determined. It has to be pointed out that the measured energy values are of excitonic nature.[19,20] The measured values correspond, in other words, to the energy difference between he quantized electron and hole states in the wells minus the binding energy of the exciton.

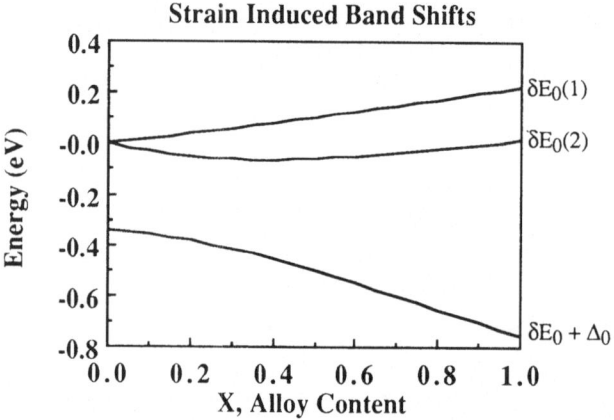

Fig. 2   Strain induced shifts of the valence bands $\delta E_0(1)$, $\delta E_0(2)$ and $\delta E_0 + \Delta_0$ in InGaAs alloys as a function of alloy content. The energy shifts are calculated assuming all strain is incorporated in the thin InGaAs layer. The hydrostatic contribution to the band shifts is determined from the pressure dependence of the spin orbit band relative to the energy gap. T = 300K.

For sample G63, the two transitions 1C-1H and 1C-2H are indicated in Figure 3. Excellent agreement is found for all fits and errors for transition energies are found within $\pm$ 0.5 meV (absolute error of the critical point energy fit). Table II lists the results for other samples including the 6K photoluminescence value for the 1C-1H transition and associated calculated values. Since no reliable data for the exciton binding energy in InGaAs/GaAs quantum well systems are available, the calculated transition energies were not corrected for exciton binding energy at low temperature. This effect is believed to be rather small, inducing an additional possible error of only a few (0 to 8) meV[22] for the thin and low alloy content SSQWs. Excellent agreement is found for each sample and both temperatures. In the calculations, a band offset ratio of 85:15 ($\Delta E_{CB}:\Delta E_{VB}$) is taken in accordance with the values determined experimentally by Kowalczyk et al[5] and Hwang et al[6] and that calculated by Potz and Ferry[8]. Unacceptable agreement was obtained utilizing a value of 0.70, contrasting to other reports[7,21].

For the low temperature calculations, we used the formula of Goetz et al[23] to find the alloy band gap energy at 6K. We calculated the low temperature transition energies using temperature independent strain and material contributions. Typical low temperature

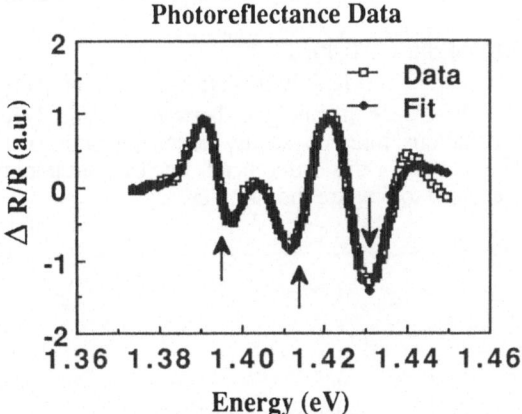

Photoreflectance Data

Fig. 3   Photoreflectance spectrum for the $In_xGa_{1-x}As$/GaAs SSQW with x= 0.07 and a well width of 5.5nm. Excitation is a 1mW 633nm He-Ne laser. Solid symbols indicate the associated theoretical curve with transition energies marked with arrows.

Table II. Results for Photoreflectance (300K) and Photoluminescence (6K) for $In_xGa_{1-x}As$ Single Strained Quantum Wells and Corresponding Theoretical Values for the 1C-1HH Quantum Transitions.

Sample	Transition Energy (eV)			
(Structure) Temp.	300K[a]	Calc	6K[b]	Calc
G44	1.355	1.355	1.442	1.441
15%, 4nm				
GaAs	1.422	1.424		
G63	1.394	1.393	1.481	1.484
7%,5.5nm				
(1C-2HH)	1.413	1.412		
GaAs	1.429	1.424		
G218	1.301	1.302	1.383	1.386
15%, 20nm				
(2C-2HH)	1.344	1.334		
GaAs	1.421	1.424		
G186	1.260	1.256	1.335	1.336
30%, 3.8nm				
(1C-2HH)	1.345	1.327		
GaAs	1.422	1.424		

a) Values determined by Non Linear Least Square Fitting of Photoreflectance Spectra.
b) Values determined from peak positions in Photoluminescence Spectra.

photoluminescence spectra are shown in Figure 4. The linewidths, 6 meV for G63 at 6K and 7meV for G44 at 77K are indicative of the high quality SSQW. Good agreement between theoretical and experimental values at 6K was obtained. From experiments performed at 77K, we have calculated the 77K alloy band gap as a function of In fraction. We determine the equation to be

$$E_g^0(77K) = 1.508 - 1.580x + 0.496x^2 \qquad (6)$$

The fitted coefficients do not vary considerably from the previous bulk crystal values of Leu et al[24]. However, when the difference in band gaps between 6K and 77K is calculated and compared to the previous equations, more physically reasonable and experimental verifiable behaviour is predicted. The reason for this difference is unclear, but may be due to the different growth techniques used to prepare the samples.

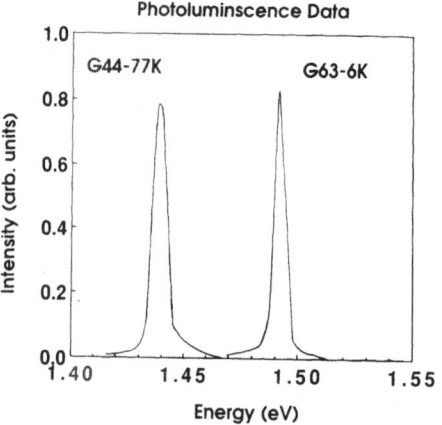

Fig. 4   Low Temperature Photoluminescence spectra of sample G63 and G44. Linewidths (FWHM) are 6meV and 7meV for G63 (6K) G44 (77K), respectively. A comparison of relative intensity is not implied.

From the above data, we have analyzed the temperature dependence of the band gap of the InGaAs alloy. The InGaAs alloy, at least at the compositions studied, behaves most like GaAs. Further indication of this is seen in Fig 5 where we have plotted the normalized percent change to $E_g^0(300K)$ in the band gap between 6K and 77K vs the In mole fraction. The two end points of this curve are GaAs and InAs. The behaviour observed is not obvious at first reasoning whereas one would expect a linear extrapolation between InAs and GaAs. Clearly, at low In concentrations, the GaAs behaviour dominates, possibly governed by the GaAs substrate and the sandwich construction of the SSQW. The implications of this finding are clear when considering structures for novel, high speed electronic or optoelectronic devices to be used in low temperature environments. The behaviour of InGaAs/GaAs heterojunctions on both GaAs and InAs substrates is currently being investigated.

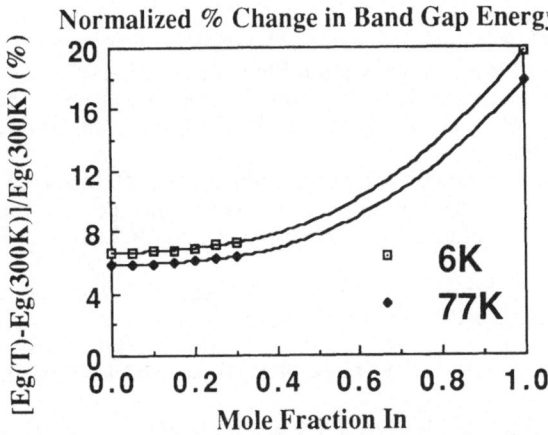

Fig. 5 Percent Change in the Band Gap Energy normalized to $E_g^o(300K)$ at 6K and 77K vs In Mole Fraction. The line drawn is for visual aid.

V. SUMMARY AND CONCLUSIONS

Strain induced band shifts in InGaAs/GaAs single quantum wells have been calculated by a theory incorporating a shared hydrostatic influence based upon the pressure dependence of the fundamental band gap and the split off band. A conduction band offset percentage of 85% is utilized and excellent agreement is found between observed and calculated transition energies. Based on room temperature photoreflectance measurement and low temperature photoluminescence, an equation is derived for the 77K InGaAs alloy band gap as a function of In mole fraction. The excellent agreement between theory and experiment found at lower temperatures suggests that the strain induced band energy changes are temperature independent. We find for low In fractions, $x < 0.3$, that the behaviour of the InGaAs alloy band gap with temperature resembles that of GaAs not InAs. This behaviour is not linearly dependent on the In concentration suggesting that device designs utilizing GaAs substrates should utilize GaAs- related properties and not those extrapolated between InAs and GaAs.

DJA acknowledges the support of a NATO grant awarded in 1988.

REFERENCES

1. J.W. Matthews and A.E. Blakeslee, J.Cryst.Growth *27*, 118, 1974.
2. See for example I Bar-Joseph, G. Sucha, D.A.B. Miller, D.S. Chemla, B.I.Miller, and U. Koren, Appl.Phys.Lett. *52*, 51, 1988.
3. Y.J. Yang, K.Y. Hsieh and R.M. Kolbas, Appl.Phys.Lett. *51*, 215, 1987, and references therein.
4. J.-Y. Marzin, M. N. Charasse, and B. Sermage, Phys.Rev.B, *31*, 8298, 1985.
5. S. P. Kowalczyk. W.J. Schaffer, E.A. Kraut, and R.W. Grant, J.Vac. Sci. Tech.B *20*, 705, 1982.
6. J. Hwang, P. Pianetta, C.K. Shih, W.E. Spicer, Y.-C. Pao, and J.S. Harris, Jr., Appl. Phys. Lett. *51*, 1632, 1987.
7. G. Ji, D. Huang, U.K. Reddy, H. Unlu, T.S. Henderson, and H. Morkoc, J.Vac.Sci.Tech.B *5*, 1346, 1987.
8. W. Potz and D.K Ferry, J.Vac.Sci.Tech.B *4*, 1006, 1986.
9. J. Menendez, A. Pinczuk, D.J. Werder, A.Y. Cho, and D.L.Sivco, J.Vac.Sci.Tech.B *5*, 1256, 1986.
10. I.J. Fritz, S.T. Picraux. L.R. Dawson, W.D. Laidig, and N.G. Anderson, Appl.Phys.Lett. *46*, 967, 1985.
11. I.J. Fritz, P.L. Gourley, and L.R. Dawson, Appl.Phys.Lett. *51*, 1004, 1987.

12. M. Gal, P.C. Taylor, B.F.Usher, and P.J. Orders J.Appl.Phys. *62*, 3898, 1987.
13. P.M.J. Maree, J.C. Barbour, J.F. van der Veen, K.L. Kavanagh, C.W.T.Bulle-Lieuwma, and M.P.A. Viegers,.Appl.Phys. *62*, 4413, 1987.
14. P.L. Gourley, I.J. Fritz, and L.R. Dawson, Appl.Phys.Lett. *52*, 337, 1988.
15. T.P. Pearsall, F.H. Pollak, J.C. Bean, and R. Hull, Phys.Rev.B *33*, 6821, 1986, and references therein.
16. J.Y. Marzin, in Heterojuntions and Semiconductor Superlattices, ed. G.Allan, G. Bastard, N. Boccara, M. Lannoo, and M. Voos, Springer, Berlin. 1986, p.161.
17. R.E. Nahorey, M.A. Pollack, W.D. Johnston, Jr, and R.L. Barns, Appl.Phys.Lett. *33*, 659, 1978.
18. D.E. Aspnes, in Handbook on Semiconductors, ed. M. Balkanski, North Holland, New York, 1980, Vol.2, p.109, and references therein.
19. B.V. Shanabrook, O.J. Glembocki, and W.T. Beard, Phys.Rev.B. *35*, 2540, 1987.
20. W.M. Theis, G.D. Sanders, C.E. Leak, K.K. Bajaj, and H. Morkoc, Phys.Rev.B. *37(6)*, 3042, 1988.
21. G. Ji,D. Huang, U.K. Reddy, T.S. Henderson, R. Houdre, andH. Morkoc, J.Appl.Phys. *62*, 3366, 1987.
22. See for example calculations for AlGaAs/GaAs quantum well systems, R.L Greene, K.K. Bajaj, and D.E. Phelps Phys.Rev.B. *29(4)*, 1807, 1984; G.D.Sanders and Y.C. Chang, Phys.Rev.B. *32(8)*, 5517, 1985.
23. K.H. Goetz, D. Bimberg, H. Jur, J. Selders, A.V. Solomonov, G.F. Glinksii, M. Razeghi, and J.J. Robin, J.Appl.Phys. *54*, 4543, 1983.
24. Y.T. Leu, F.A. Thiel, H. Scheiber, B.I. Miller, and J. Bachmann, J.Electron.Mater. *8*, 663, 1979.

EXCITONS IN QUANTUM WELL STRUCTURES

K. K. Bajaj
US Air Force Wright Aeronautical Laboratories,
Avionics Laboratory, Wright-Patterson AFB Ohio, 45433, USA

G. D. Sanders
Universal Energy Systems, Inc., 4401 Dayton-Xenia Rd.
Dayton, Ohio 45432, USA

R. L. Greene
Department of Physics, University of New Orleans
New Orleans, Louisiana 70148, USA

In this paper we provide a brief overview of the properties of
excitons in GaAs-AlGaAs quantum well structures. We first briefly review
the various calculations of the binding energies of excitons in quantum
wells as a function of well size and compare these results with the
available experimental data. We then outline briefly the calculations of
the binding energies, oscillator strengths and absorption spectra of
excitons in quantum wells in the presence of an externally applied electric
field and again compare these results with the experimental data. We also
point out the applications of these results to a class of opto-electronic
devices.

INTRODUCTION

There has been a great deal of interest in both the theoretical and
experimental investigations of the behavior of Wannier excitons in semiconductor quantum well structures in recent years. This work has been motivated both from the point of view of basic understanding and applications.
Excitonic transitions have been observed in a variety of III-V and II-VI
semiconductor quantum well structures both in absorption and in emission
and their behavior has been studied in the presence of external perturbations such as electric and magnetic fields. A great deal of useful information concerning their properties has been obtained from these studies.

In this paper we briefly review the properties of excitons in quantum
well structures. Though we focus most of our attention on GaAs/AlGaAs
systems, the discussion presented here is also applicable to other similar
systems where the electrons and holes are confined in the same semiconductor component. We shall briefly review various calculations of the binding
energies of excitons in quantum wells and compare these results with the
available experimental data. We shall then outline briefly the theoretical
investigations of the behavior of excitons in quantum wells in the presence
of an applied electric field and again compare these results with the

experimental data. We shall also point out the applications of these results to opto-electronic devices such as spatial light modulators.

During the past decade it has become possible to grow systems consisting of alternate layers of two different semiconductors with controlled thicknesses and relatively sharp interfaces using well developed epitaxial crystal growth techniques such as molecular beam epitaxy (MBE) and metal-organic chemical vapor deposition (MOCVD). Though a variety of systems have been grown, the most extensively studied structure is the one consisting of alternate layers of GaAs and $Al_xGa_{1-x}As$. Depending on the Al concentrations in $Al_xGa_{1-x}As$ its bandgap can be made considerably larger than that of GaAs thus leading to discontinuities of the conduction and valence band edges at the interfaces. Early work[1] suggested that the conduction and valence band offsets were 85 and 15 percent respectively of the bandgap difference between the two semiconductors. A variety of recent measurements, however, have indicated that these values are close to 60 and 40 percent respectively[2]. Thus electrons and holes in the GaAs matrix find themselves in potential wells whose heights depend on the Al concentration in the surrounding $Al_xGa_{1-x}As$ layers. This results in discrete energy levels for electrons and holes and in addition, the lifting of the degeneracy of the valence band leading to the formation of heavy-and light-hole subbands. The Coulomb interaction between the electrons and the holes thus leads to the formation of heavy-and light-hole excitons.

EXCITON BINDING ENERGIES: ZERO FIELD CASE

The Hamiltonian of an exciton in a structure consisting of a layer of GaAs sandwiched between two semi-infinite layers of $Al_xGa_{1-x}As$ grown along the (001) direction can be expressed, using an effective mass approximation as

$$h = \frac{\hbar^2}{2m_e} + T_h(-i\nabla_h) - \frac{e^2}{\varepsilon_0|\vec{r}_e - \vec{r}_h|} + V_e(z_e) + V_h(z_h) \tag{1}$$

here the first term denotes the kinetic energy of the conduction electron with effective mass m_e and the second term is the kinetic energy of the hole as first described by Luttinger[3]. In the present case we can ignore the split off valence band and express this as

$$T_h(\vec{k}) = \begin{bmatrix} L_+ & M & N & 0 \\ M^* & L_- & 0 & N \\ N^* & 0 & L_- & -M \\ 0 & N^* & -M^* & L_+ \end{bmatrix} \tag{2}$$

where

$$L_\pm = \frac{\hbar^2}{2m_0}\left[(\gamma_1 \pm \gamma_2)(k_x^2 + k_y^2) + (\gamma_1 \mp 2\gamma_2)k_z^2\right] \tag{3}$$

$$M = -\frac{\hbar^2}{m_0}\sqrt{3}\gamma_3 k_z(k_x - ik_y) \tag{4}$$

and

$$N = \frac{\hbar^2}{2m_0}\sqrt{3}[\gamma_2(k_x^2 - k_y^2) - 2i\gamma_3 k_x k_y] \tag{5}$$

Here m_o is the free electron mass and γ_1, γ_2 and γ_3 are three material parameters which describe the valence band structure. The positions of the electron and the hole are designated by $\vec{r}_e$ and $\vec{r}_h$ respectively and ϵ_o is the static dielectric constant of the material assumed to be the same for GaAs and $Al_xGa_{1-x}As$. The potential wells for the conduction electron $V_e(z_e)$ and for the holes $V_h(z_h)$ are assumed to be square wells of width L,

$$V_e(z_e) = \begin{cases} 0 & |z_e| \quad <L/2 \\ \\ V_e & |z_e| \quad >L/2 \end{cases} \tag{6a}$$

and

$$V_h(z_h) = \begin{cases} 0 & |z_h| \quad <L/2 \\ \\ V_h & |z_h| \quad >L/2 \end{cases} \tag{6b}$$

here we have chosen, without any loss of generality, the origin of the coordinate system to be the center of the GaAs well. The values of the potential heights V_e and V_h are determined from the Al concentration in $Al_xGa_{1-x}As$, using the following expression for the total-band-gap discontinuity:

$$\Delta E_g = 1.155x + 0.37x^2 \tag{7}$$

in units of electron volts. The values of V_e and V_h are assumed to be 60 and 40% of ΔE_g respectively. The first attempt to calculate the binding energies of the ground state (1s-like, hereafter referred to as 1s) and of the first excited state (2s-like, referred to as 2s) of excitons associated with the lowest hole subbands was made by Miller et al[5]. They assumed that the heavy- and the light-hole subbands were completely decoupled, namely they ignored the contributions of the off-diagonal terms of the Luttinger hamiltonian [Eq.(2)]. This leads to the formation of two types of excitons, one associated with the heavy hole subband and the other with the light-hole subband. With this approximation, the hamiltonian of a heavy (light) exciton in a quantum well reduces to

$$H = -\frac{\hbar^2}{2\mu_\pm} \left[\frac{1}{\rho} \frac{\partial}{\partial \rho} \rho \frac{\partial}{\partial \rho} + \frac{1}{\rho^2} \frac{\partial^2}{\partial \phi^2} \right] - \frac{\hbar^2}{2m_e} \frac{\partial^2}{\partial z_e^2}$$
$$- \frac{\hbar^2}{2m_\pm} \frac{\partial^2}{\partial z_h^2} - \frac{e^2}{\epsilon_o |\vec{r}_e - \vec{r}_h|} + V_e(z_e) + V_h(z_h) \tag{8}$$

where we have used cylindrical coordinates and have used the following definitions

$$\frac{1}{\mu_\pm} = \frac{1}{m_e} + \frac{1}{m_o} (\gamma_1 \pm \gamma_2) \tag{9}$$

and

$$\frac{1}{m_\pm} = \frac{1}{m_o} (\gamma_1 \mp 2\gamma_2) \tag{10}$$

In Equations 9 and 10 the upper sign refers to the $J_z = \pm 3/2$ (heavy-hole) band and the lower sign to the $J_z = \pm 1/2$ (light-hole band). Fairly accurate values of the Luttinger parameters γ_1, γ_2 and γ_3 for GaAs are now available[6]. An exact solution of the Schrödinger equation corresponding to the exciton Hamiltonian [Eq. (6)] is not possible. They therefore followed a variational approach and used the following form of the trial wave function:

$$\psi = f_e(z_e) \, f_h(z_h) \, g(\rho,z,\phi) \tag{11}$$

where $z = z_e - z_h$, $f_e(z_e)$ and $f_h(z_h)$ are the ground-state solutions of the electron and the hole respectively in the quantum well and $g(\rho,z,\phi)$ describes the internal motion of the exciton. In order to further simplify their calculations, they assumed infinite potential barriers and used for g wave functions which are appropriate for thin wells (<200A). Using the known values of the dielectric constant and the various mass parameters they minimized the expectation values (E) of the Hamiltonian [Eq. (8)] with respect to the variational parameters of the trial wave function [Eq. (11)]. The binding energy of the exciton say for the 1s state was then obtained by subtracting E_{1s} from the total ground-state-energy of the electron and the hole in the wells. They found that the binding energies of the 1s and the 2s states of both the heavy- and the light-hole exciton increased as the well size (L) was reduced and reached their respective two dimensional values as the well size went to zero. Bastard et al[7] also calculated the binding energies of the ground state of these excitons as a function of well size assuming isotropic hole masses and infinite potential barriers. Following a variational approach and using a trial wave function which was appropriate for an arbitrary size well, they obtained results similar to those derived by Miller et al[5]. Greene and Bajaj[8] were the first to calculate the binding energies of the 1s state of both the heavy- and the light-hole excitons as a function of the well size using a more realistic case of finite potential barriers for electrons and holes. They solved the Hamiltonian [Eq. (8)] following a variational approach using a trial wave function of the form given by Eq. (11) where $f_e(z_e)$ and $f_h(z_h)$ were now the solution of the particle in a box problem with finite barriers. They used the following expression for the function g

$$g(\rho,z,\phi) = (1 + \alpha z^2) \, e^{-\sigma(\rho^2 + z^2)^{1/2}} \tag{12}$$

where α and σ were the variational parameters. The unnormalized function $f_e(z_e)$, for instance, is given as

$$f_e(z_e) = \begin{cases} \cos(k_e z_e), & |z_e| < L/2 \\ B_e \, e^{-\kappa_e |z_e|} & |z_e| > L/2 \end{cases} \tag{13}$$

where the parameter k_e is determined from the energy of the first electron subband and, B_e and κ_e are obtained from k_e by requiring continuity of f_e and its first derivative at the interface. The hole wave function f_h is obtained in a similar fashion. In order to improve the accuracy of their results Greene et al[9] later used the following expression for g

$$g(\rho,z,\phi) = \rho^{|m|} e^{im\phi} \sum_j a_j g_j (\rho,z) \tag{14}$$

where m is the projection of the angular momentum along the z-axis and the basis functions g_j are taken to be

$$g_1 (\rho,z) = e^{-\alpha(\rho^2 + z^2)^{1/2}} \tag{15a}$$

$$g_2 (\rho,z) = z^2 \, e^{-\alpha(\rho^2 + z^2)^{1/2}} \tag{15b}$$

$$g_3 (\rho,z) = \rho e^{-\beta(\rho^2 + z^2)^{1/2}} \tag{15c}$$

314

The quantities α and β are nonlinear variational parameters and are adjusted to minimize the total energy. The co-efficients a_j are determined in the usual fashion by solving the matrix eigenvalue equation

$$h\psi = EU\psi \qquad (16)$$

The Hamiltonian and the overlap matrices are formed using the complete basis described above,

$$\psi_j = f_e(z_e) \, f_h(z_h) \, \rho^{|m|} \, e^{im\phi} \, g_j \, (\rho, z) \qquad (17)$$

For $m = 0$ (1s and 2s states) all three g_j functions were used; for $m = \pm 1$ (2p$_\pm$ states) only the first two g_j functions were used. The variational binding energies of the 1s, 2s and 2p$_\pm$ states were obtained by subtracting from the lowest electron and hole subband energies (E_e and E_h) the eigenvalues of Eq. (16). These subband energies are determined by numerically solving the transcendental equations for the finite square well[10]

$$\left[\frac{E_e}{V_e} \right]^{1/2} = \mathrm{Cos} \left[\left(\frac{m_e E_e}{2\hbar^2} \right)^{1/2} L \right] \qquad (18a)$$

and

$$\left[\frac{E_h}{V_h} \right]^{1/2} = \mathrm{Cos} \left[\left(\frac{m_\pm E_h}{2\hbar^2} \right)^{1/2} L \right] \qquad (18b)$$

Greene et al.[9] calculated the values of the binding energies of 1s, 2s and 2p$_\pm$ states of the heavy-hole exciton and the light-hole exciton as a function of L for values of Al concentrations x = 0.15 and 0.30. In order to compare their results with those of Miller et al.[5] and Bastard et al.[7] they also calculated the binding energies of these levels for the case of an infinite potential barrier. The values of the various physical parameters used in this calculation are given in Ref. (9). The variation of the binding energies of the 1s state of a heavy-hole exciton E_{1s}(h) (solid lines) and light-hole exciton E_{1s}(l) (dashed lines) as a function of the well-size L for three different values of the potential-barrier heights are displayed in Fig. 1. One finds that for a given value of x, the value of E_{1s}(h) increases as L is reduced until it reaches a maximum and then drops quite rapidly. The value of L at which E_{1s}(h) reaches a maximum is smaller for larger x. Essentially the same behavior is exhibited by E_{1s}(l). The reason for this is quite simple. As L is reduced the exciton wave function is compressed in the quantum well, leading to increased binding. However, below a certain value of L, which depends on the Al concentration, the spread of the exciton wave function into the surrounding $Al_x Ga_{1-x} As$ layers becomes more important. This forces the binding energy of the exciton to go over to the value typical of bulk $Al_x Ga_{1-x} As$ as L is reduced further. In the case of the infinite potential barriers the values of E_{1s}(h) and E_{1s}(l) increase monotonically as L is reduced and go over to their respective two dimensional values, i.e. four times their bulk values, as L goes to zero. Similar behavior is found for the 2s and 2p$_\pm$ states and is discussed in Ref. (9).

Greene et al.[9] in their calculations assumed the same values of the conduction electron mass and the Luttinger mass parameters in both the well and the barrier material. This is a fairly good approximation for commonly used values of the Al concentration (<0.4) and well sizes (>70A). Jiang[11] and Priester et al[12] have calculated the values of E_{1s}(h) and E_{1s}(l) as a function of L assuming different values of these mass parameters in the well and in the barriers following a variational - perturbation and variational approach respectively. They both find that the values of

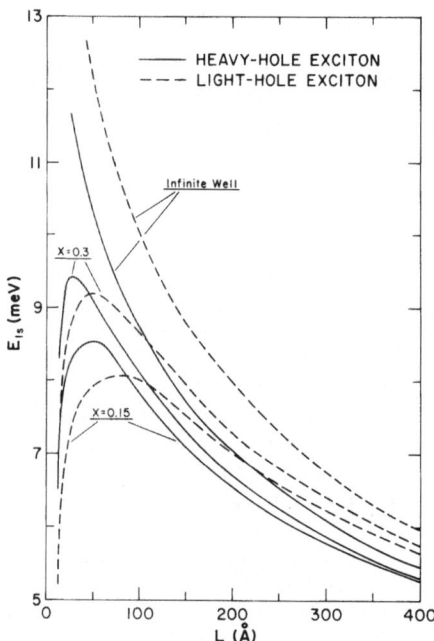

Fig. 1. Variation of the binding energy of the ground state (E_{1s}) of a
heavy-hole exciton (solid lines) and a light-hole exciton (broken
lines) as a function of the GaAs quantum well size (L) for Al
concentration x = 0.15, x = 0.3 and for an infinite potential
well.

$E_{1s}(h)$ and $E_{1s}(l)$ increase slightly (<1 meV) for well sizes smaller than 100Å.

In the foregoing calculations the effect of the off-diagonal terms in the exciton Hamiltonian [Eq. (1)] on its binding energy has been ignored. In the case of an exciton in bulk GaAs, the contribution of the off-diagonal terms to its binding energy is known to be small (<5%)[13] This may, however, not be the case in quantum wells. It is known[14,15] that in quantum wells, even in the absence of the Coulomb term, the inclusion of the off-diagonal terms results in strong mixing between the heavy- and the light-hole states of both parities, thus leading to dipole allowed transitions between all pairs of valence and conduction subbands. In addition, this hybridization leads to hole subband structure which is highly anisotropic in the transverse direction. Several of the hole subbands can have even negative zone-center masses. These features, when included in the calculations of the exciton binding energies, can lead to results quite different from those obtained by using decoupled parabolic valence subbands. In this picture, the heavy-hole subband states, for instance, have admixture of light-hole states and vice versa. Even though the heavy-hole and the light-hole subbands are no longer purely heavy-hole or light-hole in character, we still refer to them as heavy-hole and light-hole subbands for the sake of convenience. Sanders and Chang[16] and Broido and Sham[17] have independently calculated the contribution of the off-diagonal terms to the values of $E_{1s}(h)$ and $E_{1s}(l)$ using somewhat different approaches. Both these groups find that the inclusion of the off-diagonal terms leads to enhancement of the values of $E_{1s}(h)$ and $E_{1s}(l)$ by about 1-2 meV. Recently Ekenberg and Altarelli[18] have also calculated the values of $E_{1s}(h)$ and $E_{1s}(l)$ using the complete exciton Hamiltonian [Eq. (1)] using an approach similar to that of Jiang[11]. They also find increase in the values of the binding energies which however, is somewhat less than that calculated by Sanders and Chang[16] and Broido and Sham[17]. In all these calculations, the effect of the off-diagonal terms on the binding energies of excitons is taken into account through the changes these terms produce in the valence subband structure. Zhu and Huang[19] have recently pointed out that the inclusion of the off-diagonal terms also affects the Coulomb interaction between the electron-hole pair, because the different spinor components of the hole-subband wave function correspond to different in-plane angular momenta and lead to characteristically different radial distributions in the exciton structure. This has the effect of reducing the strength of the Coulomb interaction between the electron-hole pair. Zhu and Huang[19] calculate the values of $E_{1s}(h)$ and $E_{1s}(l)$ as a function of L for GaAs-Al$_{0.25}$Ga$_{0.75}$As quantum well system and find that the contribution of this effect to the binding energy increases as the well size is increased. For instance, for L = 100Å the values of $E_{1s}(h)$ and $E_{1s}(l)$ they obtain are very close to those calculated by Greene et al.[9] using decoupled band approximation. For L = 200Å their values are even smaller than those calculated using decoupled bands.

We shall now compare the results of these calculations with the available experimental data. A number of attempts have been made to measure the exciton binding energies in quantum wells during the past decade. We shall here, however, discuss only those experiments in which the excitonic transitions associated with both the 1s and the 2s states are observed. Miller et al.[5] were the first to observe the excitonic transitions associated with 1s and 2s states of both the heavy-hole and the light-hole excitons in single and multi-quantum well structures using photoluminescence excitation spectroscopy (PLE) at liquid helium temperature. The transitions associated with the 1s states were sharp and intense whereas those associated with 2s states appeared as shoulders on the high energy side. Dawson et al.[20], again using PLE, were able to resolve these shoulders into sharp peaks and thus were able to determine the energy differences between the 1s and the 2s states of both excitons. The values they found were only slightly higher than those determined by Miller et al.[5]. Koteles and Chi[21] have also observed distinct excitonic peaks associated with 1s and 2s states in GaAs-Al$_{0.22}$Ga$_{0.78}$As quantum wells using PLE at 5K. They determine the variation of the 1s-2s energy splittings for both exciton systems as a function L for well sizes ranging from 38 to 218A. They find that their values are very close to those determined by Dawson et al.[20]. In addition, they derive the values of $E_{1s}(h)$ and $E_{1s}(l)$ by adding the theoretical values of the binding energies of 2s states to the 1s-2s splitting energies. The values of $E_{1s}(h)$ and $E_{1s}(l)$ thus determined are somewhat higher than those calculated by Greene and Bajaj[8,9] using decoupled valence bands and are in good agreement with those calculated by Sanders and Chang[16] and Broido and Sham[17] using coupled valence bands. For narrow wells (L<70A) the experimental values of $E_{1s}(h)$ and $E_{1s}(l)$ also agree rather well with those calculated by Zhu and Huang[19] but for wider wells, the theoretical values are considerably smaller. It is not clear why the results of the calculations of Zhu and Huang[19] which are presumably more complete do not agree with experiment. Recently Reynolds et al.[22] have observed excitonic transitions associated with 2s and 3s excited states of heavy- and light-hole excitons in a 225A GaAs-Al$_{0.35}$Ga$_{0.65}$As multiple quantum well structure using PLE. Their values of the 1s-2s energy splittings agree with those of Koteles and Chi[21].

EXCITONS IN ELECTRIC FIELDS

In the past few years several groups have investigated the electronic and optical properties of single-quantum-well and multiple-quantum-well structures in the presence of electric fields. This work has been motivated strongly by its potential applications in a variety of electro-optic devices. Most of the attention has been focussed on the case where the field is applied perpendicular to the plane of the quantum well interfaces. Both emission and absorption studies have been done in these systems and the spectra associated with both the heavy-hole and the light-hole excitons have been observed. All these studies show that the transition energies of the excitons associated with the lowest subbands decrease as the applied field is increased. In addition, the lines become broader and the intensity decreases with increasing field. The conduction- and hole-subband energies and the binding energies of excitons in GaAs-AlGaAs quantum well structures in the presence of an electric field have been calculated by several groups,[23-29] assuming decoupled heavy- and light-hole subbands. Recently Sanders and Bajaj[29] and Hong and Singh[30] have examined this problem including the effects of off-diagonal terms.

Bastard et al.[23] were the first to calculate the energies of the lowest conduction and hole subbands in a quantum well with finite potential barriers in the presence of an electric field applied perpendicular to the interfaces, using a variational approach. They found that the confinement energies decreased as a function of the electric field. Brum and Bastard[25] calculated the binding energies of excitons associated with the lowest subbands as a function of the electric field following a variational approach. They found that the values of the binding energies decreased as the electric field was increased due to electron and hole charge separation. Miller and his co-workers[24,27] have also studied this problem and have arrived at essentially the same results. Matsuura and Kamizato[28] have calculated the binding energies and oscillator strengths of excitons associated with the two lowest conduction and two lowest heavy- and light-hole subbands in quantum wells with infinite barriers as a function of the electric field using a variational approach. It should be pointed out that the problem of eigenvalues of a particle in an infinite potential well in the presence of an electric field is exactly solvable. The solutions are linear combinations of two independent Airy functions.

Sanders and Bajaj[29] were the first to extend earlier studies of electro-optic properties in quantum wells with finite barriers to the higher lying excitons. The nth conduction subband will be referred to as CBn and the nth heavy- and light-hole subbands will be referred to as HHn and LHn respectively. An effective mass model neglecting valence subband coupling was used similar in spirit to the calculations of Brum and Bastard[25] for the lowest lying heavy- and light-hole quantum well excitons (HH1-CB1 and LH1-CB1) in an electric field. In agreement with the previous studies Sanders and Bajaj found that the binding energies and oscillator strengths of the lowest lying heavy- and light-hole excitons decreased monotonically with the applied field due to the separation of the electron and hole charge densities. For HH1-CB1 the electron and hole wavefunctions in the growth direction have single maxima at the center of the well in the absence of the field. Application of the field results in a separation of electron and hole charge densities and a monotonic decrease in the corresponding binding energies and oscillator strengths which both depend to a large extent on the degree of electron-hole overlap as described in Ref. 29. For higher lying excitons the picture is much more complicated with the excitons growing and decaying with the applied field. Figures 2(a) and 2(b) show the results obtained for the exciton binding energies in the uncoupled valence bands approximation as a function of the applied electric field for a 200A GaAs-Al$_{0.25}$Ga$_{0.75}$As quantum well for several exciton transitions to the first two conduction subbands. Figures 3(a) and 3(b) display the corresponding oscillator strength information. In the case of hh3-CB1 exciton binding energy, for example, the hh3 charge density has three maxima and the largest contribution to the potential energy comes from the interaction of the electron and hole charge densities centered at the center of the well. When an electric field F is applied, the potential energy initially decreases as the electron and hole charge densities in the center of the well separate but then increases again before beginning a uniform decline as the peak in the CB1 charge cloud overlaps with one of the secondary peaks in the hh3 charge density. Similar trends are seen in the oscillator strength as a result of the variation of electron-hole overlap with electric field and in particular forbidden lines ($n \neq 0$) at zero field can gain appreciable oscillator strength at finite fields as is evident in the figure.

Sanders and Bajaj[29] in the same paper, also investigated the effect of an applied electric field F on the valence subband structure of GaAs-Al$_x$Ga$_{1-x}$As quantum wells including effects of the hybridization of heavy- and light-holes. The valence subbands in the presence of an applied electric field were found to be rather complicated due to the strong interactions between different subbands at nonzero values of $\vec{k}$ as a result of the mixing of heavy- and light-hole states by the off-diagonal components of $T_h(\vec{k})$. At $k=0$, these off-diagonal components are zero and heavy- and light-hole states are decoupled leading to pure heavy- and light-hole states at the zone center. Away from the zone center, the off-diagonal components increase giving rise to strong nonparabolicities in the valence subband structure. It is convenient to label the subbands after the pure heavy- and light-hole states at $\vec{k}=0$. The valence subband structures of a 200A GaAs-Al$_{0.25}$Ga$_{0.75}$As quantum well for electric field strengths of F=0 and 50kV/cm are shown in Figure 4. In the absence of an applied electric field, the inversion symmetry of the square-well potential causes the valence subbands to exhibit a twofold Kramer's spin degeneracy. In this case, the two sets of degenerate states are obtained by changing the signs of the spin indices and the states are seen to exhibit parity symmetry (even 3/2, odd 1/2, even -1/2, odd -3/2) and (odd 3/2, even 1/2, odd -3/2, even -3/2). When an electric field is applied the inversion symmetry of

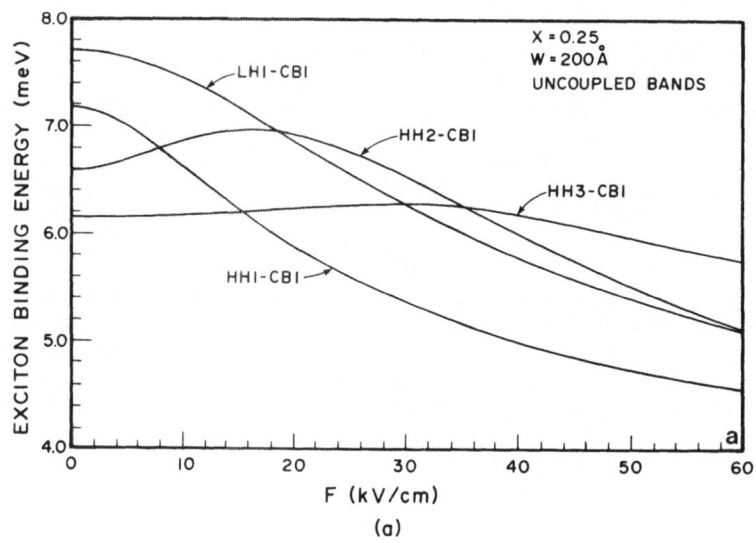

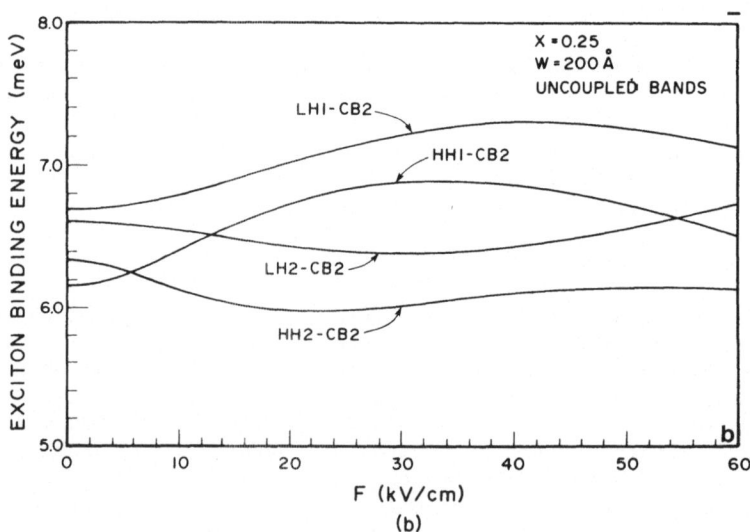

Fig. 2. Exciton binding energies in the uncoupled valence band approximation as a function of applied electric field F for a 200Å
GaAs-Al$_{0.25}$Ga$_{0.75}$As quantum well for excitonic transitions to (a)
the first and (b) the second conduction subband.

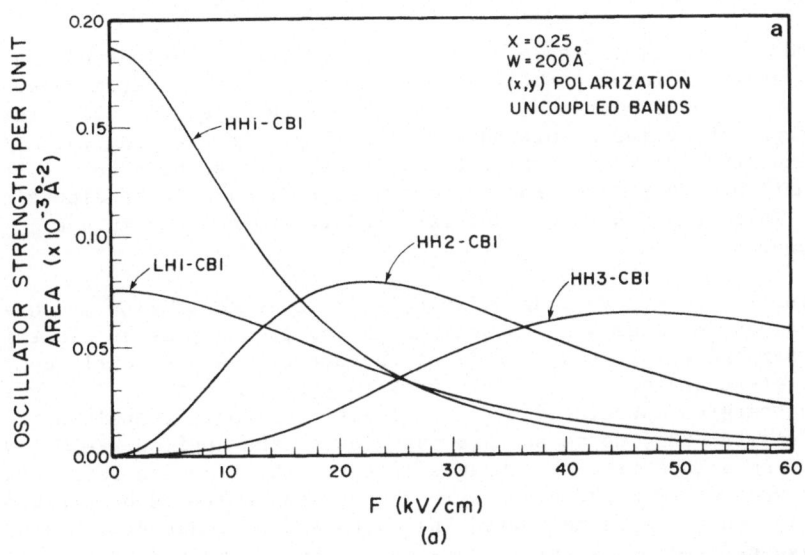

(a)

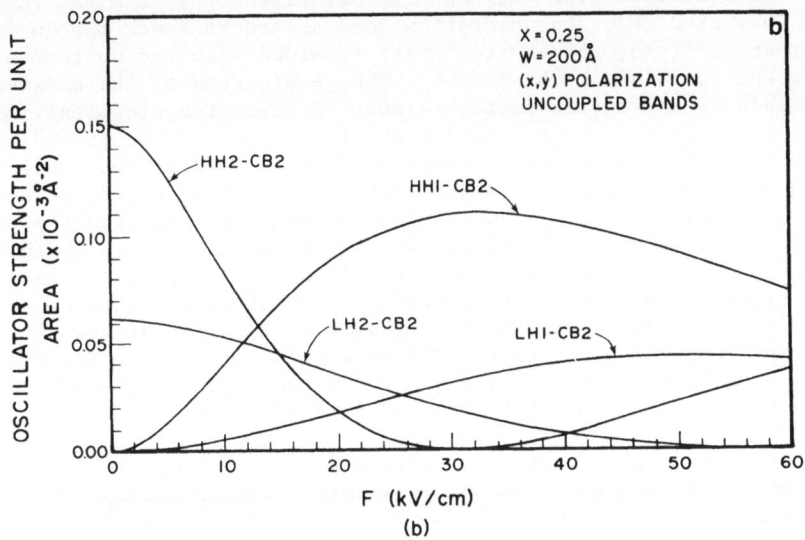

(b)

Fig. 3. Exciton oscillator strengths per unit area in the uncoupled valence subband model for several prominent excitons as a function of applied field in a 200A GaAs–Al$_{0.25}$Ga$_{0.75}$As quantum well for excitonic transitions to (a) the first and (b) the second conduction subband.

the hole potential is lost and the twofold degeneracy at nonzero values of $\bar{k}$ is lifted due to spin orbit coupling. This can be seen explicitly by block diagonalizing the 4x4 Luttinger Hamiltonian into two 2x2 subblocks via a unitary transformation as pointed out by Ekenberg and Altarelli[31] and Broido and Sham[32]. Thus, the upper and lower branches of the spin-split valence subbands do not mix as evidenced by the numerous crossings seen in Figure 4.

Sanders and Bajaj[29] have also calculated the variations of the binding energies and oscillator strengths of excitons associated with several conduction and hole subbands as a function of the electric field taking into account the valence subband mixing. They find that the variations thus calculated are very similar to those obtained using decoupled bands approximation. They also find exciton splittings at finite electric fields. This is not physical and is a result of an error in the derivation.

Several groups have made emission as well as absorption measurements in GaAs-AlGaAs quantum well structures in the presence of an electric field applied perpendicular to the plane of the structures. A fairly complete list of references describing this work is given in Ref. 29. They all find that the energies associated with the lowest excitonic transitions, namely HH1-CB1 and LH1-CB1, decrease as a function of the applied field. Most of the decrease arises from the changes in the conduction and hole subband energies and not from the changes in the binding energies of excitons. This is in general agreement with the calculations. Recently Miller et al.[27] have carried out a fairly detailed study of the effects of electric field on the optical absorption spectrum of a GaAs-Al$_{0.32}$Ga$_{0.68}$As multi-quantum-well structure with well and barrier sizes of 95 and 98A respectively. They find that the absorption edge shifts to lower photon energies with increasing field; the exciton peaks remained resolved up to fields of 100 kV/cm and shift almost by 25 meV. The persistence of the excitonic peaks at such high electric fields (almost 50 times the classical ionization field) is due to reduced tunnelling of the confined particles out of the wells and the presence of the strong Coulomb interaction. Collins et al.[33] and Yamanaka et al.[34] have also studied the behavior of the excitonic transitions in GaAs-Al$_x$Ga$_{1-x}$As multi-quantum wells in electric fields using photocurrent spectroscopy. Both groups find a rich structure in their spectra and observe as many as eight transitions. As the value of the electric field is increased, these transitions shift toward lower energies and their relative strengths change. The behavior of these transitions is in agreement with the predictions of the calculations of Sanders and Bajaj.[29]. A more detailed discussion of the comparison between theory and experiments is contained in Reference 29.

As mentioned earlier, the interest in electro-absorption effects associated with excitons in quantum well structures is strongly motivated by their applications in a variety of electro-optic devices[35] such as spatial light modulators, self-linearized modulators, wave length selective detectors and optically bistable switches. Reference 35 provides an excellent review of this area of research. The behavior of excitons in quantum wells in the presence of an electric field applied perpendicular to the interfaces forms the basis of operation, for instance, of a spatial light modulator. This device consists of a p-i-n diode where the undoped region i consists of a multi-quantum well structure. In the absence of an electric field, for a photon energy, say E_p, corresponding to HH1-CB1 transition, the system exhibits maximum absorption. As the electric field is applied HH1-CB1 transition moves to lower energy, thus decreasing the absorption at E_p. This leads to the modulation of the optical signal and forms the basis of operation of the spatial light-modulators.

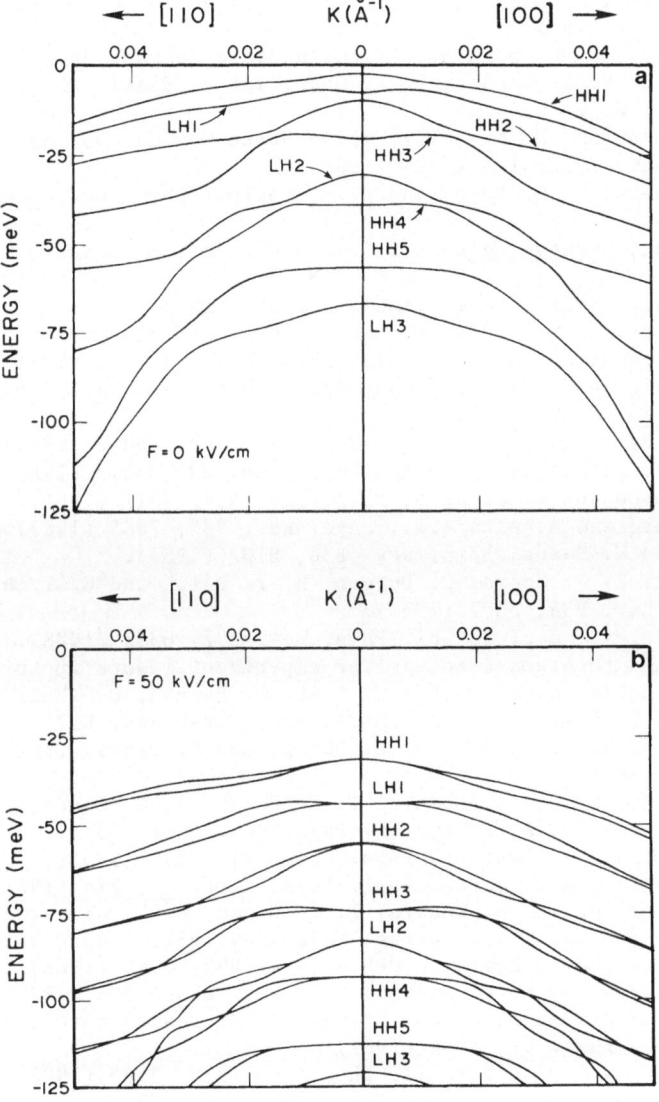

Fig. 4. Valence subband structures for a 200Å GaAs–Al$_{0.25}$Ga$_{0.75}$As quantum well for (a) F = 0 and (b) F = 50 kV/cm.

REFERENCES

1. R. Dingle, in Advances in Solid State Physics, edited by H. J. Queisser (Pergamon/Vieweg, Braunschweig, 1975) Vol. XV, p.21.
2. G. Duggan, J. Vac. Sci. Technol. B3, 1224 (1985).
3. J. M. Luttinger, Phys. Rev. 102, 1030 (1956).
4. h. J. Lee, L. Y. Juravel, J. C. Wolley and A. J. Springthorpe, Phys. Rev. B21, 659 (1980).
5. R. C. Miller, D. A. Kleinman, W. T. Tsang and A. C. Gossard, Phys. Rev. B24, 1134 (1981).
6. K. K. Bajaj and C. H. Aldrich, Solid State Commun. 35, 163 (1980).
7. G. Bastard, E. E. Mendez, L. L. Chang and L. Esaki, Phys. Rev. B26, 1974 (1982).
8. R. L. Greene and K. K. Bajaj, Solid State Commun. 45, 831 (1983); J. Vac. Sci. Technol. B1 391 (1983).
9. R. L. Greene, K. K. Bajaj and D. E. Phelps, Phys. Rev. B29, 1807 (1984).
10. A. Messiah, Quantum Mechanics (North-Holland, Amsterdam, 1961) Chap. 3, Vol 1.
11. T. F. Jiang, Solid State Commun. 50, 589 (1984).
12. C. Priester, G. Allan and M. Lanno, Phys. Rev. B30, 7302 (1984).
13. A. Baldereschi and N. O. Lipari, Phys. Rev. B3, 439 (1971).
14. Y. C. Chang and J. N. Schulman, Appl. Phys. Lett. 43, 536 (1983); Phys Rev. B31, 2069 (1985).
15. G. D. Sanders and Y. C. Chang, Phys. Rev. B31, 6892 (1985).
16. G. D. Sanders and Y. C. Chang, Phys. Rev. B32, 5517 (1985).
17. D. A. Broido and L. J. Sham, Phys. Rev. B34, 3917 (1986).
18. U. Ekenberg and M. Altarelli, Phys. Rev. B35, 7585 (1987).
19. B. Zhu and K. Huang, Phys. Rev. B36, 8102 (1987).
20. P. Dawson, K. J. Moore, G. Duggan, H. I. Ralph and C. T. B. Foxon, Phys. Rev. B34, 6007 (1986).
21. E. S. Koteles and J. Y. Chi, Phys. Rev. B37, 6332 (1988). This paper contains references to earlier experimental work in this area.
22. D. C. Reynolds, K. K. Bajaj, C. Leak, G. Peters, W. Theis, P. Yu, K. Alavi, C. Colvard and I. Shidlovsky, Phys. Rev. B37, 3117 (1988).
23. G. Bastard, E. E. Mendez, L. L. Chang, and L. Esaki, Phys. Rev. B28, 3241 (1983).
24. D. A. B. Miller, D. S. Chemla, T. C. Damen, A. C. Gossard, W. Wiegmann, T. H. Wood and A. C. Burrus, Phys. Rev. Lett. 53, 2173 (1984).
25. J. A. Brum and G. Bastard, Phys. Rev. B31, 3893 (1985).
26. E. J. Austin and M. Jaros, Appl. Phys. Lett. 47, 274 (1985).
27. D. A. B. Miller, D. S. Chemla, T. C. Damen, A. C. Gossard, W. Wiegmann, T. H. Wood and C. A. Burrus, Phys. Rev. B32, 1043 (1985).
28. M. Matsuura and T. Kamizato, Phys. Rev. B33, 8385 (1986).
29. G. D. Sanders and K. K. Bajaj, Phys. Rev. B35, 2308 (1987).
30. S. Hong and J. Singh, Superlattices and Microstructures, 3, 645 (1987); J. Appl. Phys. 62, 1994 (1987).
31. U. Ekenberg and M. Altarelli, Phys. Rev. B30, 3569 (1984).
32. D. A. Broido and L. J. Sham, Phys. Rev. B31, 888 (1985).
33. R. T. Collins, K. v. Klitzing and K. Ploog, Phys. Rev. B33, 4378 (1986).
34. K. Yamanaka, T. Fukunaga, N. Tsukada, K. L. I. Kobayashi and M. Ishii, Appl. Phys. Lett. 48, 840 (1986).
35. D. A. B. Miller, D. S. Chemla, T. C. Damen, T. H. Wood, C. A. Burrus, A. C. Gossard and W. Wiegman, IEEE J. Quantum Electron, Vol. QE21, 1462 (1985).

FOURIER DETERMINATION OF THE HOLE WAVEFUNCTIONS IN P-TYPE MODULATION DOPED QUANTUM WELLS BY RESONANT RAMAN SCATTERING

G. Fasol, T. Suemoto*, U. Ekenberg and K. Ploog**

Cavendish Laboratory, Madingley Road, Cambridge CB3 OHE, England

*Research Institute for Scientific Measurements, Tohoku University, Katahira, Sendai 980, Japan

**Max-Planck-Institut für Festkörperforschung, Heisenbergstrasse 1, D-7000 Stuttgart 80, Fed. Rep. of Germany

ABSTRACT

We determine the electronic structure of p-type modulation doped quantum wells by resonant Raman scattering and using self consistent envelope function calculations. We demonstrate that closely spaced electronic energy levels in wide quantum wells can be determined with high precision from resonant Raman measurements. We determine the conduction band non-parabolicity experimentally and theoretically. We show that the height of sharp resonant peaks in the Raman scattering cross section as a function of laser energy is related to the transition matrix elements and yields the Fourier components of the hole wavefunction in terms of the electron wavefunctions. The hole wavefunction determined experimentally agrees reasonably well with the results of the calculations.

1. INTRODUCTION

In order to use "wavefunction engineering" for new devices it is necessary to determine the wavefunctions in a given artificial semiconductor structure. It is a straight forward procedure to calculate the energy levels and the wavefunctions in modulation doped heterojunctions and in modulation doped quantum wells by solving the Schrödinger equation and Poisson's equation selfconsistently. It is difficult to determine the wavefunctions directly experimentally. Such information is usually only very indirectly contained in experiments. In the present paper we discuss how resonant Raman scattering can be used to determine the shape of wavefunctions in modulation doped structures. We have introduced this new technique recently [1]. We show that the scattering cross section as a function of laser energy for Raman scattering from hole intersubband excitations and from LO phonons in p-type modulation doped GaAs/AlGaAs quantum wells shows series of many sharp and pronounced peaks. We show that their position can be used to determine electron binding energies in the conduction band well, and the conduction band non-parabolicity[1].

The resonant Raman method[1,2,3,4,5] of exploring the conduction subband structure is fundamentally different from the method of luminescence spectroscopy. Figure 1 shows that the underlying electronic processes are quite different. The absorption and the luminescence emission processes are both one step processes, where a photon is converted into an exciton (or an excitonic polariton) or vice-versa. Phase and energy relaxation take place between luminescence excitation and luminescence emission. Raman scattering processes, on the other hand, are higher order processes which are described theoretically by a series of quantum mechanical perturbation

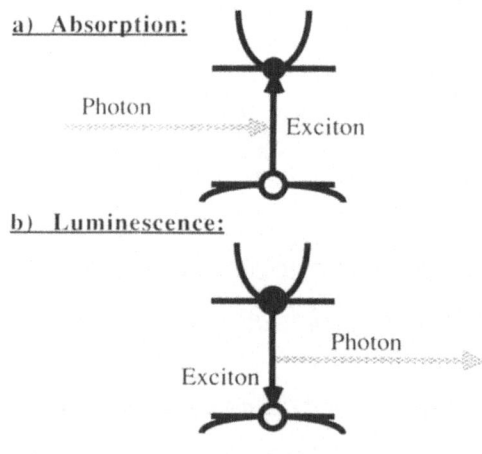

a) Absorption:

Photon — Exciton

b) Luminescence:

Exciton — Photon

c) Raman Light Scattering:

Excitation (Phonon, Plasmon,....)

Laser Photon ω_L

Scattered Photon $\omega_L - \omega_{exc}$

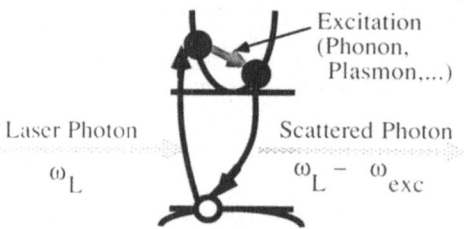

Figure 1. Schematic view of (a) absorption, (b) luminescence, and (c) Raman processes. Note that the Raman mechanism is a higher order process involving several transitions.

expansion terms. One or more terms in such a perturbation expansion have to be taken into account in particular cases. Figure 1c shows schematically a typical Raman scattering process[5]. The transitions take place coherently on a very short time scale. The expressions for the Raman intensity contain resonant denominators which diverge when the laser energy (or for outgoing resonance, the laser energy minus the energy of the phonon or the electronic excitation) coincides with electronic energies of the semiconductor.

Figure 2 demonstrates schematically the difference between determining the electronic structure of a quantum well by luminescence spectroscopy, luminescence excitation spectroscopy and by Raman resonance spectroscopy as described in the present article. The luminescence spectra of high quality quantum wells usually show a single narrow line due to the recombination of the lowest exciton. The luminescence excitation spectra are essentially proportional to the optical absorption - however, for a complete interpretation of excitation spectra carrier relaxation and diffusion effects have to be considered as well. In a two-dimensional system the optical absorption is proportional to the sum of a series of step functions. As the energy increases every new subband adds one new step function. In addition, there is a substantial enhancement of the absorption just below the absorption steps due to the excitonic interaction, which makes luminescence excitation spectroscopy useful to study energy levels in quantum wells. In modulation doped quantum wells the excitonic interaction is screened by the free carriers, which reduces the height of the excitonic absorption peaks and therefore reduces the sensitivity of luminescence excitation measurements for the determination of the electronic structure. The Raman scattering intensity, on the other hand, as a higher order process[5] is proportional to the sum of a series of resonant denominators which are expected to diverge whenever the laser light and/or the Raman scattered light is equal to the spacing of energy levels in the structure[1,2,3,4]. It has been demonstrated, that the spacing of energy levels can be determined with high precision[2,3,6,7,8] and up to high quantum numbers[1,4] by observing the sharp peaks in the resonant Raman profile - we observe levels up to n = 19 in wide wells. This method will be discussed below in this article. We will show that these measurements can determine the conduction band non-parabolicity in the well.

The present work studies and exploits a second important feature of the Raman resonance intensity: we show that the Raman scattering efficiency is a function of the overlap integral between the hole wave function and the electron wave function of the states participating in the Raman process, using a few plausible approximations. Therefore by measuring the height of the various resonant peaks in the Raman resonance profile we can determine the components of the Fourier expansion of the hole wave function in terms of the electron wave functions. Thus we can use the Raman resonance profile to determine the wavefunctions of bound hole states in the valence band well. In the present article we demonstrate this technique by studying experimentally the hole wavefunction in asymmetric p-type modulation doped AlAs/GaAs/AlGaAs quantum wells.

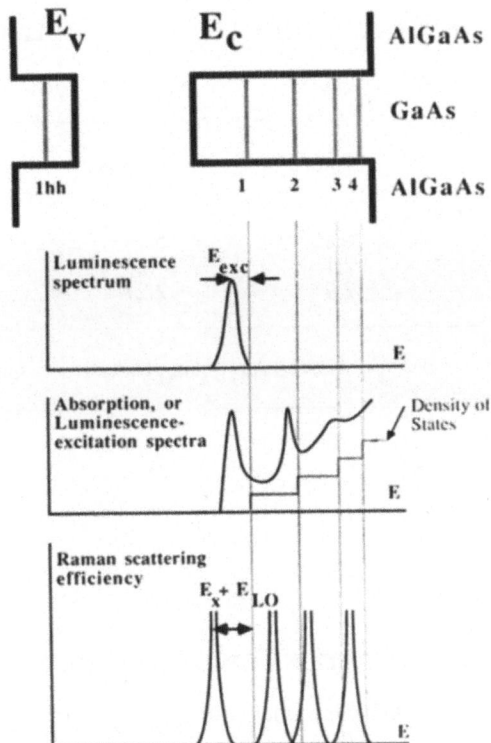

Figure 2. Schematic view of luminescence spectrum, luminescence excitation spectrum and Raman resonance profile from a quantum well.

"Wave function engineering" makes it desirable to be able to determine the shape of wavefunctions. This is usually done by solving the Schrödinger equation and Poisson's equation selfconsistently. We show the results of such calculations for the present quantum wells. We show that resonant Raman scattering allows a fairly direct comparison of the calculated wavefunction with experimental results.

TABLE I   SAMPLE PARAMETERS AND FITTING PARAMETERS

Sample	x	ρ (cm^{-2})	μ (cm^2/Vs)	d_{tot} (Å)	d (Å)	ϵ (meV)	F (meV/Å)	z (Å)	δ (Å)	Interface
A (4984)	0.43	1×10^{11}	3850	1279	745	-3	0.15	65	9.3	GaAlAs
							0.15	72	16.3	AlAs
B(4982)	0.42	7×10^{10}	3300	1214	670	-18	0.08	71.5	2.8	GaAlAs
							0.08	78.5	9.8	AlAs

TABLE II   INPUT PARAMETERS FOR THE CALCULATIONS

	GaAs	$Al_{0.43}Ga_{0.57}As$	AlAs
Electron mass	0.0665	0.1024	0.15
Valence-band parameters:			
γ_1	6.85	5.39	3.45
γ_2	2.1	1.49	0.68
γ_3	2.9	2.21	1.29

Interface:	GaAs - $Al_{0.43}Ga_{0.57}As$	GaAs - AlAs
ΔE_c (meV)	350	1040
ΔE_v (meV)	202	469

2. STRUCTURE OF P-TYPE MODULATION DOPED QUANTUM WELLS

We study several p-type modulation doped multi-quantum well samples prepared by molecular beam epitaxy on GaAs substrates. Each period of these multi-quantum wells consists of the following sequence of layers: a 250 Å wide layer of Be-doped AlGaAs, followed by a 50 Å AlAs spacer layer, an up to 745 Å wide GaAs well, another 150 Å wide AlGaAs spacer layer, followed by the next Be-doped AlGaAs layer. The basic period of our sample structures is shown schematically in Figure 3. The samples consist of twenty such periods. The thicknesses per period for our samples as determined by X-ray diffraction, the concentration and the mobility of the holes from Hall effect measurements are tabulated in Table I. We investigated sample A (No. 4984) most carefully, because it had the clearest resonance behaviour.

We have performed self-consistent subband structure calculations for comparison with the experimental results described below. The transfer of holes from the barriers to the quantum well sets up a potential which is displayed in Figure 4, where we also have plotted the wave functions for the two highest hole sublevels and the two lowest electron sublevels. It is seen that the holes are concentrated at the interface regions but the electrons are distributed over the whole quantum well. First the valence-band structure is calculated self-consistently in the multi-band envelope function approximation. We use the Luttinger-Kohn Hamiltonian[9] for the kinetic energy of the holes:

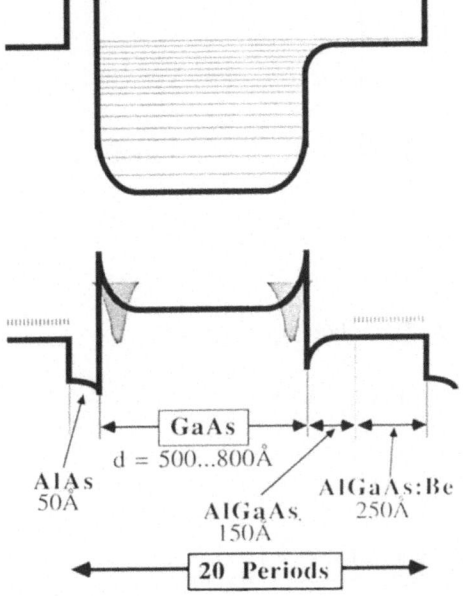

Figure 3. Schematic view of one period of the p-type modulation doped quantum wells studied in the present article.

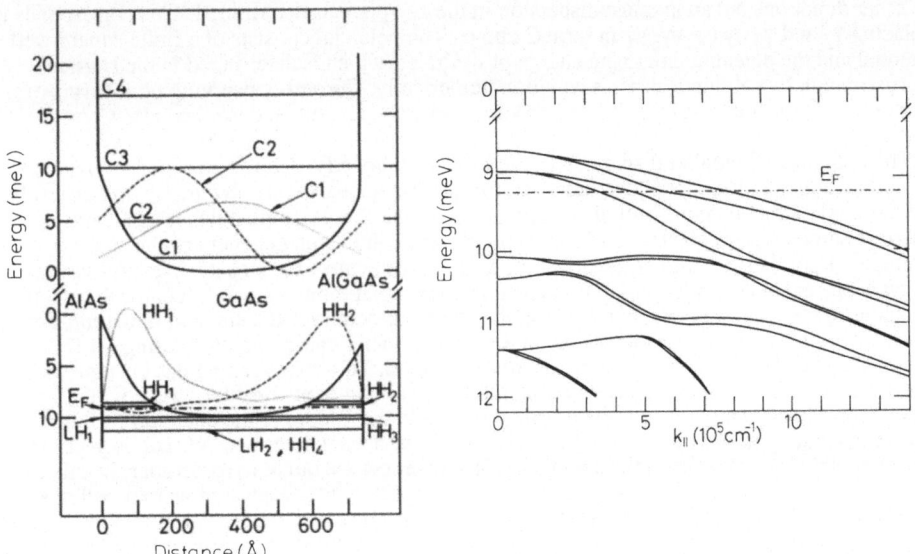

Figure 4. Potential in the conduction and the valence band for a 745 Å p-type modulation-doped quantum well. The AlAs layer is to the left and the $Al_{0.43}Ga_{0.57}As$ is to the right. The energy levels at $k_{//} = 0$ are shown as horizontal solid lines and the dash-dotted line denotes the Fermi energy. Wave functions for the lowest electron sublevels and for the uppermost hole sublevel (dotted lines) and those for the second sublevels (dashed lines) are drawn with the zero level at the relevant sublevel.

Figure 5. Energy dispersion parallel to the interfaces for the upper hole subbands. The splitting at finite $k_{//}$ is due to the asymmetry of the quantum well. The Fermi level is shown by the dash-dotted line. The near-degeneracy of the fifth sublevel at $k_{//} = 0$ is accidental.

$$\begin{bmatrix} A_+ & B & C & 0 \\ B^* & A_- & 0 & C \\ C^* & 0 & A_- & -B \\ 0 & C^* & -B^* & A_+ \end{bmatrix}$$

[1]

where

$$A_\pm = -\frac{1}{2}(\gamma_1 \; 2\gamma_2) k_z^2 - \frac{1}{2}(\gamma_1 \pm \gamma_2)(k_x^2 + k_y^2)$$

[2]

$$B = \sqrt{3}\;\gamma_3\;(k_x - i\,k_y)\,k_z$$

[3]

and

$$C = \frac{\sqrt{3}}{2}[\,\gamma_2\,(k_x^2 - k_y^2) - 2\,i\,\gamma_3\,k_x\,k_y]$$

[4]

The valence band structure is described by three material parameters, γ_1, γ_2 and γ_3 which are given in Table 2 for GaAs, for $Al_{0.43}Ga_{0.57}As$ and for AlAs. We use units such that $h = m_0 = 1$.

We choose the z-direction to be perpendicular to the interfaces, replace k_z by the operator $- i$ d/dz and add the potential along the diagonal. The filled hole states occupy a very small region in k-space so we can safely apply the axial approximation, in which the z-direction is inequivalent to the other directions but an average dispersion in the x-y-plane is determined. This corresponds to replacing γ_2 and γ_3 by $(\gamma_2+\gamma_3)/2$ in term C above. The potential consists of a finite square well potential and the potential due to the charge of the holes, which is determined in the Hartree approximation by solving Poisson's equation numerically. Current-conserving boundary conditions are fulfilled with the use of a modified variational method described elsewhere[10].

The complicated parallel dispersion of the valence subbands is shown in Fig. 5 . To understand the main features of the subband structure it is important to analyze the symmetry properties. In a narrow, symmetric quantum well, the potential due to the charge of the holes is small and can for practical purposes be neglected. Each subband has a two-fold spin degeneracy due to the inversion symmetry around the middle of the quantum well[11]. (We neglect the lack of inversion symmetry which is present also in bulk GaAs, because it is known that it has a small effect for the small $k_{||}$-values relevant in Fig. 5)[12]. For the roughly triangular potential at a single modulation-doped heterojunction the strong violation of the inversion symmetry causes a spin splitting for finite values of $k_{||}$ but the two-fold degeneracy remains at $k_{||}$=0. This spin splitting has been predicted theoretically[12,13,14,15] and verified experimentally[16,17]. If we next consider a very wide symmetric modulation-doped quantum well, the potential barrier in the middle of the quantum well is so large that we essentially have two independent hole gases, one at each interface. We can expect the same type of subband structure as in the single interface case, but we ought to reconsider the labelling of the subbands. It is essential that the potential wells at the two interfaces are inverted with respect to each other. If we choose the z-axis as the direction for quantization of angular momenta we can say that if the upper band corresponds to "spin up" at the left interface, it corresponds to "spin down" at the right interface. Since we have two hole gases and inversion symmetry around the middle of the quantum well, each subband should be considered as a separate subband with a two-fold spin degeneracy. If we now imagine that the width of the quantum well is decreased so that the two hole gases start to interact with each other the four-fold degeneracy at $k_{||}$=0 is lifted into two sublevels, each with two-fold degeneracy [18]. We here have a previously unexplored mechanism for band structure engineering in which the energy separation between the first two sublevels can be made arbitrarily small. If we finally make the quantum well asymmetric, the two-fold degeneracy is lifted for finite values of $k_{||}$. As is seen in Fig. 5, we therefore have four spin subbands very close to each other and for the low hole density in this case only these four subbands contain holes.

It is worth noting that the wave functions for two of these four subbands are usually localized to the region near the left interface and the other two wave functions are localized at the right interface. In Fig. 4 we have drawn the wave functions at $k_\parallel = 0$, where each level has a two-fold degeneracy and the wave function is the same for both spin directions. For $k_\parallel = 0$ the wave functions have pure heavy-hole or light-hole character but for finite $k_\parallel$ the wave functions have both a heavy-hole and a light-hole component. The wave functions for the second and third subbands in Fig. 5 change with the wave vector $k_\parallel$ in a remarkable way. For small $k_\parallel$ the wave function of the second subband is localized near the AlAs interface while that of the third subband is at the other interface. For large values of $k_\parallel$ this is reversed and for $k_\parallel \approx 5 \times 10^5$ cm^{-1} both the wave functions are extended over both interface regions. This is a very interesting case of "wave function engineering" where the extension of the wave functions can be modified by a judicious choice of well width, spacer layer widths and dopant concentrations.

After having determined the band bending due to the transfer of holes self-consistently we use this potential to determine the energy levels and wave functions for the electrons. The numerical calculation of the electron states is performed in the approximation of parabolic subband dispersions.

3. RAMAN MEASUREMENTS ON MODULATION DOPED QUANTUM WELLS

Experimental data on the wavefunctions and the electron energies are extracted from around 150 separate Raman spectra for each sample. With a conventional single channel Raman system such a series of Raman spectra would take a prohibitive amount of time. Therefore it was essential to measure with an optical multichannel system. We used an optical multi-channel analyzer (OMA) system consisting of a position sensitive detector (ITT, Mepsicron F4146M) combined with a triple grating monochromator (Spex Triplemate 60cm focal length with 600/mm gratings).

The samples were mounted in a cryostat on a cold finger held at 4K in a He gas atmosphere and excited by an Ar ion laser pumped CW dye laser with the dyes DCM, pyridine I and pyridine II depending on the wavelength region. The spontaneous emission from the dye was carefully removed by an interference filter which was placed before the samples.

Under relatively high excitation intensity (typically 100 mW), the Raman resonance profile is influenced by the photoexcited carriers. The luminescence peak also shifts to higher energy continuously with the laser power, which may be due to the effect of the photo-created carriers. Therefore all the measurements are done with less than 25 mW which corresponds roughly to 12 W/cm^2. The Raman resonance profiles were corrected for the response of the detection system, for absorption in the sample and for reflection at the sample interfaces.

Figure 6 shows a series of Raman spectra for sample A, taken for parallel polarisations at various laser frequencies. Each spectrum shows firstly a peak at 0 meV due to the attenuated elastically scattered laser light, secondly a broad feature between 5 and 8 meV assigned to hole intersubband excitations and a Raman signal near 36 meV due to GaAs LO phonons, confined in the quantum wells. Since the present quantum wells are so thick, this phonon is practically a bulk LO phonon. AlAs like LO phonons cannot be seen using the present laser energies. In Figure 6 the low energy side of the band is cut off at around 3.5 meV (vertical line in Figure 6) by the window defined by the filter stage monochromator. Thus the maximum of the hole intersubband excitation spectrum may be hidden below this experimental cut-off. The vertical dashed lines in Figure 6 mark the hole intersubband excitations.

4. RAMAN RESONANCE PROFILES OF THE LO PHONON BAND AND OF THE HOLE INTERSUBBAND EXCITATIONS

Figures 7a and 7b show the Raman resonance profiles for sample A from which we determine the electron subband energies, the conduction band well non-parabolicity and the Fourier components of the hole wavefunction. In Figure 8 we show the resonance profile for the LO phonon for sample B. The Raman resonance profiles show the intensity of the Raman scattered light at constant Raman shift as a function of dye laser energy. Each resonance profile curve is extracted from around 150 or more Raman spectra. In the case of the phonons we plot the integral over the Raman line after background subtraction, while in the case of the hole intersubband excitation we simply plot the integral between 4 and 10 meV. The Raman resonance profiles shown in Figure 7a, 7b and 8 show a series of very pronounced sharp peaks of varying heights. Clearly, a lot of information is contained in the position and the height of these peaks. The present section 4 and the following two sections 5 and 6 will show how we extract this information: The resonant peaks correspond to the resonant denominators in the expressions for the Raman intensity, while their height reflects the transition matrix elements, involving the wavefunctions in the well.

We note immediately that the spacing of the peaks increases with increasing energy - i.e. the peaks labeled 8 and 9 in Fig. 7a and 7b are closer spaced than the peaks labelled 18 and 19, for example. This fact finds its natural explanation in the assignment of the resonance peaks to the electron levels in the conduction band well - since the energies E_n of the bound levels in a square well are proportional to n^2. The varying spacing between levels allows to establish a one to one correspondence between the resonance peaks of the LO phonon (Figure 7a) and the resonance peaks of the hole intersubband excitation (Figure 7b). Thus by comparing Figures 7a and 7b it is easily proven that those peaks in Figures 7a and 7b labelled with the same number belong to the same electronic subband level.

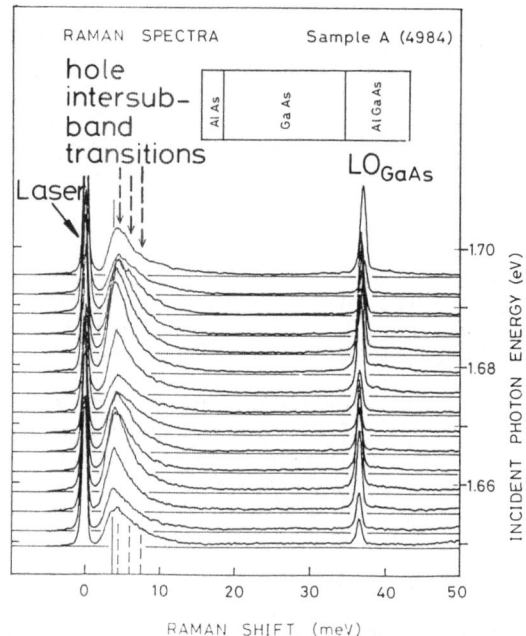

Figure 6   *Raman spectra for sample A taken at various laser photon energies. The curves are shifted vertically according to the incident photon energy, which is shown on the right hand side of the Figure. The full vertical line shows the position of the edge of the filter monochromator window. The insert shows one period of the superlattice.*

We will first discuss the hole intersubband excitations in detail. The resonance curve of this signal is shown in Figure 7b. The structures in this Figure are assigned to Raman processes due to transitions between hole levels and the conduction subband levels. (There is one exception: the peak at 1.87 eV does not correspond to an electron subband but it is due to luminescence superimposed onto the Raman band). The underlying Raman process is sketched in the inset of Figure 7b: the incoming light excites an electron from a valence subband state to an intermediate state, from where the electron recombines with a hole state higher than the original state. The net effect of the process is the emission of a Raman photon with an energy lower than the incident laser photon and the creation of a hole subband excitation. Whenever the intermediate, virtual electron state falls close to a conduction band well state, resonance occurs. The peaks in the Raman profile (except the one at 1.87 eV) are assigned to transitions to the conduction subband levels. Because of the very large well width, the hole wavefunctions are attracted towards the ionized acceptors in

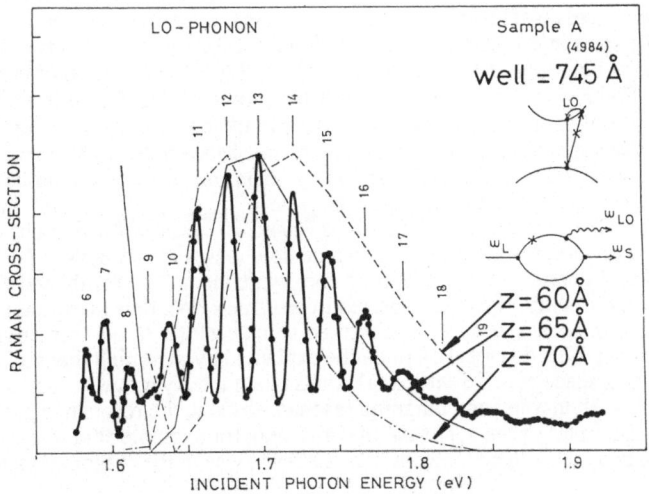

Figure 7a Resonance profile of the LO phonon Raman peak in sample A. The thick full line is meant to guide the eye. The vertical bars indicate the calculated energies of the resonance peaks. The chained, solid and dashed curves show the envelope functions calculated for three different hole wavefunctions. The thin full curve shows the best fit (see text).

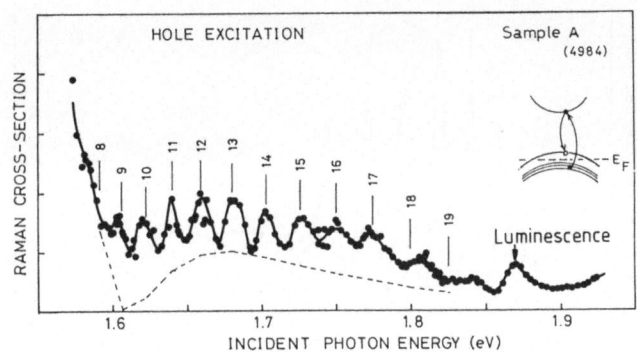

Figure 7b Resonance profile of the hole intersubband excitation near 5 meV in sample A. The vertical bars with numbers indicate the calculated energies of the resonance peaks. The dashed curve shows a calculated envelope for the height of the resonance peaks (see text).

the barrier by the electrostatic forces (see Figure 4). Since the doping density is low the Fermi level is very close to the top of the highest valence subbands as shown in Figure 5. The initial state of the Raman transition is therefore close to the top of the highest four valence bands whereas the final states are assumed to be closely spaced subbands, situated 4 and 10 meV from the top of the highest valence bands.

As pointed out above there is a unique one-to-one correspondence of the peaks in the resonance profile of the LO phonon (shown in Figure 7a) and the resonance peaks of the hole intersubband excitations (shown in Figure 7b). All peaks in the LO phonon resonance profile occur exactly 17.5

meV higher than the corresponding peaks in the resonance profile of the hole excitation band. This fact means that the processes relevant for the Raman resonance of the hole band and of the LO phonon involve the same intermediate states. Raman resonance can occur in principle at the ingoing photon energy, at the outgoing photon energy, or at both. From examination of the resonance profiles shown in Figures 7a , 7b and 8 and also those shown in Ref. 4 it follows, that we have resonance predominantly either at the ingoing or at the outgoing energy but not at both. If the resonance of the hole intersubband excitations and the resonance of the LO phonon Raman signals are both outgoing resonances, the two corresponding resonance profiles should be shifted by the difference $\Delta\omega$ of the two Raman signals: $\Delta\omega = \omega_{LO} - \omega_{\text{hole intersubband exc.}} = 36.4$ meV - 7 meV $\approx$ 29 meV. Comparing the peak positions in Figures 7a and 7b reveales a difference of the peak positions of the LO phonon resonance peaks and the hole intersubband peaks of 17.5 meV. If the resonance of the LO Raman signal is an ingoing one, while the hole intersubband excitation is an outgoing resonance, this shift would be expected to be -7meV. We thus interpret the resonance peaks as outgoing resonances, because the observed value is closer to the former value, although not in exact agreement. This discrepancy might be explained by assuming that the transitions occur at different points in k space at different spatial points along the layer due to spatial inhomogeneities. If both ingoing and outgoing resonances exist, the resonance peaks should occur in pairs, i.e. the counterpart of the 1.721 eV (n = 14) peak for example should appear at 1.685 eV. The fact that no peak appears there indicates that the resonance at the outgoing photon energy dominates.

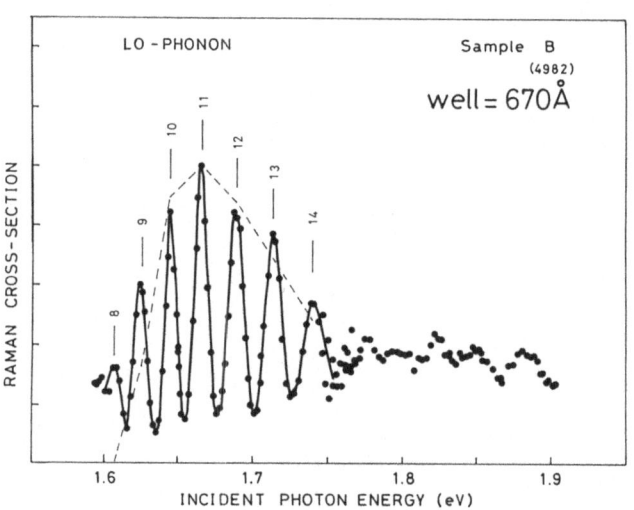

Figure 8 *Resonance profile of the LO phonon peak in sample B. The vertical bars indicate the calculated energies and the dashed curve shows the calculated envelope of the resonance peaks.*

The results for the resonance profiles of the LO phonon and the hole subband resonances for sample B are exactly analogous to the results for sample A. In particular, the peaks in the LO phonon resonance profile also occur exactly 17.5 meV higher than the peaks in the hole resonance profile. We only show the resonance profile of the LO phonon Raman signal for sample B here in Figure 8. Note that the heights of the resonance peaks show a different envelope for sample B than for sample A. This fact will find its explanation in section 6.

5. ELECTRON SUBBAND ENERGIES AND CONDUCTION BAND NONPARABOLICITY

Our previous discussion has shown that the peaks in the resonance profiles in Figures 7a, 7b and 8 arrise from resonance of the outgoing Raman photon with an excitonic energy of the quantum well. Because only a narrow range of hole energies contributes, the series of peaks in the resonance curves is due to the subband structure of the conduction band well.

The insert of Figure 9 shows the situation: the electronic subbands in the deep and wide conduction band well are very close to those of a square well of the appropriate depth. The holes on the other hand are localized close to the edges of the well - the holes are attracted by the ionized acceptors in the barrier toward the edge of the well. In Figure 9 we plot the experimental values for the conduction band levels, determined from the peaks in the resonance profiles, against the k-values determined by solving the square well problem for the conduction band well in a simple particle-in-a-box model. The experimental points show a clear deviation from the parabolic behaviour (dashed line). Thus our experiment determines directly the nonparabolicity of the conduction band in our well.

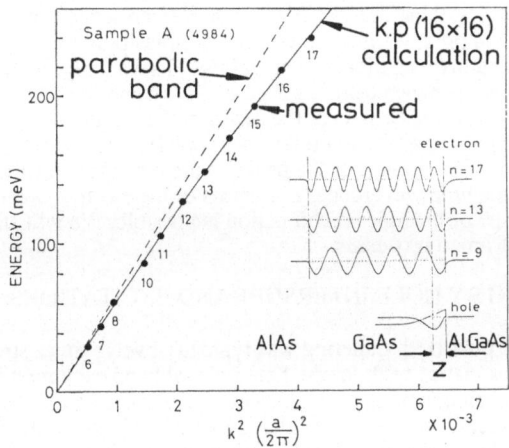

Figure 9   *Experimental conduction subband energies (full points) as a function of squared wavevector k, as obtained from solutions of the square well problem. The dashed line corresponds to a parabolic band with a mass of 0.0665 and the solid curve corresponds to binding energies calculated using the bulk effective masses from a 16x16 k.p calculation.*

The wavefunctions of the electron can be approximated by the solutions of the square well problem. The positions of the outgoing resonance peaks of the LO phonon can be written as,

$$E = E_g + E_{el} + \omega_{LO} + \varepsilon \qquad [5]$$

where E_g is the band gap of bulk GaAs, E_{el} is the energy of the levels in the quantum well measured from the bottom of the conduction band, $\omega_{LO} = 36.4$ meV is the LO phonon energy. In the case of the hole intersubband excitation, ω_{LO} should be replaced by the corresponding hole excitation energy. ε involves the contributions from (a) the exciton binding energy, (b) the hole subband energy, (c) the kinetic energy of the holes and of the electrons in the direction parallel to the layers at the Fermi wavevector of the 2D hole gas and (d) the band bending. The exciton binding energy should be less than the bulk value of - 5 meV, because the electrons and the holes are spatially separated. There is an additional slight reduction because of the screening by the holes. The contribution (c) is estimated to be around 5 meV for the present doping level of the hole. The band bending is calculated to be 10 meV (see Fig. 4). We calculate the energy levels in the quantum well in terms of the effective mass approximation. To include the non-parabolicity in our calculation we use the (100) dispersion of the lowest conduction band in GaAs calculated by Fasol[19] using a 16×16 k p Hamiltonian, based on the calculation described by Rössler[20]. After adjustment of the well width d and the constant ε, we obtained the positions of the outgoing resonance which are shown by vertical bars in Figures 7a, 7b and 8. The best parameters are listed in Table I. A more detailed analysis of non-parabolicity effects based on calculational methods of Ref. 21 will be published shortly.

The energies resulting from this calculation are shown as the full line in Figure 9. The agreement of the calculated values with the experimental results is excellent (Figures 7a and 7b) up to $n = 17$. For sample B we obtained equally good agreement up to $n = 14$ as indicated in Figure 8. The best parameters for d and ε are shown in the column 6 and 7 of Table I. Thus the positions of the peaks in the Raman resonance curve yield the conduction band non-parabolicity.

6. FOURIER DETERMINATION OF THE HOLE WAVEFUNCTION

Since several initial hole subbands contribute to the Raman scattering and since the hole wavefunctions also depend on k, we treat here only a simple approximation with a single initial hole state. A more detailed investigation will be published separately. In the present section we interpret the height of the resonance peaks in Figures 7a, 7b and 8 based on this simple approximation. We show that the heights of the peaks in the LO phonon resonance profiles are determined by the overlap integrals between the hole wavefunction and the relevant electron wavefunctions in the well. The heights of the peaks in the resonance profile therefore measure the Fourier components of the hole wavefunction in terms of the electron wavefunctions. Thus a determination of the shape of the hole wavefunction is possible. We will start this section by discussing the relevant Raman processes.

RAMAN SCATTERING BY HOLE INTERSUBBAND EXCITATIONS

In back scattering the polarized scattering was typically twelve times stronger than the depolarized one at the resonance peak for the hole intersubband excitation band. This suggests that scattering by spin density fluctuations is less important than that due to charge density fluctuations. The proposed Raman process for the hole excitation band is shown in the insert of Figure 7b. The intensity of the resonant Raman scattering due to electronic excitations in the quantum wells has been discussed by Abstreiter et. al.[5] and is given by:

$$I \sim |\sum_{m,n,k,i} \frac{<h_m \mid H_{eR} \mid n\,k><n\,k \mid H_{eR} \mid h_i>}{E(n,k) - \omega_L}|^2 \qquad [6]$$

where H_{eR} is the Hamiltonian for the electron radiation interaction, and ω_L is the laser frequency. $<h_i \mid$ and $<h_m \mid$ stand for the initial (i-th) and the final (m-th) hole state in the well respectively and $<n\,k \mid$ stands for the intermediate state in the n-th conduction subband with a wavevector k parallel to the layer. $E(n,k)$ is the energy of the electron in the n-th subband measured from the initial hole state. We applied a summation over i and m because the initial and the final states involve several subbands. In back scattering geometry, the electric vector involved in H_{eR} lies in a plane parallel to the layer. Therefore the integration $<h_m \mid H_{eR} \mid n\,k>$ in this plane is independent of m and n. Then the relative intensity of the resonant enhancement for a peak at $\omega_L = E(n,k)$ is proportional to the overlap integral,

$$I \sim |\sum_{m,i} <h_m \mid n><n \mid h_i>|^2 \qquad [7]$$

RAMAN SCATTERING BY LO PHONONS

LO phonon Raman scattering can occur by the deformation potential mechanism and by Fröhlich interaction and it can also be defect induced. In our scattering geometry, where the electric vector of the incident light is parallel to the (110) axis of the crystal within the layer of the MQW, the polarized scattering was 13 times stronger than the depolarized one at the resonance peak. The interaction induced by the electric field in the inversion layer might cause also polarized scattering. However, the strong outgoing resonance relative to the ingoing one can be better understood in terms of the impurity assisted Fröhlich mechanism, as discussed by Menéndez and Cardona[22] for bulk GaAs. This process is shown diagrammatically in k-space in Figure 7a, where the cross indicates the impurity scattering which breaks momentum conservation. Under these conditions the outgoing resonance can be typically twice as large as the ingoing one in GaAs[23]. The intensity for scattering by LO phonons is given by

$$I_n \sim |\sum_{k,k'k''} \frac{<h,k\,|H_{eR}|\,n,k><n,k\,|H_{ei}|\,n,k'><n,k'\,|H_{eL}|\,n,k''><n,k''\,|H_{eR}|\,h,k>}{(E(n,k'')-\omega_L+\omega_{LO})\,(E(n,k')-\omega_L)\,(E(n,k)-\omega_L)}|^2 \qquad [8]$$

Here, the integral $<n\,k'\,|H_{eL}|\,n,k''>$ denotes the intraband scattering of an electron by the LO phonon within the 2D space and it can be assumed to be independent of n. Similarly the integral $<n,k\,|H_{ei}|\,n,k'>$, which denotes the elastic scattering by impurities or imperfections will be assumed to be approximately independent of n. Thus the relative intensity of resonance for different n can be written as

$$I \sim |\sum_i <h_i\,|\,n><n\,|\,h_i>|^2 \qquad [9]$$

where the matrix elements $<h_i\,k\,|H_{eR}|\,n\,k>$ again have been assumed to be independent of n, as in the case of Equation 7. Using the solution of the square well problem for electrons, we evaluated the overlap integrals of Equation 9.

The wavefunctions of the electrons or holes in the inversion layer have been discussed by Stern[24] and reviewed by Ando et. al.[25]. The Airy function, which is an exact solution of a triangular well, is known to be a reasonable approximation, although more accurate wavefunctions may be obtained from self-consistent calculations. We adjusted the strength of the field F in the inversion layer and the distance of the centre of gravity of the hole wavefunction from the interface z as variable parameters so as to obtain the best fit for the envelope of the experimental resonance intensities. The hole will have the possibility to be localized either at the normal interface or at the inverted interface or at both. The calculated envelopes for z = 60 Å, 65 Å and 70 Å are shown in Figure 7a by dashed, solid and chained lines, respectively, where we assumed the holes to be localized near the GaAlAs side. The maximum and the minimum positions of the envelopes depend on the parameter F, but less sensitively. We obtained the best fit for z = 65 Å and F = 0.15 meV/Å (solid curve in Figure 7a) in the case of sample A. The best parameters are listed in Table I. The two main features (i) the largest amplitude at n = 13 and (ii) the minimum at n = 9 are well reproduced by this calculation. The physical situation is shown in Figure 9. Since the electron wavefunction for n = 13 has a large overlap with the hole wavefunction, the transition probability is large. On the other hand, for n = 9 and for n = 17, the electron wavefunctions have a node at the centre of the hole wavefunction, therefore the transition probability is small. As shown in Figure 8, the amplitude maximum appears at n = 11 for sample B at an energy lower than that for sample A. This implies that the hole wavefunction is further away from the interface. The optimum z for sample B is 71.5 Å for the AlGaAs side which is larger than the corresponding value for sample A as shown in Table I. This result is supported by the fact that the doping level is lower for sample B. This simple analysis is in fair agreement with the selfconsistent calculations.

The Raman intensity due to hole intersubband excitations (Equations 6 and 7) is more difficult to evaluate because the wavefunctions of the higher hole states cannot be approximated by Airy functions in the same triangular well as that assumed for the highest states. Furthermore, the light hole - heavy hole interaction makes the dispersion of the holes very complicated and the integration in the 2D space difficult. The results using the full band calculations of the valence subband will be published separately. Here we assume that in Equation 7

$$\sum_{m \neq 1} <h_m \mid n> = \text{const.} \qquad [10]$$

because the contributions from different m's can be averaged out to give a smooth function when summing up. Then the hole band intensity should be proportional to

$$\mid \sum_i <n \mid h_i> \mid^2 \qquad [11]$$

which is the square root of Equation 9. This function is shown by a dashed line in Figure 7b; it is flatter than the envelope of Figure 7a but reproduces the observed behaviour qualitatively.

In conclusion, Equation 9 shows that the height of the peaks in the Raman resonance profiles for LO phonon scattering are proportional to the fourth power of the overlap integral of the components perpendicular to the well of the uppermost hole wavefunctions with the relevant electron subband wavefunction. Therefore the peak heights are proportional to the fourth power of the Fourier component of the hole wavefunction in terms of the conduction band subband wavefunctions. The envelopes for sample A (Figure 7a) and sample B (Figure 8) differ, because the overlap integrals in these two wells are different due to the different well width and carrier density.

In our case of very broad wells in principle the 18 Fourier components corresponding to the 18 bound electron wavefunctions can be determined by Raman scattering. In addition overlap integrals with higher electron wavefunctions can also be determined. Once this is done the hole wavefunction can in principle be synthesized by summing up the Fourier series. There are unfortunately two obstacles to do this. The first obstacle is in our case, that we could not determine the Fourier components below $n = 6$, because strong luminescence at the lowest exciton energy inhibits Raman measurements in this region. The second obstacle is that we are only able to determine the amplitude and not the phase of the Fourier components and in addition the continuum states should also be included.

7. CONCLUSIONS

We study the electronic structure of p-type modulation doped quantum wells by resonant Raman scattering and with self-consistent envelope function calculations. We show that resonant Raman scattering provides a technique to determine densely spaced electronic levels in quantum wells with high precision and to obtain direct experimental information on the hole wavefunctions.

The resonance profiles of the LO phonon and the hole intersubband Raman scattering shows a large number of very pronounced sharp peaks. The position of these peaks correspond to excitonic transitions in the quantum wells. The intensity of the peaks in the Raman resonance profiles is related to the transition matrix elements. With a few plausible assumptions the Fourier components of the hole wavefunction in terms of the conduction band wavefunction can be extracted from these peak heights. The shape of the hole wavefunctions determined experimentally agrees reasonably well with the calculated shapes.

Using detailed analysis of the spectra we determine densely spaced electron binding energies up to high levels (up to $n = 18$). From these energies we determine the conduction band non-parabolicity.

ACKNOWLEDGEMENTS

The authors are extremely grateful to M. Cardona for valuable discussion, support and encouragement of this work. We are indebted to A. Fischer for sample growth and to H. Hirt, M. Siemers and P. Wurster for expert technical assistance. Financial support of T. S. by the Alexander von Humboldt Stiftung and of U. E. by Trinity College, Cambridge, is gratefully acknowledged.

REFERENCES

1. T. Suemoto, G. Fasol and K. Ploog, Phys. Rev. B 37 , 6397 (1988)
2. J. E. Zucker, A. Pinczuk, D. S. Chemla, A. Gossard and W. Wiegmann, Phys. Rev. Lett. 51, 1293 (1983)
3. J. E. Zucker, A. Pinczuk, D. S. Chemla, A. Gossard and W. Wiegmann, Phys. Rev. 29, 7065 (1984)
4. T. Suemoto, G. Fasol and K. Ploog, Phys. Rev. $B34$, 6034 (1986)
5. G. Abstreiter, M. Cardona and A. Pinczuk, in Light Scattering in Solids IV, Topics in Applied Physics Series, Vol. 54, edited by M. Cardona and G. Güntherodt (Springer Verlag, Berlin, 1984), p. 5
6. A. Pinczuk, H. L. Störmer, A. C. Gossard and W. Wiegmann, in Proceedings of the 17th International Conference on the Physics of Semiconductors, ed. by J. D. Chadi and W. Harrison (Springer Verlag, New York, 1985), p. 329
7. A. Pinczuk, D. Heiman, R. Sooryakumar, A. C. Gossard and W. Wiegmann, Surf. Science 170, 573 (1986)
8. D. Heiman, A. Pinczuk, A. C. Gossard, A. Fasolino and M. Altarelli, in Proceedings of the 18th International Conference on the Physics of Semiconductors, ed. by O. Engström, (World Scientific, Singapore, 1987), p. 617
9. J.M.Luttinger and W.Kohn, Phys.Rev. 97, 869 (1955) and J.M.Luttinger, Phys.Rev. 102, 1030 (1956).
10. M.Altarelli, Phys.Rev.B 28, 842 (1983).
11. See for example: M.Altarelli, U.Ekenberg and A.Fasolino, Phys.Rev.B 32, 5138 (1985).
12. D.A.Broido and L.J.Sham, Phys.Rev.B 31, 888 (1985).
13. E.Bangert and G.Landwehr, Superl. Microstr. 1, 363 (1985).
14. T.Ando, J.Phys.Soc.Jpn. 54, 1528 (1985).
15. U.Ekenberg and M.Altarelli, Phys.Rev.B 32, 3712 (1985).
16. J.P.Eisenstein, H.L.Störmer, V.Narayanamurti, A.C.Gossard and W.Wiegmann, Phys.Rev.Lett. 53, 2579 (1984).
17. Y.Iye, E.E.Mendez, W.I.Wang and L.Esaki, Phys.Rev.B 33, 5854 (1986).
18. U.Ekenberg, to be published.
19. G. Fasol, to be published
20. U. Rössler, Solid State Commun. 49, 943 (1984)
21. U.Ekenberg, Phys.Rev.B 36, 6152 (1987).
22. J. Menéndez and M. Cardona, Phys. Rev. $B31$, 3696 (1985)
23. W. Kauschke and M. Cardona, Phys. Rev. 33, 5473 (1986)
24. F. Stern, Phys. Rev. $B5$, 4891 (1972)
25. T. Ando, A. B. Fowler and F. Stern, Rev. Mod. Phys. 54, 437 (1982)

OPTICAL PROPERTIES OF SEMICONDUCTOR SUPERLATTICES

Yia-Chung Chang and Hanyou Chu

Department of Physics
University of Illinois at Urbana-Champaign
Urbana, Illinois 61801

G. D. Sanders

Universal Energy Systems
4401 Dayton-Xenia Road, Dayton, Ohio 45432

INTRODUCTION

Semiconductor quantum wells and superlattices[1] have received growing interest in recent years. Optical measurements including photoabsorption, photoluminescence, and Raman scattering are widely adopted for probing the electronic states of these heterostructures. Using a simple quantum mechanical model which contains essentially a particle in a one-dimensional square-well potential (particle-in-a-box model)[2], one can obtain a fairly accurate description of the energy levels in a GaAs-$Al_xGa_{1-x}As$ quantum well with well size between 50 Å and 300 Å. This model predicts a selection rule for the inter-band optical transitions which requires the difference in principal quantum numbers of the initial hole state and the final electron state in a quantum well to be zero, i.e. $\Delta n = 0$. Indeed, most experimental data indicate that $\Delta n = 0$ transitions are at least an order of magnitude stronger than the other transitions which violate this selection rule.

Recent studies[3,4] of the electronic and optical properties of semiconductor quantum wells (or superlattices) have revealed that the mixing of heavy and light hole components (valence band mixing) in the quantum well (or superlattice) states can lead to $\Delta n \neq 0$ (forbidden) interband transitions with strengths much larger than those expected from the simple particle-in-the-box model. The most pronounced $\Delta n \neq 0$ transition is associated with the exciton involving the first light hole subband and the second conduction subband. This is because the first light-hole subband interacts strongly with the second heavy hole subband. Good agreement between theoretical predictions and experimental data for the absorption spectra of GaAs-Al_xGa_{1-x}As quantum wells have recently been reported[4].

There are added complications for the excitonic effect in semiconductor superlattices. At the mini-zone boundary ($q = q_{max}$), the lowest subband energy is a maximum along z , but a minimum along x or y. Thus, we have an M_1 saddle point there. In the past, excitonic effects associated with the M_1 saddle point in bulk semiconductors have attracted a great deal of interest both theoretically and experimentally[5-10]. Although qualitative understanding of this phenomenon can be obtained via a contact-potential model[6], quantitative calculations for the Coulombic potential are desired. Chu and Chang[11] have recently performed quantitative calculations for the line shapes of photoabsorption associated with saddle-point excitons in a tight-binding model for bulk semiconductors and in a Kronig-Penney model for semiconductor superlattices, including the electron-hole Coulomb interaction. It is found that when the width of superlattice band dispersion is comparable to the exciton binding energy, some prominent structures appear which can be interpreted as saddle-point exciton resonances.

ELECTRONIC STRUCTURES OF GaAs-Al_xGa_{1-x}As SUPERLATTICES

Since GaAs-Al_xGa_{1-x}As superlattices are made of direct semiconductors, the conduction-band states of interest are mainly derived from the Γ valley. In this case it is often sufficient to use the simple effective-mass method[2] to calculate the energy dispersion and envelope functions. In this method, the conduction subband state of a quantum well can be written as

$$
\psi^e_{n,\vec{k}_\parallel} = \sum_{k_z} e^{i\vec{k}.\vec{r}} f_{n,\vec{k}_\parallel}(k_z) |c,\vec{k}>,
$$

(1)

where $|c, \vec{k} >$ is the cell-periodic function of the conduction band state at $\vec{k}$. The Fourier transform of $f_{n,\vec{k}_\parallel}(k_z)$ satisfies the simple effective-mass equation:

$$\left[\frac{\hbar^2}{2m_e(z)}(k_{\parallel}^2 - \partial^2/\partial z^2) + V_e(z)\right] f_{n,\vec{k}_{\parallel}}(z) = E_n^e(\vec{k}_{\parallel}) f_{n,\vec{k}_{\parallel}}(z). \tag{2}$$

where $V_e(z_e)$ is the superlattice potential for the electron and $m_e(z)$ equals the effective electron mass in GaAs or $Al_xGa_{1-x}As$ depending on where z is located. It is easy to see that in this approximation, the envelope function f is independent of $\vec{k}_{\parallel}$ and we shall drop the $\vec{k}_{\parallel}$ index for f from now on.

The valence band structures are much more complicated. The valence subband structures are highly nonparabolic even at wave vectors very close to the zone center (within 1% of the superlattice Brillouin zone along directions parallel to the interface) and the valence band states contain substantial admixtures of heavy and light hole components.[3,4,14] Such a "valence band mixing" phenomenon has nontrivial and very interesting effects on the optical properties. A number of theoretical methods have been developed to treat the valence-band states in superlattices. These include the tight-binding method[12], the multi-band effective mass method[4], the envelope function $(\vec{k} \cdot \vec{p})$ method[13-15], and the bond-orbital method.[16] All these methods produce almost identical results, provided that the band parameters used give the same bulk heavy and light hole effective masses. The tight-binding method is the most complete, but it requires rather heavy computation. The multi-band effective mass method is only suitable for superlattices with thick barriers (ie. in the quantum well region). The envelope function method is the most convenient for calculating superlattice states, but it is somewhat clumsy if one wishes to take into account the boundary conditions properly with different Luttinger parameters in the two different materials which comprise the superlattice. The bond-orbital method has advantages over the above methods, except that it can not be reduced to a simple spherical model as the envelope-function method can. For studying the optical properties of $GaAs-Al_xGa_{1-x}As$ quantum wells, we shall use the multi-band effective-mass method for calculating the valence subband structures.

The m-th valence subband state can be written as

$$\psi_{m,\vec{k}_{\parallel}}^h = \sum_{v,k_z} e^{i\vec{k}\cdot\vec{r}} g_{m,\vec{k}_{\parallel}}^v(k_z)|v,\vec{k}\rangle, \tag{3}$$

where $|v,\vec{k}\rangle$'s are the cell-periodic functions of the well-material valence-band states correct to the first order in the $(\vec{k} \cdot \vec{p})$ perturbation[17]. The envelope function $g^v_{m,\vec{k}_{\parallel}}(k_z)$ satisfies a multi-band effective mass equation in $\vec{k}$-space, viz.

$$\sum_{v'} [H^{(0)}_{v,v'}(\vec{k}_{\parallel},k_z) - E^h_m(\vec{k}_{\parallel})\delta_{v,v'}]g^{v'}_{m,\vec{k}_{\parallel}}(k_z) + \sum_{k'_z} < k_z|V_h(z)|k'_z>g^v_{m,\vec{k}_{\parallel}}(k_z) = 0, \quad (4)$$

where $V_h(z)$ is the quantum well potential seen by the hole and $H^{(0)}_{v,v'}(\vec{k}_{\parallel},k_z)$ are matrix elements of the Luttinger-Kohn Hamiltonian[18] for describing the bulk valence band structure. Here we have ignored the mismatch between the valence band parameters for the well and barrier materials. Because the wave functions of interest are mostly confined in the well material, the approximation used is justified. Both Eq. (1) and Eq.(2) can be solved by a variational method in which the real-space envelope functions are written in terms of linear combinations of Gaussian-type orbitals[4]. Detailed results for the valence subband structures can be found in Ref. 4. The material parameters for the bulk band structures and valence-band offset used are listed in Table I. As an illustration we show in Fig. 1 the valence-subband structure calculated with the multi-band effective mass method for a number of GaAs-$Al_{0.25}Ga_{0.75}As$ quantum wells.

Table I. Material parameters used in the calculation.

Substance	m_C/m_0	γ_1	γ_2	γ_3	$E_G(eV)$	$E_v(eV)$
GaAs	0.0665	6.85	2.1	2.9	1.518	0
AlAs	0.15	4.04	0.78	1.57	3.1	-0.46

PHOTOABSORPTION OF GaAs-$Al_xGa_{1-x}As$ QUANTUM WELLS

In the discussion of absorption coefficients, it is important to understand the electronic structures of excitons. Green and Bajaj[19] have studied the exciton states in a simple variational calculation in which the valence-band mixing is ignored. Sanders and Chang[4] and Broido and Sham[15] have considered the effect of band mixing on the excitonic states and their contributions to the photoabsorption.

In a two-band approximation, the excitonic states associated with the n-th conduction subband and the m-th valence subband can be written as

$$\psi_x^{mm} = \sum_{\vec{k}_\parallel} G_{mm}(\vec{k}_\parallel) \psi_{n,\vec{k}_\parallel}^e \psi_{m,\vec{k}_\parallel}^h.$$ (5)

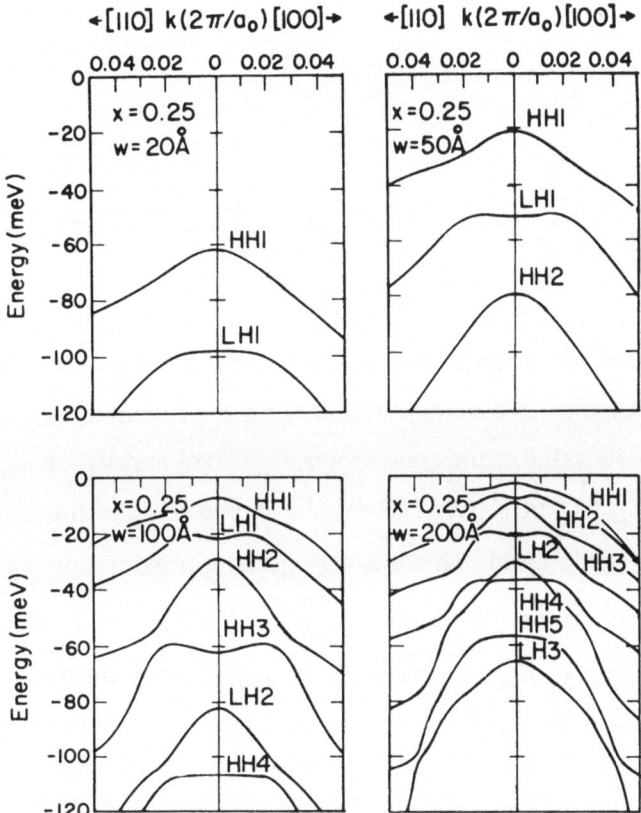

Fig. 1. Valence-subband structures of a number of GaAs-Al$_{0.25}$Ga$_{0.75}$As quantum wells.

Here we have ignored the interaction between excitonic states associated with different pairs of electron and hole subbands. This approximation is valid when the energy separation between excitons derived from different pairs of subbands is large compared to the exciton binding energy. Substituting the expansion into the Schrödinger equation yields an effective-mass equation for the exciton envelope function:

$$[E_n^e(\vec{k}_{\parallel}) - E_m^{ex}]G_m(\vec{k}_{\parallel}) - \sum_{\vec{k}'_{\parallel}} V_{mm}(\vec{k}_{\parallel} - \vec{k}'_{\parallel})G_{mm}(\vec{k}'_{\parallel}) = 0. \tag{6}$$

with

$$V_{mm}(\vec{k}_{\parallel} - \vec{k}'_{\parallel}) = -\frac{4\pi e^2}{\varepsilon} \sum_q \frac{F_n(q)G_m(\vec{k}'_{\parallel}, \vec{k}_{\parallel}, q)}{|\vec{k}_{\parallel} - \vec{k}'_{\parallel}|^2 + q^2}, \tag{7}$$

where

$$F_n(q) = \sum_{k_z} f_n^*(k_z) f_n(k_z - q),$$

$$G_m(\vec{k}'_{\parallel}, \vec{k}_{\parallel}, q) = \sum_{k_z, v} (g_{m,\vec{k}'_{\parallel}}^v(k_z))^* g_{m\vec{k}_{\parallel}}^v(k_z - q),$$

and ε is the static dielectric constant. In order to simplify the calculation, Sanders and Chang have made the following approximations: (1) $G_m(\vec{k}'_{\parallel}, \vec{k}_{\parallel}, q)$ is replaced by its value at the zone center, $G_m(0,0,q)$ and V_{nm} becomes a function of $|\vec{k}_{\parallel} - \vec{k}'_{\parallel}|$ only. (2) The angular dependence of the expression $E_n^e(\vec{k}_{\parallel}) - E_m^h(\vec{k}_{\parallel})$ is ignored. This is a very good approximation since $E_n^e(\vec{k}_{\parallel})$ has a circular symmetry and it dominates $E_m^h(\vec{k}_{\parallel})$ for all cases of interest. The angular deviation in $E_m^h(\vec{k}_{\parallel})$ is found to be around 20 % for GaAs-Al$_x$Ga$_{1-x}$As quantum wells[3,4]. The problem is then converted into that of a simple quasi 2-d exciton with circular symmetry. Variational methods were used to find the binding energies and oscillator strengths of several prominent excitons in GaAs-Al$_x$Ga$_{1-x}$As quantum wells.

With the excitonic effects included the absorption coefficient is given by[20]

$$\alpha(\hbar\omega) = \frac{C}{\omega} \sum_{mn} \lg_{mn} \Delta_{mn}(\hbar\omega - E_{mn}^{ex})$$

$$+ \sum_{\vec{k}_{\parallel}} h_{mn}(\vec{k}_{\parallel}) |\hat{\varepsilon} \cdot \vec{P}_{nm}(\vec{k}_{\parallel})|^2 \delta(E_n^e(\vec{k}_{\parallel}) - E_m^h(\vec{k}_{\parallel}) - \hbar\omega)], \tag{8}$$

where $C = 4\pi^2 e^2 / n_0 cm_c^2 V$ and $g_{nm} = (2/m_0) |\Sigma_{\vec{k}_{\parallel}} \hat{\varepsilon} \cdot \vec{P}_{nm}(\vec{k}_{\parallel})| 2 \cdot m_0$ is the free electron mass, V is the volume of the crystal, and $\hat{\varepsilon}$ indicates the direction of polarization of the incident radiation. $h_{nm}(\vec{k}_{\parallel})$ is the Coulomb enhancement factor due to the final-state electron-hole interaction. This factor is proportional to the probability of finding the electron and hole at the same point in space, i.e. $h_{nm}(\vec{k}_{\parallel}) \propto |\psi^h_{X,\vec{k}_{\parallel}}(0)|^2$, where $\psi^h_{X,\vec{k}_{\parallel}}(\vec{r})$ is the $\vec{k}_{\parallel}$-th exciton continuum state associated with subbands (n,m). Note that in the presence of the electron-hole Coulomb interaction, $\vec{k}_{\parallel}$ is no longer a good quantum number, but merely a label of the continuum states.

$\Delta_{nm}(E) = (\Gamma_{nm}/\pi)(E^2 + \Gamma^2_{nm})^{-1}$ is a Lorentzian function of half width Γ_{nm}. It was found that the experimental data can be fit reasonably well with the empirical rule $\Gamma_{nm} = \Gamma_0 nn'$, where n and n' are the principal quantum numbers for the electron and heavy- or light- hole quantum well states. Γ_0 is an empirical parameter selected to be around 1 meV. The theoretically predicted absorption spectrum for a 102-Å GaAs-Al$_{0.27}$Ga$_{0.73}$As quantum well for unpolarized light propagating along the growth direction (z) is shown in Fig. 2. The experimental data of Miller et. al.[21] is superposed for comparison. The agreement between the theoretical and experimental results is fairly good. The excitonic effect in conjunction with the valence-band mixing accounted for the strong $\Delta n \neq 0$ forbidden transitions identified as HH2-CB1, HH3-CB1and LH1-CB2.

The theory of Sanders and Chang, however, omitted the angular dependence of the mixing coefficient. This turns out to be an important consideration for the parity-forbidden LH1-CB2 transition. A theory which incorporates the angular dependence into the excitonic states was recently developed by Zhu and Huang[22] and the oscillator strengths of excitonic transitions were analyzed by Zhu.[23] It was shown that the four components of the envelope function in a valence band state (say m) have different angular dependence given explicitly by

$$g^v_{m,\vec{k}_{\parallel}}(k_z) = \bar{g}^{-v}_{m,k_{\parallel}}(k_z) e^{i(1/2-v)\phi} \quad ,$$

where v=3/2,1/2,-1/2,-3/2, and ϕ is the angle between $\vec{k}_{\parallel}$ and the x-axis (the azimuthal angle for $\vec{k}$). Thus in an exciton state, different components of the valence-band envelope function must be associated with different angular momenta. For example, an s-like v = 3/2 (heavy-hole like) envelope function must be coupled to a p-like v =1/2 (light-hole like) envelope function. As a consequence, the parity-forbidden LH1-CB2 excitonic transition must have an orbital angular momentum l=1, as it is coupled to the s-like HH2-CB2 excitonic transition. This calls for a revised calculation of the

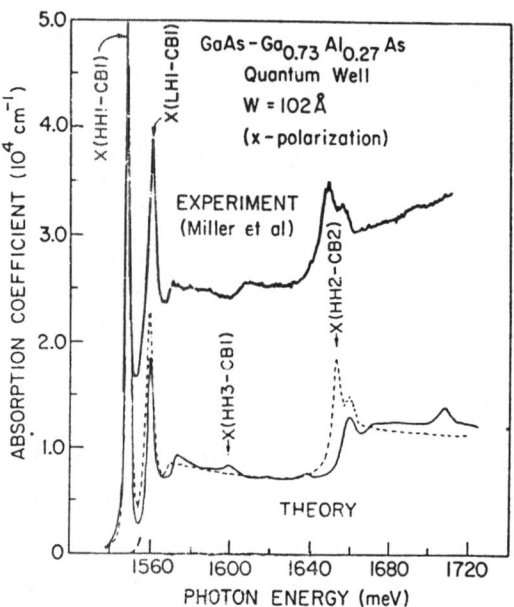

Fig. 2. Theoretical absorption spectrum and experimental excitation spectrum (from Ref. 21) for a 102 Å GaAs-Al$_{0.27}$Ga$_{0.63}$As quantum well.

absorption spectra. Using oscillator strengths for the prominent excitons similar to those obtained by Zhu[23] and the same band-to-band transitions obtained by Sanders and Chang[4], we obtain a revised absorption coefficient for the 102-Å quantum well (with x=0.27). The result is shown as the solid curve in the lower part of Fig. 3 together with the experimental data of Miller et. al.[21] (shown in the upper part). It is interesting to note that the revised absorption spectrum is in much worse agreement with the data. In particular, the parity-forbidden LH1-CB2 (2p) excitonic transition becomes too weak to be noticeable. This is because the 2p state has much weaker binding energy, and

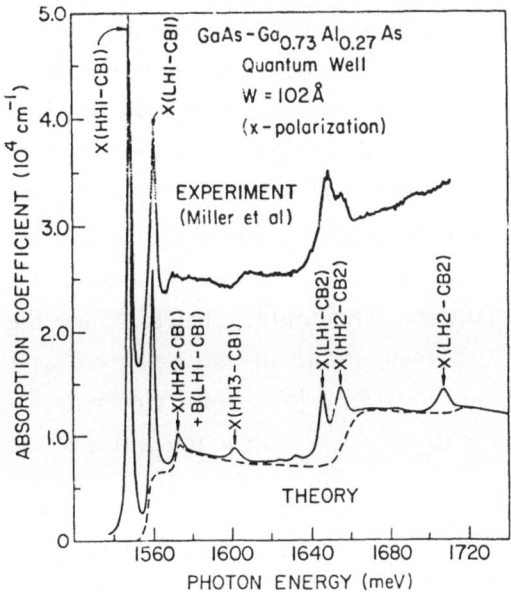

Fig. 3. Theoretical absorption spectra and experimental excitation spectrum (from Ref. 21) for a 102 Å GaAs-Al$_{0.27}$Ga$_{0.63}$As quantum well. Lower solid curve: calculations with the use of Zhu's (Ref. 23) oscillator strengths; dashed curve: calculations including coupling with continuum states.

thus much smaller oscillator strength than one would expect from a 1s state, if they have equally strong mixing with the HH2-CB2 (1s) exciton. Therefore, the doublet structure of the HH2-CB2 transition observed by Miller et. al. can not be satisfactorily explained. We further notice that the oscillator strength of the allowed LH1-CB1(1s) transition calculated by Zhu[23] is systematically lower than the available experimental result for a number of quantum wells by about 50%. All these discrepancies indicate that Zhu's theory for excitons in quantum wells still needs improvement.

We feel that in order to calculate the excitonic transitions properly, one must go beyond the two-band approximation. Several theoretical calculations [4,15,22,23] have attempted to go beyond the two-band approximation by writing the exciton state as a linear combination of several two-band exciton states. The results are not substantially different for quantum wells with well widths less than 200 Å. However, such an expansion ignores the possibility that one particular two-band exciton can be coupled to many excited states and continuum states of other two-band excitons. In fact since the LH1-CB1 (1s) exciton typically sits very close to the onset of the HH1-CB1 continuum, a strong mixing between them is expected. In the next section we describe a new method developed by Chu and Chang[24] for calculating the absorption coefficient of GaAs-Al$_x$Ga$_{1-x}$As superlattices which takes into account the interaction of a two-band exciton with the continuum of other two-band excitons. If we use the theory described in the next section (in a three-band model, including couplings among HH1, LH1, and HH2 states) to calculate the absorption coefficient of the 102-Å GaAs-Al$_{0.27}$Ga$_{0.73}$As quantum well, we obtain the dashed curve in the lower part of Fig. 3. We found that the oscillator strength of the LH1-CB1 transition is now in much better agreement with the experimental data. The doublet structure near the HH2-CB2 transition is also clearly seen, and its line shape agrees well with the data. By examining the symmetry-resolved joint density of states[24], we found that the first peak associated with the doublet structure is due to the LH1-CB2 p-like transition.

SUPERLATTICE EFFECT ON OPTICAL PROPERTIES

Superlattices can be viewed as multiple quantum wells if the barrier material in each superlattice unit cell is sufficiently wide so that interactions between electronic states associated with one well and those with adjacent wells are negligible. For superlattices with narrow barrier width, the energy dispersion for wave vectors along the growth direction (z) is large and its effect on the absorption coefficient is important. One expects the line shape of the absorption spectrum changes gradually from a 3-d character in the ultra-thin barrier case to a quasi 2-d character in the wide-barrier case, and for intermediate barrier widths it is difficult to predict what the absorption spectrum is like, unless a realistic theoretical calculation is done.

We denote the superlattice wave vector $\vec{k}$ as $(\vec{k}_{\|}, q)$ where q is the projection of the wave vector in the growth direction. We define the n-th conduction subband state as

$$\psi_{n,\vec{k}}^{e} = \sum_{k_z} e^{i\vec{k}\cdot\vec{r}} f_n(q+k_z)|c,\vec{k}>.$$

We further define the m-th eigenstate of the v-th diagonal component of the hole Hamiltonian (the zero-th order hole Hamiltonian) at $\vec{k}$ as

$$\psi_{m,v,\vec{k}}^{h0} = \sum_{k_z} e^{i\vec{k}\cdot\vec{r}} g_{m,v}^{0}(q+k_z)|v,\vec{k}>.$$

The corresponding energy eigenvalues are denoted by $E_n^{(e)}(\vec{k})$ and $E_{m,v}^{0}(\vec{k})$ Note that due to the superlattice periodicity in the z direction, k_z in the above summations must be equal to a superlattice reciprocal lattice vector, i.e.

$$k_z = s(\pi/d); \; s = \text{integer},$$

where d is the length of the superlattice unit cell.

In the calculation of absorption coefficient, only excitonic states of zero total momentum needs to be considered. For simplicity, we ignore the coupling of excitonic states associated with different conduction subbands, but keep the coupling of excitonic states associated with different valence subband states. This is justified because the energy separation between conduction subbands is much larger than that between valence subbands. The excitonic states associated with conduction subband n can be written as

$$\psi_X^n = \sum_{m,v} \sum_{\vec{k}} G_{m,v}^n(\vec{k}) e^{i(1/2-v)\phi} \psi_{n,\vec{k}}^{e} \psi_{m,v,\vec{k}}^{h0}, \tag{9}$$

where ϕ is the azimuthal angle of $\vec{k}$. It can be shown that the coefficients G_{mv} $(\vec{k})$ satisfy the following effective-mass equation

$$[E_n^e(\vec{k}) - E_{m,v}^{0}(\vec{k}) - E_n^{ex}]G_{m,v}^n(\vec{k}) + \sum_{m',v',k'} [V_{v,m,m'}^n(\vec{k};\vec{k'})\delta_{v,v'}$$

$$+ \sum_{k_z} H_{v,v'}^{(0)}(\vec{k}_{||},k_z) g_{m,v}^{0}(q+k_z) g_{m',v}^{0}(q+k_z) e^{i(v-v')\phi}\delta_{\vec{k},\vec{k'}}]G_{m',v'}^n(\vec{k'}) = 0, \tag{10}$$

The Coulomb matrix element is given by

$$V_{v,m,m}^{n}(\vec{k};\vec{k}') = \frac{4\pi e^{2}}{\varepsilon} \sum_{s} F_{n}^{e}(q, q' - sK)F_{v,m,m}^{h}(q, q' - sK)\frac{e^{iv(\phi - \phi')}}{|\vec{k}_{\parallel} - \vec{k}'_{\parallel}|^{2} + (q - q' + sK)^{2}} \quad (11)$$

where

$$F_{n}^{e}(q, q') = \sum_{s} f_{n}(q + sK)f_{n}(q' + sK) \quad \text{and} \quad F_{v,m,m}^{h}(q, q') = \sum_{s} g_{m,v}^{0}(q + sK)g_{m,v}^{0}(q' + sK)$$

Here $K = \pi/d$, and s runs through all integers.

We first consider the lowest two-band exciton (HH1-CB1) and ignore the valence-band mixing. In this case, the indices n,m,v can be dropped, and we denote the electron-hole product state at $\vec{k}$ simply $|\vec{k}\rangle$. To solve the Schrödinger equation (10), we expand the wave function Ψ_{X} in terms of linear combinations of a set of basis states defined by

$$\beta_{j} = \sum_{\vec{k} \varepsilon \Delta_{j}} |\vec{k}\rangle / \sqrt{\Delta_{j}}, \quad (12)$$

i.e. $\Psi_{X} = \Sigma_{j} G(\vec{k}_{j}) \beta_{j}$. where Δ_{j} denotes a small volume in $\vec{k}$-space centered at $\vec{k}_{j}$.

Substituting this expansion into the Schrödinger equation for Ψ_{X} immediately leads to a simple eigenvalue problem:

$$\sum_{j'} \overline{H}_{j,j'} G(\vec{k}_{j'}) - EG(\vec{k}_{j}) = 0, \quad (13)$$

with

$$\overline{H}_{j,j'} = \frac{\sum_{k \varepsilon \Delta_{j}} \sum_{k' \varepsilon \Delta_{j'}} H(\vec{k}, \vec{k}')}{\Delta}, \quad (14)$$

where $H(\vec{k}, \vec{k}')$ is the exciton effective-mass Hamiltonian in $\vec{k}$ space. Eq.(13) can then be diagonalized to give the energies and corresponding wave functions for a few low-lying discrete exciton states and a good sampling of the continuum states. The absorption coefficient for interband transitions can be written as:

$$\alpha(\hbar\omega) = \frac{C}{\omega} \sum_{i} |\psi_{X,i}(0)|^{2}\delta(E_{i} - \hbar\omega), \quad (15)$$

where C is a constant and $\Psi_{\chi,i}(\vec{r})$ is the wave function of the i-th excitonic states with an energy eigenvalue E_i. Here the label i runs through discrete states as well as the continuum states. Within the effective-mass approximation, we have $\Psi_\chi(0) = \sum_{\vec{k}} G(\vec{k})$.

To minimize the size of the matrix to be diagonalized, while maintaining the high precision, the symmetry of the system must be fully exploited. Note that the absorption spectrum only depends on $\Psi_\chi(0)$, which is nonzero only for states with full symmetry of the system. Within the approximations used, the superlattice has a circular symmetry in parallel directions and a reflection symmetry in the growth direction. We then use symmetrized basis states labeled by the radial component of $\vec{k}_\parallel$ and for each fixed q. A cut-off $\wedge$ is introduced for the sampling of $k_\parallel$. The final results for energies near the saddle point are insensitive to the choice of the cut-off, as long as $\wedge$ is large enough.

In order to obtain a smooth absorption spectrum, we replace the delta function in Eq.(15) by a Lorentzian function with a half-width Γ. The magnitude of Γ must match the size of Δ_j.

Fig. 4 shows the calculated absorption spectra associated with the saddle-point exciton in GaAs-$Al_{0.25}Ga_{0.75}As$ superlattices for a number of $Al_{0.25}Ga_{0.75}As$ layer thicknesses (L_B). The width of the GaAs layer (L_W) is kept at 80 Å. All spectra are broadened by a Lorentzian function with a width of 1 meV. The dashed curves are the absorption spectra in the absence of the electron-hole Coulomb interaction. Prominent structures associated with saddle-point exciton resonances can be seen in this figure for intermediate values of L_B. At large and small L_B's the spectra are similar to the corresponding quantum well and bulk results.

We have also considered the effect of band mixing on the saddle-point excitons. To make the computation feasible, the axial approximation introduced by Altarelli[14] was used. This allows one to combine the v component with the $-v$ component so that the number of valence subbands are reduced by a factor of two (each valence subband becomes doubly degenerate). Furthermore, the valence subband structure becomes spherical in the x-y plane. With the specific angular-dependent phase factor introduced in the exciton envelope function, the coefficients $G_{m,v}(\vec{k})$ become independent of ϕ. Since the number of states are substantially larger in this situation, the brute force matrix diagonalization is impractical. From Eq. (15) we see that the absorption coefficient can be written as the imaginary part of the green's function at $\vec{r}=0$. Thus we adopted the recursion method[25] to calculate projected density of states.

The problem is therefore substantially simplified. We shall consider excitonic states with orbital angular momenta l=0 and 1 (i.e. s-like and p-like) only. The higher angular momentum states are much less important in the calculation of the absorption coefficient[23]. Due to symmetry the two sets of excitonic states described by {(s-like, $|\nu|=3/2$), (p-like, $|\nu|=1/2$)} and {(p-like, $|\nu|=3/2$), (s-like, $|\nu|=1/2$)} are decoupled. To obtain accurate results, many

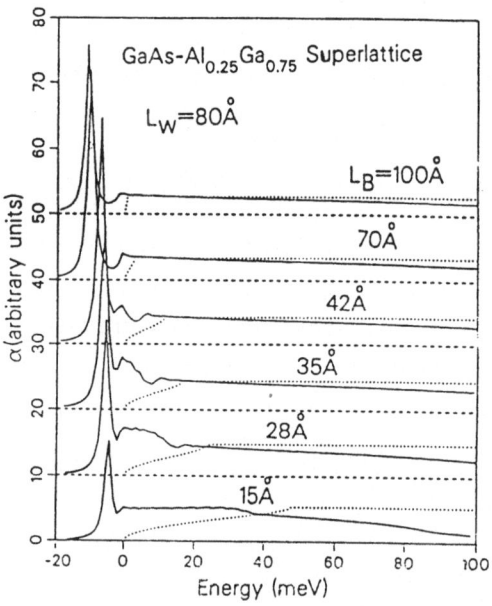

Fig. 4 Theoretical absorption spectra for GaAs-Al$_{0.25}$Ga$_{0.75}$As superlattices with well width L$_W$ = 80 Å and barrier widths L$_B$ = 15 Å, 28 Å, 35 Å, 42 Å, 70 Å, and 100 Å.

zero-th order valence subband states of different m values are needed in the expansion of the exciton states. Below we show results of a preliminary calculation which includes both s-like and p-like excitonic states associated with HH1, HH2, and LH1 valence subbands only.[25] We refer to this as a three-band model. To test the adequacy of this model, we compare in Fig. 5 the valence subband structure obtained by this model (dashed curves) and those obtained by solving the full $\vec{k} \cdot \vec{p}$ Hamiltonian (the `exact' model) (solid curves) for a (L$_W$=75 Å, L$_B$=105 Å) GaAs-Al$_{0.18}$Ga$_{0.82}$As superlattice. Here L$_W$ and L$_B$ denote the well width and barrier width, respectively in each superlattice period. The comparison shows that with only three zero-th order valence subband states included, the valence subband structures are reasonably accurate for small wave vectors where the excitonic effect is most important.

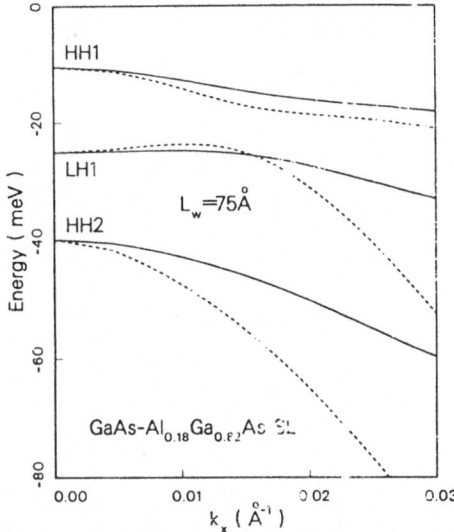

Fig. 5 Valence subband structures of a (L_W=75 Å, L_B=105 Å) GaAsAl$_{0.18}$Ga$_{0.82}$As superlattice with q=0. Solid: exact model; Dashed: three-band model.

The absorption coefficients for excitonic states associated with the first conduction subband of a (L_W = 75 Å, L_B=105 Å) GaAs-Al$_{0.18}$Ga$_{0.82}$As superlattice and a (L_W = 75 Å, L_B = 60 Å) GaAs-Al$_{0.18}$Ga$_{0.82}$As superlattice obtained in the three-band model are shown in Fig. 6. We choose a small Al mole fraction (x=0.18), so that the conduction subband has substantial dispersion even for a decent barrier width. Experimentally, one can grow larger width barrier material with smaller fractional error.[26] In this figure, the dashed curve is due to the set of excitonic transitions involving predominantly HH1-CB1 s-like states and the dotted curve is due to the set of excitonic transitions involving predominantly LH1-CB1 s-like states. The solid curve is the sum of the two. Comparing the two absorption spectra, we see a clear difference in the line shape as we change the well width from 105 Å (the uncoupled quantum well case) to 60 Å(the coupled case).

High resolution excitation spectroscopy measurements for a large number of GaAs-Al$_x$Ga$_{1-x}$As superlattices (including the two cases shown in Fig. 6) have recently been performed[26]. Variation of the line shape of the absorption spectra due to the change of barrier thickness is apparent. Good agreement between our theoretical predictions and the experimental data are obtained.

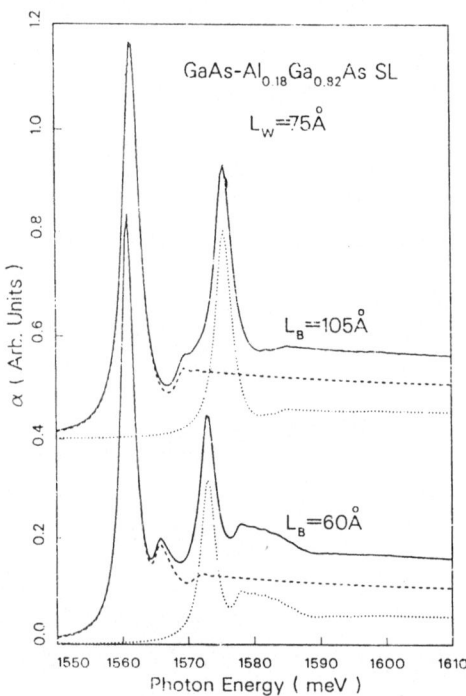

Fig. 6 Theoretical absorption spectra for GaAs-Al$_{0.18}$Ga$_{0.82}$As superlattices with well width L$_W$ = 75 Å and barrier widths L$_B$ = 105 Å and 60 Å. Dashed: HH1-CB1 contribution; dotted: LH1-CB1 contribution; Solid: total.

ACKNOWLEDGMENTS

This work was supported by the University of Illinois Materials Research Laboratory under the National Science Foundation (NSF) NSF-DMR-86-12860 and the U. S. Office of Naval Research (ONR) under Contract No. N00014-81-K-0430.

REFERENCE

1. L. Esaki, R. Tsu, IBM J. Res. Develop. 14, 61 (1970).

2. A. C. Gossard, P. M. Petroff, W. Wiegman, R. Dingle, and A. Savage, Appl. Phys. Lett. **29**, 323 (1976); E. E. Mendez, L. L. Chang, C. A. Chang, L. F. Alexander, and L. Esaki, Surf. Sci. **142**, 215 (1984).

3. Y.-C. Chang, J. N. Schulman, Appl. Phys. Lett **43**, 536 (1983); Phys. Rev. B. **31**, 2069 (1985).

4. G. D. Sanders, Y. C. Chang, Phys. Rev. B **31**, 6892 (1985); Phys. Rev. B **32**, 4282 (1985); Phys. Rev. B **35**, 1300 (1987).

5. J. C. Phillips, Phys. Rev. **136A**, 1705 (1964).

6. B. Velicky, J. Sak, Phys. Status Solidi **16**, 147 (1966).

7. H. Kamimura, K. Nakao, J. Phys. Soc. of Japan, V.24, No.6, 1313(1968).

8. E. O. Kane, Phys. Rev. **180**, 852 (1969).

9. J. E. Rowe, F. H. Pollak, M. Cardona, Phys. Rev. Lett. **22**, 933 (1969).

10. S. Antoci, E. Reguzzoni, G. Samoggia, Phys. Rev. Lett. **24**, 1304 (1970).

11. H. Chu, Y. C. Chang, Phys. Rev. B **36**, 2946 (1987).

12. J. N. Schulman and Y. C. Chang, Phys. Rev. B **24**, 4445 (1981).

13. C. Mailhiot, D. L. Smith, and T. C. McGill, J. Vac. Sci. Technol. **B2(3)**, 371 (1984).

14. A. Fasolino and M. Altarelli, in *Two-Dimensional Systems, Heterostructures, and Superlattices*, edited by G. Bauer, F. Kucher, and H. Heinrich (Springer-Verlag, New York, 1984); M. Altarelli, Phys. Rev. B **32**, 5138 (1985).

15. A. Broido and L. J. Sham, Phys. Rev. B **34**, 3917 (1986).

16. Y. C. Chang, Phys. Rev. B (in press).

17. E. O. Kane, J. Phys. Chem. Solids, **1**, 82 (1956).

18. J. M. Luttinger, W. Kohn, Phys. Rev. **97**, 869 (1956).

19. R. L. Greene, K. K. Bajaj, Solid State Commun. **45**, 831 (1983).

20. See for example, F. Bassani and C. P. Parrasvacini, Electronic States and Optical Properties in Solids (Pergammon, New York, 1975).

21. R. C. Miller, A. C. Gosard, G. D. Sanders, Y. C. Chang, J. N. Schulman, Phys. Rev. B **32**, 8452 (1985).

22. B. Zhu and K. Huang, Phys. Rev. B **36**, 8102 (1987).

23. B. Zhu, Phys. Rev. B **37**, 4689 (1988).

24. H. Chu and Y. C,. Chang (unpublished).

25. G. Grosso and G. Pastori Parravicini, in *Memory Function approaches to Stochastic Problems in Condensed Matter*, Advances in Chemical Physics, V.62.

26. J. J. Song, private communications.

AB-INITIO CALCULATED OPTICAL PROPERTIES

OF [001] (GaAs)ₙ-(AlAs)ₙ SUPERLATTICES

R. Eppenga and M.F.H. Schuurmans

Philips Research Laboratories, P.O.Box 80.000
5600 JA Eindhoven, The Netherlands

Introduction

Recently, the gap between experimentally and theoretically accessible [001] (GaAs)ₙ-(AlAs)ₙ superlattices has been bridged. Following the pioneering work of Ishibashy et al., [1] several [2-5] (GaAs)ₙ-(AlAs)ₙ superlattices of high quality have been grown and characterized down to $n=1$. Initiated by the LMTO calculation of Christensen et al. [6] for the (GaAs)₁-(AlAs)₁ superlattice, *ab-initio* bandstructure methods have been applied to these superlattices up to $n=4$. [7-11]

From photoluminescence and excitation photoluminescence experiments on MBE grown superlattices [2-3] and MOCVD grown superlattices [1,5] the following picture evolves concerning the electronic properties: the $n=1$ superlattice appears to be indirect and from $n=2$ on, the superlattices are direct. The lowest Γ-point conduction band state of the thin ($n=1$-8) superlattices is a band-folded X_z-point conduction band state.
We note that these results should be viewed with caution: down to $n=3$ the structural quality of the (GaAs)ₙ-(AlAs)ₙ superlattices is reasonable. X-ray diffraction analysis shows that the average Al contents of the superlattice and the superlattice periodicity can be obtained within a few percent and half of a monolayer, respectively. [3] The structural quality of the superlattices is questionable for $n=1$ and $n=2$. [3]

The electronic structure that evolves from the bandstructure calculations of Christensen et al., [6] Nakayama et al., [7] Bylander et al., [8] Gilbert et al., [9] Nelson et al. [10] and Ciraca et al. [11] is not unique for $n\geq2$. For $n=1$ the minimum of the conduction band is found to be at [6-9] R, making the $n=1$ superlattice indirect. For $n=2$ the minimum of the conduction band is found to be at Γ by Nakayama et al., [7] Nelson et al. [10] and Ciraci et al. [11] and in between Γ and Z by Gilbert et al.. [9] Theoretically it is thus unclear whether the $n=2$ superlattice is direct or indirect. Moreover, Nakayama et al. find the minimum of the conduction band to be a band-folded Γ state, whereas Nelson et al. and Ciraci et al. find it to be a band-folded X_z state. For $n=3$ and $n=4$ the superlattices are found to be direct. [7-11] However, again Nakayama et al. [7] find the minimum of the conduction band to be a band-folded Γ state whereas Nelson et al. [10] and Ciraci et al. [11] find it to be a band folded X_z state. We also note that the very thin (GaAs)ₙ-(AlAs)ₙ superlattices are found to be unstable against disproportionation into their constituents. [8,12,13] This may be related to the poor quality of the actually grown $n=1$ and $n=2$ superlattices.

In view of the differences between the various theoretical predictions there is a clear need for knowledge of the strength of the optical transitions across the direct and indirect gap. In this paper we present the calculated oscillator strengths of Γ-point across-gap transitions for the very thin ($n=1,2,3,4$) [001] (GaAs)ₙ-(AlAs)ₙ superlattices using the Augmented Spherical Wave (ASW) [14] *ab-initio* bandstructure method. We note that such calculations cannot be done using semi-empirical (for example envelope function type) [15] approaches since for the thin superlattices considered here the superlattice crystal potential must be determined selfconsistently. The strength of the indirect transition is estimated. The implications for the interpretation of (excitation) photoluminescence experiments are briefly discussed.

Method

A self-consistent potential for the [001] $(GaAs)_n$-$(AlAs)_n$ superlattices was generated by solving the Kohn and Sham equation [16] iteratively within the ASW basis set, [14] using the local density approximation (LDA) for the exchange and correlation functional. [16] Scalar-relativistic effects were ignored and "empty" spheres were placed at the interstitial sites. [17] The ASW basis set consisted of s,p and d orbitals centered at each atomic and "empty" sphere site. All muffin-tin sphere radii were taken to be equal. The lattice constant was set equal to 5.653 Å. The number of special k-points in the irreducible wedge of the Brillouin zone was taken to be 18 (6) for the tetragonal structure and 27 (8) for the orthorhombic structure (the numbers in parenthesis apply to the n = 2 SL). Selfconsistency

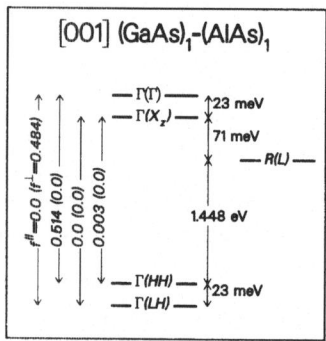

Fig. 1   Schematic representation of the energy levels at high symmetry points of the [001] $(GaAs)_1$-$(AlAs)_1$ superlattice. $\Gamma(HH)$ and $\Gamma(LH)$ denote superlattice valence band states and $\Gamma(X_z)$, $\Gamma(\Gamma)$ and $R(L)$ denote conduction band states. The *fcc* origin of the superlattice symmetry points is indicated in parentheses. The calculated oscillator strengths f between a conduction band state Γ_c and a valence band state Γ_v for light polarization in plane $(f^{||} \equiv f^{xx} + f^{yy})$ and along [001] $(f^{\perp} \equiv f^{zz})$ are in units of the calculated direct-gap oscillator strength $f^{GaAs} \equiv f^{xx} + f^{yy} + f^{zz}$ (= 13.8) [18] of bulk GaAs. For comparison: the calculated result of f^{AlAs} is 7.0 for bulk AlAs. [18]

of the superlattice crystal potential was achieved at a level of 1 mRy. The across-gap oscillator strength between a valence band state ϕ_v and a conduction band state ϕ_c induced by light of polarization $\mathbf{e}$ is equal to $\frac{2}{m} | < \phi_v | \mathbf{p}.\mathbf{e} | \phi_c > |^2 / \Delta E$; here m is the electron mass, $\mathbf{p}$ is the momentum operator and ΔE is the energy difference between the states ϕ_v and ϕ_c. We have shown before that calculated oscillator strengths are typically accurate to within [18] 20% despite the fact that the gap is typically calculated to be 30-50% too small. [18] Details concerning the calculation of the oscillator strengths from the selfconsistently obtained wavefunctions can be found elsewhere. [18] We estimate the accuracy of the relative positions of the energy levels in the conduction band to be around [19] 30-100 meV.

Our results for the optical properties of the superlattices are as follows; results for ground state properties are summarized in footnote 20. For the [001] $(GaAs)_n$-$(AlAs)_n$

superlattice we find Fig. 1. The bottom of the conduction band is at $R(L)$, which is the folded L point of the underlying *fcc* lattice. The lowest superlattice conduction-band Γ-state is X_z-derived, i.e. $\Gamma(X_z)$, and is 71 meV higher in energy than $R(L)$. The Γ-derived state $\Gamma(\Gamma)$ is still 23 meV higher in energy. The heavy-hole *(HH)* and light-hole *(LH)* derived top of the valence band states $\Gamma(HH)$ and $\Gamma(LH)$ are split by 23 meV, the $\Gamma(HH)$ state being higher in energy. In accordance with the experimental result [1] and other bandstructure calculations [6-10] the $(GaAs)_1$-$(AlAs)_1$ superlattice is thus found to be *energywise* indirect. We find that the direct-gap oscillator strength $\Gamma(X_z) - \Gamma(HH)$ of the $(GaAs)_1$-$(AlAs)_1$ superlattice is very small, 0.3% of that of bulk GaAs and 0.6% of the $\Gamma(\Gamma) - \Gamma(HH)$ superlattice oscillator strength.

For the $n = 2$ [001] superlattice we arrive at a more complicated picture; see Fig. 2. The indirect gap $M(X_x) - \Gamma(HH)$ is barely smaller (1 meV) than the direct gap $\Gamma(X_z) - \Gamma(HH)$. We find the lowest Γ-derived superlattice-conduction-band state $\Gamma(\Gamma)$ to be 135 meV higher in energy than the state $\Gamma(X_z)$. The $\Gamma(X_z) - \Gamma(HH)$ oscillator strength in the [001] $(GaAs)_2$-$(AlAs)_2$ superlattice is found to be 6% of that of bulk GaAs and 13% of the $\Gamma(\Gamma) - \Gamma(HH)$ superlattice oscillator strength.

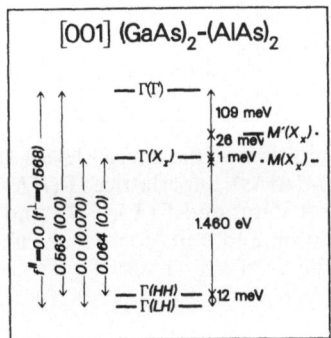

Fig. 2 Schematic representation of the energy levels at high symmetry points of the [001] $(GaAs)_2$-$(AlAs)_2$ superlattice. $\Gamma(HH)$ and $\Gamma(LH)$ denote superlattice valence band states and $\Gamma(X_z)$, $\Gamma(\Gamma)$, $M(X_x)$ and $M'(X_x)$ denote conduction band states. Notation and units for the calculated oscillator strengths $f^{\parallel}$ and $f^{\perp}$ are the same as in Fig. 1.

The results for the $n = 3$ ($n = 4$ in parentheses) [001] superlattice are depicted in Fig. 3 (4). The superlattice has a direct gap $\Gamma(X_z) - \Gamma(HH)$ which is 31 meV (54 meV) smaller than the smallest indirect gap $M(X_x) - \Gamma(HH)$. The direct-gap oscillator strength is 0.0004% (0.02%) of that of bulk GaAs. The second superlattice Γ conduction band state is the band-folded Γ state $\Gamma(\Gamma)$. It lies 174 meV (266 meV) higher in energy than the lowest superlattice conduction band state $\Gamma(X_z)$ and the corresponding $\Gamma(\Gamma) - \Gamma(HH)$ direct-gap oscillator strength is 60% (52%) of that of bulk GaAs.

The values for the calculated oscillator strengths are related to the origin of the corresponding conduction band state; they are large for Γ derived conduction band states and small for any other (e.g. X_z) derived conduction band state.

The ratio r of the rates for emission across the direct and indirect gap can be estimated as follows. We first note that the direct and indirect gaps of the [001] $(GaAs)_n$-$(AlAs)_n$

(n = 2-4) superlattices have comparable magnitude ($E_{direct} - E_{indirect} \sim$ 0-54 meV; see Figs. 2-4). This is not surprising since the corresponding conduction band states are in each case band-folded X-states.. Elementary calculation shows that the ratio r of the radiative transition rates for the direct $\Gamma(X_z)$-$\Gamma(HH)$ and the indirect $M(X_x)$-$\Gamma(HH)$ transition is given by [21] $r \simeq 0.01 (S/\Delta E)^2$. Here S is an energy measure of the strength of the electron-phonon coupling involved in the phonon-assisted indirect rate and ΔE is the difference in energy between the direct and the indirect transitions involved in the perturbative calculation of that rate or, if the relevant phonon energy is larger, ΔE is that phonon energy (10-30 meV). [22] We estimate [23] $S \simeq 10 - 100$ meV and therefore $r \simeq 1$: direct and indirect transitions are equally strong.

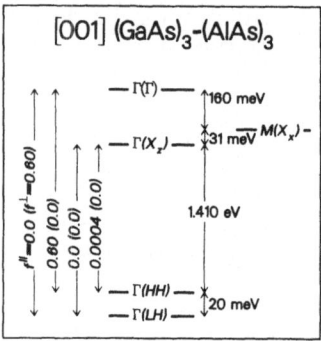

Fig. 3   Schematic representation of the energy levels at high symmetry points of the [001] $(GaAs)_3$-$(AlAs)_3$ superlattice. $\Gamma(HH)$ and $\Gamma(LH)$ denote super-lattice valence band states and $\Gamma(X_z)$, $\Gamma(\Gamma)$ and $M(X_x)$ denote conduction band states. Notation and units for the calculated oscillator strengths f and $f^\perp$ are the same as in Fig. 1.

We now discuss the consequences of the foregoing for the principal across-gap transitions. The $(GaAs)_1$-$(AlAs)_1$ superlattice is indirect. However, one should realize that the direct ($\Gamma(X_z)$-$\Gamma(HH)$)) and the indirect transitions ($R(L)$-$\Gamma(HH)$) have comparable strength and are extremely weak. The oscillator strength of the direct transition ($\Gamma(X_z)$-$\Gamma(HH)$)is only 0.6% of that of the $\Gamma(\Gamma)$-$\Gamma(HH)$ transition. The situation for the $(GaAs)_2$-$(AlAs)_2$ superlattice is complicated. We find energy differences $\Delta E(\Gamma(X_z) - M(X_x)) \simeq 1$ meV and $\Delta E(\Gamma(\Gamma) - \Gamma(X_z)) \simeq 130$ meV, i.e. energywise the superlattice is barely indirect. However, the direct and indirect transitions are equally strong ($r \simeq 1$) as we have shown in the preceding paragraph. The $(GaAs)_2$-$(AlAs)_2$ superlattice can therefore be qualified neither as direct nor as indirect from a spectroscopic point of view. The n = 3 and n = 4 $(GaAs)_n$-$(AlAs)_n$ superlattices are direct. However, the across-gap oscillator strength is only 0.0004% (0.02 for n = 4) of that of GaAs. The indirect transitions are close in energy and have strengths comparable to those of the direct transitions. In view of the small values for our calculated oscillator strength the direct recombination would probably also have a substantial contribution due to non-radiative decay.

We stress that the issue of a superlattice being energywise direct or indirect cannot be fully settled by ab-initio bandstructure calculations since they are known to be able to produce relative positions of energy levels in conduction band spectra only to an accuracy of [19] 30-100 meV. The $(GaAs)_2$-$(AlAs)_2$ superlattice is particularly difficult in this respect since the relevant energy difference $E(\Gamma(X_z)) - E(M(X_x))$ is so small ($\simeq 1$ meV). This is unfortunate since the ordering of the X_{xy} and X_z derived states is subject to debate. [4,24]

Table 1   Comparison between measured and calculated values for several energy gaps of the [001] $(GaAs)_n$-$(AlAs)_n$ superlattices. A constant value of 0.6 eV has been added to the calculated energy gaps.

		Exp. (eV)	Theory (eV)
$n=1$	$R(L) - \Gamma(HH)$	1.93-2.06	2.05
$n=2$	$M(X_x) - \Gamma(HH)$	1.97-2.07	2.06
$n=3$	$\Gamma(X_x) - \Gamma(HH)$	2.04	2.01
$n=4$	$\Gamma(X_x) - \Gamma(HH)$	1.93-2.02	2.00
$n=1$	$\Gamma(\Gamma) - \Gamma(HH)$	-	2.14
$n=2$	$\Gamma(\Gamma) - \Gamma(HH)$	2.21	2.20
$n=3$	$\Gamma(\Gamma) - \Gamma(HH)$	-	2.20
$n=4$	$\Gamma(\Gamma) - \Gamma(HH)$	2.16-2.17	2.18

However, we note that the accuracy of this calculated energy difference is on a considerably better level (a few meV) than the 30-100 meV mentioned before since both states originate from in the bulk equivalent X states. But even the ordering of $\Gamma(X_z)$ and $\Gamma(\Gamma)$ is subject to debate. We find $\Gamma(X_z)$ to be 135 meV lower in energy than $\Gamma(\Gamma)$. Nakayama and Kamimura [7] find the opposite result: from Fig. 4 of their paper we estimate $\Gamma(\Gamma)$ to be $\simeq 100$ meV lower in energy. These authors have adjusted the α-parameter of the exchange-correlation potential in order to obtain calculated bandstructures of bulk GaAs and AlAs which agree better with the experimental bandstructures. Nevertheless differences between experimental and theoretical conduction bandstructures are still on the level of 100-200 meV. This is close to the level of inaccuracy of the relative positions of conduction band energy levels expected anyway for true ab-initio bandstructure calculations. [19] Similar remarks apply to the $n=3$ and $n=4$ $(GaAs)_n$-$(AlAs)_n$ superlattices.

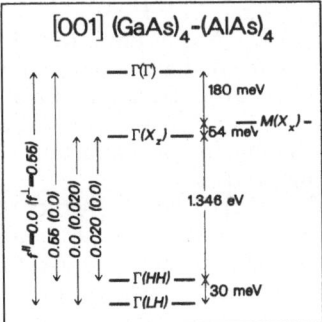

Fig. 4   Schematic representation of the energy levels at high symmetry points of the [001] $(GaAs)_4$-$(AlAs)_4$ superlattice. $\Gamma(HH)$ and $\Gamma(LH)$ denote superlattice valence band states and $\Gamma(X_z)$, $\Gamma(\Gamma)$ and $M(X_x)$ denote conduction band states. Notation and units for the calculated oscillator strengths $f^\parallel$ and $f^\perp$ are the same as in Fig. 1.

In Table I we present measured and calculated values for the direct and indirect gaps of the $(GaAs)_n$-$(AlAs)_n$ superlattices. Since energy gaps that are calculated within the local density approximation are usually 30-50% too small, [19] we have added a constant value of 0.6 eV to all calculated gaps. The calculated gaps thus obtained agree with the measured gaps within the experimental error.

In conclusion, we have calculated the radiative rates of direct transitions and estimated the phonon-assisted radiative rates of indirect transitions in [001] $(GaAs)_n$-$(AlAs)_n$ superlattices for n = 1-4. The rates in [001] superlattices are at least two orders of magnitude smaller than the across-gap direct transitions in GaAs. These superlattices therefore do not hold many prospects for light emission applications. Our calculations show that the radiative rates of direct and indirect transitions, and possibly also non-radiative rates, are of comparable magnitude. We have also shown that from a spectroscopic point of view the $(GaAs)_2$-$(AlAs)_2$ superlattice can neither be called direct nor indirect since the corresponding transitions are equally strong. The $(GaAs)_3$-$(AlAs)_3$ and $(GaAs)_4$-$(AlAs)_4$ superlattices are direct but have very small direct-gap oscillator strengths.

References

1. A. Ishibashi, Y. Mori, M. Itabashi and N. Watanabe, J. Appl. Phys. 58, 2691 (1985)
2. T. Isu, De-Sheng Jiang and K. Ploog, Appl. Phys A43, 75 (1987)
3. K.J. Moore, G. Duggan, P. Dawson and C.T.B. Foxon, to be published.
4. E. Finkman, M.D. Sturge and M.C. Tamargo, Appl. Phys. Lett. 49, 1299 (1986);
5. N. Kobayashi and Y. Horikoshi, Appl. Phys. Lett. 50, 909 (1987)
6. N.E. Christensen, E. Molinari and G.B Bachelet, Solid State Comm. 56, 125 (1985)
7. T. Nakayama and H. Kamimura, J. Phys. Soc. Japan 54, 4726 (1985)
8. D.M. Bylander and L. Kleinman, Phys. Rev. B 34, 5280 (1986)
9. T.G. Gilbert and S.J. Gurman, Superl. and Microstr. 3, 17 (1987)
10. J.S. Nelson, C.Y. Fong and I.P. Batra, Appl. Phys. Lett. 50, 1595 (1987)
11. S. Ciraci and I.P. Batra, Phys. Rev. B36, 1225 (1987); I.P. Batra, S. Ciraci and J.S. Nelson, J. Vac. Sci. Technol. B4, 1300 (1987)
12. D.M. Bylander and L. Kleinman, Phys. Rev. B 36, 3229 (1987); D.M. Bylander and L. Kleinman, Phys. Rev. Lett. 59, 2091 (1987)
13. D.M. Wood, S.H. Wei and A. Zunger, Phys. Rev. Lett. 58, 1123 (1987)
14. A.R. Williams, J. Kubler, C.D. Gelatt, Phys.Rev. B19, 6094 (1979)
15. see e.g. G. Bastard and J.A. Brum, IEEE Journ. of Quant. Elec. QE-22, 1625 (1986) and references therein.
16. P. Hohenberg and W. Kohn, Phys. Rev. 136B, 864 (1964); W. Kohn and L.J. Sham, Phys. Rev. 140A, 1133 (1965)
17. T. Jarlborg and A.J. Freeman, Phys. Lett. 74A, 349 (1979)
18. H.W.A.M. Rompa, R. Eppenga and M.F.H. Schuurmans, Physica 145B, 5 (1987)
19. G.W. Godby, M. Schluter and L.J. Sham, Phys. Rev. B35, 4170 (1987); These authors have shown that the experimental conduction bands of GaAs and AlAs can, to within 100 meV, be obtained from their ab-initio density functional calculations using a rigid shift of 0.8 eV for GaAs and 0.9 eV for AlAs. Using a similar approach we find the calculated relative energy positions in the conduction bands of GaAs and AlAs to be accurate on the level of 30 meV.
20. Our calculated results for the ground state properties of these superlattices are in accordance with the results from other ab-initio calculations. We define the GaAs/AlAs interface heat of formation as $\Delta H_n = E((GaAs)_n$-$(AlAs)_n)/n - (E(GaAs) + E(AlAs))/2$; we have calculated the total energies $E(GaAs)$ and $E(AlAs)$ under the same conditions as the SL calculation, i.e. using the same unit cell, the same number of k-points in the BZ, etc. We find $\Delta H_1 \simeq 30$ meV (cf. Bylander and Kleinman (15 meV) using relativistic pseudopotentials [12] and Wood et al. (25 meV) using both semirelativistic pseudopotentials and the LAPW method). [13] We find $\Delta H_2 \simeq 19$ meV. By shifting the bulk GaAs and AlAs potential rigidly to fit the potential of the corresponding monolayers of the GaAs/AlAs SL optimally, we find a value of 0.6 meV for the valence band offset $\Gamma(HH)^{GaAs} - \Gamma(HH)^{AlAs}$ in [001] GaAs-AlAs superlattices (cf. 446 meV in Ref. 12).
21. Note that we need not consider the virtual process involving the valence band states since the corresponding energy denominator is much larger.

22. A.S. Barker, J.L. Merz and A.C. Gossard, Phys. Rev. B17, 3181 (1978); C. Colvard, R. Merlin, M.V. Klein and A.C. Gossard, Phys. Rev. Lett. 45, 298 (1980); C. Colvard, R. Merlin, M.V. Klein and A.C. Gossard, J. de Physique C6, 631 (1981)
23. O.J. Glembocky and F.H. Pollak, Phys. Rev. Lett. 48, 413 (1982); O.J. Glembocky and F.H. Pollak, Phys. Rev. B25, 1193 (1982);
24. J. Ihm, Appl. Phys. Lett. 50, 1068 (1987)

EFFECT OF A PARALLEL MAGNETIC FIELD ON THE HOLE LEVELS IN SEMICONDUCTOR SUPERLATTICES

A.Fasolino, International School for Advanced Studies, Strada Costiera 11, I-34014 Trieste, Italy

M. Altarelli, Max Planck Institut für Festkörperforschung, Hochfeld Magnetlabor, B. P. 166X, F-38042 Grenoble Cedex, France and European Synchrotron Radiation Facility, B.P. 220, F-38043 Grenoble Cedex, France

ABSTRACT

We present a calculation of the hole levels in a GaAs-GaAlAs superlattice in a magnetic field parallel to the layers, when the magnetic length is greater or comparable with the superlattice period. A comparison with the electronic levels in the same field configuration shows noticeable analogies, but also differences related to the more complex hole subband structure. A calculation of interband magneto-optical transition intensities is also presented.

The properties of electrons in semiconductor superlattices in a magnetic field parallel to the layers have been recently studied experimentally[1,2,3]. The new interest of this field configuration lies in the fact that electrons are forced by the magnetic field to execute orbits by tunneling through the layers, in a situation where the magnetic length can range from much greater to less than the superlattice period. In a superlattice with barriers so thin as to induce coupling between neighbouring wells, the superlattice band structure is formed by subbands of finite width, corresponding to extended states, separated by minigaps where tunneling is inhibited: a different behaviour is then observed in these two types of energy regions once a magnetic field is applied parallel to the layers. A study of this situation yields information about the subband dispersion in the growth direction, not easily accessible otherwise, and about tunneling and perpendicular transport as well. For example, the magneto-optical experiments of Ref.1 show a rather regular ladder of peaks below a field independent energy threshold which is suggested to be, and indeed roughly coincides with the first subband edge. Moreover the width of such peaks, which has implications on the mobility across the layers, is only modestly larger than in a perpendicular field.

An analogous study of the hole levels is interesting at least in two respects. The first is that in a magnetic field both electrons and holes perform orbits of the same size, whereas their band structure and hence their masses and penetration in the barrier layers is very different. It is then interesting to study how the peculiarities of the hole band structure show up in this case. The second is that a quantitative interpretation of interband magneto-optical experiments, such as those of Ref.1, requires, beyond the electronic energy levels which are well understood experimentally and theoretically[1,3], a knowledge of the hole levels and matrix elements for optical transition as well. In the following we will present a calculation of the hole energy levels and of the interband magneto-optical transition intensities. In particular we wish to verify if interband absorption can measure directly the combined electron and hole subband width as suggested in Refs. 1,3, based on a simple model for holes.

Let us introduce our terminology while briefly reviewing results for electrons[1,3]. The Schrödinger equation for an electron of mass m^* moving under the action of a magnetic field $\vec{B} = (0, 0, B)$, represented by the vector potential $\vec{A} = (0, Bx, 0)$, in a superlattice grown along x can be written as a separable, $1-$dimensional differential equation

$$(\frac{1}{2m^*}p_z^2 + \frac{1}{2m^*}p_x^2 + \frac{1}{2m^*}\omega_c^2(x - x_0)^2 + V(x))\psi = E\psi \tag{1}$$

where $\omega_c = eB/m^*c$ is the cyclotron frequency, $x_0 = -l^2 p_y$ is the cyclotron orbit centre, $l = (c/eB)^{1/2}$ being the magnetic length, and $V(x)$ is the band edge profile. It should be noted that the energy dependence on x_0 clearly reflects the in-plane subband dispersion, which is more complicated for holes, and that the presence of $V(x)$ lifts the degeneracy of free electrons with respect to the position of the orbit centre x_0.

At the top of Fig. 1 the resulting levels for a $40\mathring{A}$ well, $20\mathring{A}$ barrier superlattice at $B = 10$ and 15 Tesla are shown as a function of the position of the cyclotron orbit centre with respect to the superlattice potential. One can see that the first few levels, which occur within the first miniband width, hardly depend on x_0 while they become dispersive above the subband edge. The threshold where this happens clearly does not depend on the field but only on the band structure. These results, which are explained in detail in Ref. 3 , can be understood if we think of the superlattice as a new *bulk* material, where, within the minibands, the degeneracy with respect to x_0 is recovered. Also, within a quasi-classical quantization scheme[3], flat and dispersive levels can be associated with close or open orbits respectively.

Let us now go over to the holes. We have calculated the hole magnetic levels in the field configuration described above, within the envelope function formalism, by means of a numerical method based on the reduction of the Luttinger Hamiltonian[4] to a finite difference system in real space. This method is particularly suited to this problem as the magnetic potential, which adds to the superlattice band edge profile, while breaking its periodicity ensures the required vanishing of the wave function at infinite distances from the orbit centre. Moreover no further approximations are required and in principle the full $k \cdot p$ Kane Hamiltonian can be directly solved. An alternative method, based on the extension of the oscillator formalism proposed by Luttinger[4] for the valence band of bulk semiconductors to a single quantum well has been recently worked out[5]. The extension of such a formalism to a full superlattice is, despite its formal elegance, very laborious and a direct numerical solution has been preferred. Moreover, the axial approximation[4,5] is somewhat more critical for a layered structure than for the bulk.

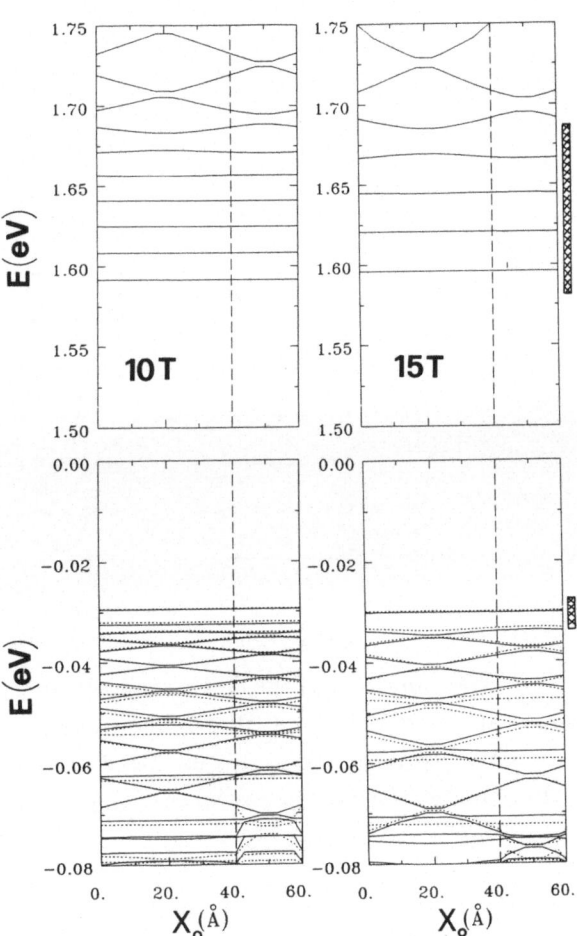

Fig.1. Electron (top) and hole (bottom) magnetic levels at 10 T (left) and 15 T (right) vs. the position of the orbit centre x_0. The dashed line in each panel indicates the boundary between the 40 Å GaAs well (left) and the 20 Å GaAlAs barrier (right). The first conduction and heavy hole subband widths are indicated on the right hand margin.

Once the k_z component of the wave vector in the direction of the magnetic field is set to zero, the Luttinger Hamiltonian[4] reduces to two inequivalent blocks for the $J_z = 3/2, -1/2$ and $J_z = -3/2, 1/2$ components of the $J = 3/2$ manifold at $k = 0$. The wavefunction is then a 2−component spinor, defined over a regular grid of N points in real space − extending from $-6l$ to $6l$, with a spacing $\delta = 0.5$Å in the present case. The superlattice potential is added to the diagonal elements. It should be noted that in principle a square well profile is only reached for $\delta \to 0$, i.e. a grading of extension δ is always present at the interfaces. The differential operators are represented as

$$\frac{d}{dx}\psi_i(j) = \frac{\psi_i(j+1) - \psi_i(j-1)}{2\delta}; \qquad \frac{d^2}{dx^2}\psi_i(j) = \frac{\psi_i(j+1) + \psi_i(j-1) - 2\psi_i(j)}{\delta^2} \tag{2}$$

i indicating the J_z component and j the point in the grid. The Hamiltonian becomes a $2N \times 2N$ asymmetric banded matrix.

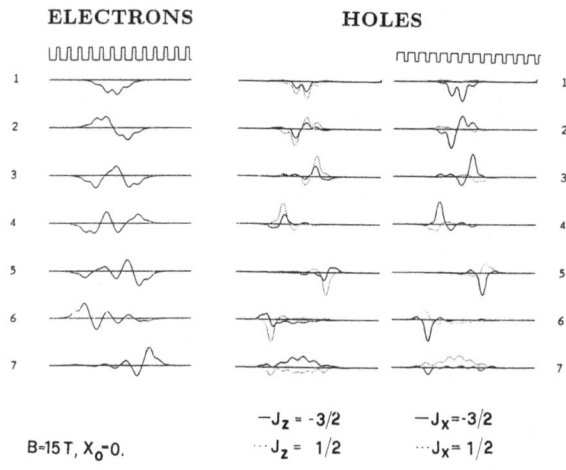

Fig.2. Wavefunctions of the first seven electron and hole levels at $x_0 = 0$ and 15 T. The superlattice confining potential is shown on top. Note that the hole wavefunctions are almost purely -3/2 or 1/2 in the J_x representation and very mixed in the J_z representation (see text).

We illustrate in the following our results for both electrons and holes and for interband transitions. In order to reduce the computer time required , we have made the following approximations: 1)band parameters are the same in the barriers and in the wells, 2) electronic and hole levels are calculated separately 3) excitonic effects ($\simeq 4.5$ meV [6]) are not included. None of these assumptions alters the qualitative picture and all of them but 3) could be easily lifted. We have used the Luttinger parameters $\gamma_1 = 6.85$, $\gamma_2 = 2.1$, $\gamma_3 = 2.9$ and $\kappa = 1.2$, $m^* = .067m_0$, a band offset of 0.375 eV of which 60 % for the conduction band.

At the bottom of Fig.1 we show results for holes at 10 and 15 T. Surprisingly enough, the overall spectrum is strongly reminiscent of that for electrons. The first levels are flat and become dispersive at lower energies. As far as they remain flat, their energy depends linearly on the field. However a second set of more widely spaced flat levels is superimposed on the dispersive levels. By inspection of the wave function we can say that these latter flat levels originate from the light hole subbands (see level n.7 in Fig. 2). The subband structure shown in Fig.3, calculated in the conventional scheme[7] for a superlattice with square wells, shows that the first set of flat levels correspond to the heavy hole miniband energy region, while the second set starts only below the upper edge of the light hole subband. The two sets of solution for $J_z = 3/2, -1/2$ and $-3/2, 1/2$, represented as solid and dotted lines respectively, show a negligible spin splitting for the heavy hole derived states. The confining superlattice potential "polarizes" the states into eigenfunction of J_z (3/2 for heavy and 1/2 for light holes), whereas the magnetic field tends to polarize the J_z component of the eigenvector. From Fig.2 one can see that indeed the first hole levels are strongly polarized in J_x, therefore mixing the J_z components, whence the small splittings. In fact the expectation values of the Hamiltonian over the eigenfunction of $J_x = \pm 3/2$ are the same. When the field is increased or the well depth decreased splittings start opening.

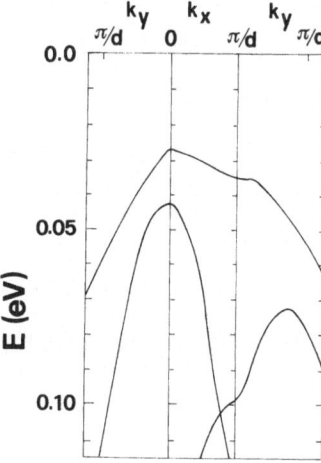

Fig. 3. Valence subband structure. $d = 60\mathring{A}$ is the superlattice period.

Finally, in Fig.4 we show the calculated intensities for interband transitions in the Voigt configuration. The vertical line at $E = 1.721$ eV indicates the combined-electron hole first subband edge, calculated at $B = 0$ for a square well superlattice – this value is underestimated by a few meV due to the finite grading of interfaces present in the calculation at $B \neq 0$. Indeed, sharp peaks are present only below this value. We have indicated in the figures where the transitions with $\Delta n = 0$, i.e. transitions from the first hole level to the first conduction level, from the second to second and so on, occur. We see that the most pronounced peaks come from the first two or three flat levels in the heavy hole subband range. However, it is interesting to notice that they always present shoulders at slightly higher or lower energies, resulting

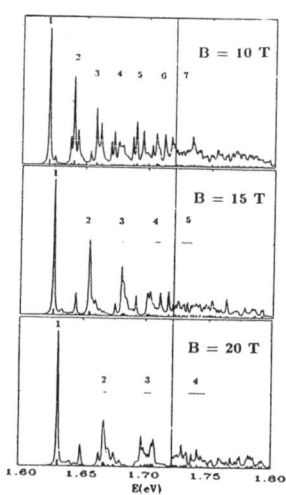

Fig. 4. Calculated interband transition intensities in the Voigt configuration. The small bars under each peak correspond to intensities at different x_0's. The sum of these intensities, dressed with 2 meV wide Lorentzians, results in the shown spectra. The vertical bar at 1.721 eV marks the combined electron hole first subband edge. $n = 1, 2, \ldots$ indicate where the transitions from the n-th hole level to the n-th conduction level occur.

from transitions from other hole levels, as a consequence of the mixed character of the wave function. These latter contributions to transitions to the same final conduction state may become comparable or predominant as the energy increases. Consequently, the calculated spectra are rather complex and even present peaks, such as the $n = 3$ peak at 20 T, with a double structure. It should be interesting to verify this prediction on high quality samples: the observation of these splittings could in fact measure the hole level structure. All these peaks become broader as soon as the hole levels become dispersive. In particular, at 15 T, where four flat conduction levels are present, the fourth peak ($\simeq 1.71$ eV) is drastically broadened by the hole level dispersion and by the less strict selection rules.

Additional sharp peaks, such as the peak at $\simeq 1.648$ eV at 20 T, come from the second set of flat levels, of light hole character. The position of such peaks is very dependent on the band offset and could contribute to its determination. They also provide the only sharp features beyond the combined subband edge. In this energy region, since the magnetic length is the same for electrons and holes (see Fig.2) and successive dispersive levels are localized for both particles in the same region of space, the overlap remains effective and the observed disappearance of sharp structures in interband absorption[1,3] is only due to a smearing out in energy.

In summary, although none of our conclusion contrasts with the first interpretation of interband absorption experiments, we feel that a detailed calculation of the hole levels and transition intensities is useful for a more quantitative interpretation of the experimental results. In particular, in a situation where the number of electron and hole levels contained in the respective subband widths is not the same, as in the present calculation, the cut off energy for transitions is not sharply defined. We hope that our results will stimulate further experimental work on this subject.

We are grateful to J. C. Maan for many useful discussions. One of us (AF) thanks Prof. P. Wyder for his hospitality at the Max Planck Institute in Grenoble,where this work was carried out with a grant from the Consiglio Nazionale delle Ricerche. Calculations have been performed with the support of the CCVR ,Palaiseau (France) and as a part of the SISSA-CINECA joint project sponsored by the Italian Ministry of Education.

References

1) G. Belle, J. C. Maan, G. Weimann Solid State Commun. **56**,65, 1985; G. Belle, J. C. Maan, G. Weimann Surf. Sci. **170**,611, 1986;

2) T. Duffield, R. Bhat, M. Koza, F. De Rosa, D. M. Hwang, P. Grabbe and S. J. Allen, Jr. Phys. Rev. Lett. **56**, 2724, 1986; T. Duffield, R. Bhat, M. Koza, K. M. Rush and S. J. Allen, Jr. Phys. Rev. Lett. **59**, 2693, 1987;

3) J. C. Maan, Festkörperprobleme **27**, ed. by P. Grosse, p. 137, 1987;

4) J. M. Luttinger, Phys. Rev **102**, 1030, 1956;

5) M. Altarelli, G. Platero Surf. Sci **196**, 540, 1988;

6) A. Chomette, B. Lambert, B. Deveaud, F. Clerot, A. Regreny, G. Bastard Europhys. Lett. **4**, 461, 1987;

7) A. Fasolino, M. Altarelli in Springer Series in Solid State Sciences, vol. 53, ed. by G. Bauer, F. Kuchar and H. Heinrich (Springer, Berlin) 1984, p. 176;

PROFIT FROM HETEROSTRUCTURE ENGINEERING

G. J. Rees

Plessey Research and Technology
Caswell, Towcester
Northants, U.K.

ABSTRACT

Several semiconductor devices employing heterojunctions and micro-structures are now either on the market, incorporated in systems or are soon likely to be so. In this note we examine the contribution of applied physics and modelling to the understanding and design of these devices, highlighting successes and failures and areas where further work is desirable.

INTRODUCTION

The use of heterojunctions in semiconductor devices for potential barriers to mobile charges is as old as the heterojunction laser. Dingle[1] demonstrated the physics of quantum confinement with luminescence experiments in MBE grown GaAs/AlGaAs heterostructures. Since that time research into quantum wells and low dimensional devices has mushroomed. The blurring of the distinction between research into device engineering and basic physics attests both to the sophistication of new device concepts under proposal and the dependence of advances in physics on the technology of microstructures and superlattices.

Low dimensional device research has had some notable successes to date. Devices such as the quantum well injection laser and modulation doped field effect transistor (MODFET) are appearing on the market place and in systems. Others, like the bipolar heterojunction transistor (BHT) and the quantum well avalanche photodiode are probably only a little further behind and offer advantages over their conventional counterparts.

Rather than explore further avenues of low dimensional device research this note will examine some of the contributions of physics to the understanding and design of these successful or promising devices. It is hoped that this will serve to highlight areas where additional research and modelling will improve our understanding and the performance of these and other heterostructure devices.

QUANTUM WELL LASERS

The most spectacular success of low dimensional device physics is probably the quantum well injection laser. Because of their wavelength tunability and low threshold power these devices are being incorporated into optical disc reader systems and also look attractive for inter satellite and optical fibre communications.

Figure 1 shows the band edge structure of a graded index quantum well laser. The central active multiple quantum well region is formed, for example, from alternate layers of GaAs and AlGaAs, although for longer wavelength operation alloys in the InGaAlAsP series, lattice matched to InP are preferred. The function of this central region is to play host to the stimulated radiative recombination process between quantum confined levels in the conduction and valence bands. The staircase density of states resulting from the quantum confinement provides more gain 'per electron' than in the bulk. The overall process turns out to be less expensive in terms of associated parasitic processes of spontaneous emission, Auger recombination and intervalence band absorption.

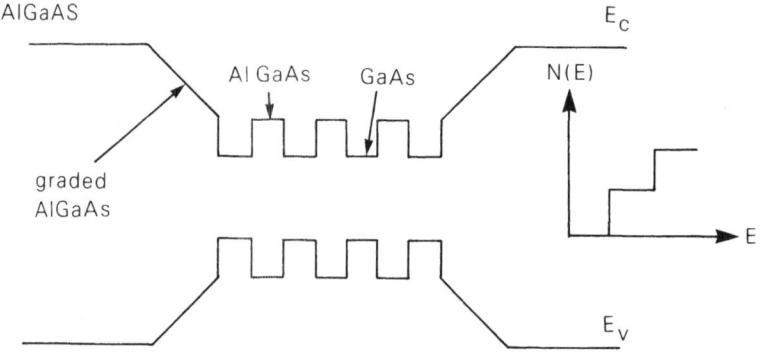

Fig.1   Band edge structure of graded index multiple quantum well injection laser. Inset shows staircase density of states in quantum well structure

The narrow average energy gap of the quantum well region leads to a higher refractive index than the surrounding medium and, as in a classical double heterojunction laser, this provides optical confinement and the enhanced overlap with the optical field necessary for stimulated emission. The graded index region surrounding the quantum well extends over several hundreds of nanometers and is intended to aid this optical confinement by providing a passive optical waveguide and a funnel for carriers injected from the widegap (low index), high Al concentration cladding.

The first step in modelling[2] for the design of a quantum well laser is to calculate the energies of the quantum confined levels and the associated two dimensional densities of states in the conduction and valence bands. These bands are broadened by the effects of electron scattering with interface roughness, phonons and other carriers[3].

Population of these bands with free carriers according to Fermi-Dirac statistics allows calculation of the spectral dependence of gain.

Theoretical work has shown how the parasitic processes of Auger[4] recombination and intervalence band absorption[5] are modified in quantum wells. Calculation of the optical field profile in the waveguide structure is straightforward and allows prediction of the strength of optical overlap in the quantum well region. Inclusion of facet reflectivity completes the picture and modelling allows prediction of threshold current for lasing and its temperature dependence for different designs of quantum well and confinement region. Other properties that can be usefully predicted include efficiency, the dependence of lasing wavelength on device current and the dependence of gain on carrier concentration, reflecting the ability of the device to respond to rapidly varying drive currents.

The modelling of quantum well lasers is particularly successful and it is instructive to examine why. Bound state energy level predictions can be checked against luminescence observations and even simple effective mass models can give reasonable agreement with measurements. Studies suggest parasitic Auger and IVBA processes are not much different from bulk rates, which have been measured. Optical recombination rates depend on matrix elements which can be checked against absorption measurements in the bulk. Modelling the confined optical field in such weakly guiding structures can be treated with simple scalar wave theory. The few weak points involve the treatment of the valence band structure, final state effects in Auger recombination and IVBA and the radiative recombination process which, in quantum well lasers appears to conserve wave vector and in conventional semiconductor lasers does not[6]. Nature seems to overlook our inadequacies in this case.

Future progress in this field is likely to be made in strained layer structures and we await confirmation of the predictions of Adams[7] and co-workers (see also Kane and Yablonovitch[8]) that the modified valence band structure with the light hole bands highest, can significantly reduce threshold current and its temperature dependence. If these predictions turn out to be correct then we shall need more detailed information on electronic structure, especially in the valence band and of band edge offsets, and also of optoelectronic processes in these systems. The resilience of these strained layer structures against defect generation in the presence of strong electron-hole recombination will then be the limiting consideration.

MODFETS

By contrast, the modulation doped field effect transistor is a difficult device to model in detail. Its low noise, high speed properties make it attractive as a front end amplifier for military systems and communications and in domestic direct broadcast satellite dishes.

The broad principles of operation are well understood. The active channel (figure 2) consists of an undoped accumulation layer in the GaAs at a GaAs/AlGaAs heterojunction interface held in place by the ionized donors in the AlGaAs epilayer. The separation of donors from electrons reduces impurity scattering and the reduction is improved by an undoped spacer layer between the dopant and the heterojunction. Source and drain contacts are made by diffusion or implantation and indeed low temperature mobilities in a device with a good interface can reach $3.10^6 cm^2 V^{-1} sec^{-1}$ [9]. The gate on the AlGaAs controls the channel carrier concentration and hence the drain current under constant drain bias.

Offstate modelling (zero drain bias) of the vertical potential and carrier profile is relatively advanced[10]. The potential profile (inset, figure 2) is found by solving the 1-D Poisson equation. This potential in turn determines the structure of the quantum confined bands in the accummulation

layer found by solving the 1-D Schrödinger equation. Population of these
bands according to Fermi-Dirac statistics determines the space charge and
completes the loop. Many body effects in the channel modify the potential
to second order and for an undoped or compensated GaAs buffer the free
carrier concentration in the bulk is negligible. Calculations along these
lines are useful for designing the combinations of alloy concentration,
thickness, doping and spacer depth to achieve adequate channel carrier
concentration and depletion at reasonable gate voltages.

Onstate modelling is a different matter. For positive drain bias
channel carriers are hot and can escape to the wide gap material with low
mobility and slow response times. At negative gate biases designed to
pinch off the device carriers reroute into the GaAs buffer where the three
dimensional electron dynamics may be dominated by space charge limited
transport, before returning to the conducting channel beyond the gate. The
potential contours around the gate are slewed towards the drain and the
vertical potential profile and hence bound state energies vary down the
channel. Transport in the important gate region is brief and for short
gates is in disequilibrium so that velocity overshoot and acceleration
effects are likely to be important.

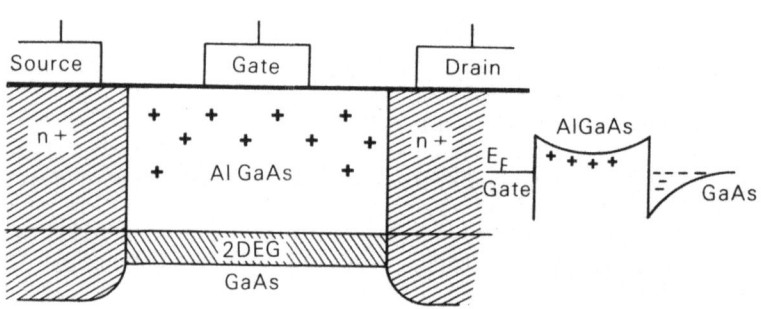

Fig. 2 Schematic of MODFET and inset showing vertical potential profile.

Valiant attempts have been made to simulate devices[11] using
hydrodynamic models[12] of transport in subbands whose spectrum is calculated
self consistently with the Schrödinger equation. These attempts serve to
highlight the enormity of the problem and the difficulty of predicting
reliable S-parameters and noise figures for a device with such complex and
ill controlled processes. It seems likely that at present design of MODFETS
and interpretation of measurements is best performed with semi analytical
"seat of pants" models[13], heavily supported by physical intuition.

Future developments are likely to see the active channel grown from
strained InGaAs on a GaAs[14] substrate or lattice matched to InP with an
InAlAs barrier. This alternative channel material has improved transport
properties[15] and the conduction band step at the new interface will help
suppress substrate current. Growth on an InP substrate also opens the way
for optoelectronic integration.

HETEROJUNCTION BIPOLAR TRANSISTOR

In the HBT the band gap difference is used to ensure electron injection from the n-type AlGaAs emitter rather than reverse hole injection from the p-type GaAs base (figure 3). This feature avoids the need for heavy doping in the emitter, associated in Si bipolar transistors with counter productive band gap narrowing. It also permits the reduction of base resistance, and so the base charging time constant, by means of high acceptor concentrations without the associated problem of inverse injection. This advantage can be traded against a shorter base with associated reduction in base transit time. It is this improvement in speed which makes the HBT attractive for fast logic and for high frequency power amplification.

There are several other features available in a heterogeneous system. A conduction band edge 'step down' injects hot, fast electrons. Alloy grading in the base provides a built in field which drives electrons from emitter to collector. 'Reverse grading' in the depleted base-collector junction helps electrons to stay near the peak of their velocity field characteristic and reduces collector transit time.

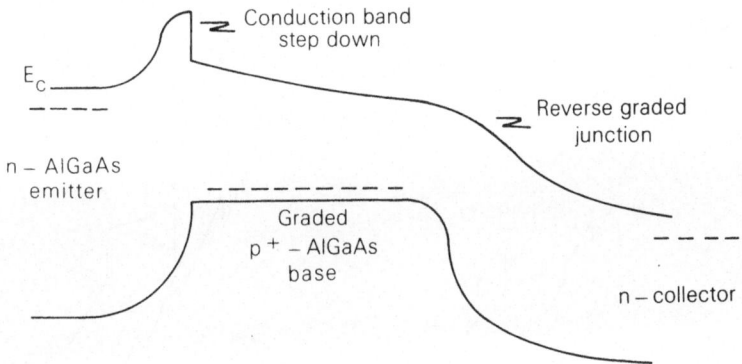

Fig.3   Band edge profile of an HBT. The 'step down' at the emitter base junction injects hot electrons, the graded base provides a built in field for drift aided electron diffusion and the reverse graded base-collector junction peaks the electron velocity.

Monte Carlo modelling of electron transport provides a useful guide to design of alloy profile to minimize electron transit times[16]. Unfortunately, the injection characteristics are dominated, at least at low bias, by the imponderable recombination through deep traps at the heterojunction interface and at isolation regions created by ion bombardment.

The simple property of a band gap difference at the emitter-base junction has produced the fastest bipolars ever and has even stimulated silicon technologists to seek analogues such as a SiC emitter or SiGe base.

QUANTUM WELL AVALANCHE PHOTODIODES

Built in gain and high sensitivity are the advantages of avalanche photodiodes which make them attractive as the detectors for long haul optical fibre communication systems. In these devices light is absorbed in the depletion region of a reverse biased p-n junction to generate

electron-hole pairs. These optically generated carriers accelerate in this
high field region, impact ionizing to produce useful gains in the region of
around 20. Unfortunately, impact ionization is intrinsically noisy but can
be made quieter if only one type of carrier initiates the process[17].

In III-V materials suitable for long wavelength detection this is not
the case: the impact ionization rates of electrons and holes are not too
different. However, by incorporating quantum wells[18] or a series of graded
steps in the multiplication region (figure 4) we can introduce strong
asymmetry into the ionization rates provided the difference in band gaps is
distributed asymmetrically between the conduction band and valence band
edges. For a materials system such as AlInAs/InGaAs lattice matched to InP,
where the conduction band edge step is the larger, electrons drifting in the
wide gap AlInAs become automatically hot when they transfer to the narrow
gap InGaAs. Heating increases the number of electrons exceeding threshold
energy for impact ionization and can increase the rate by orders of
magnitude.

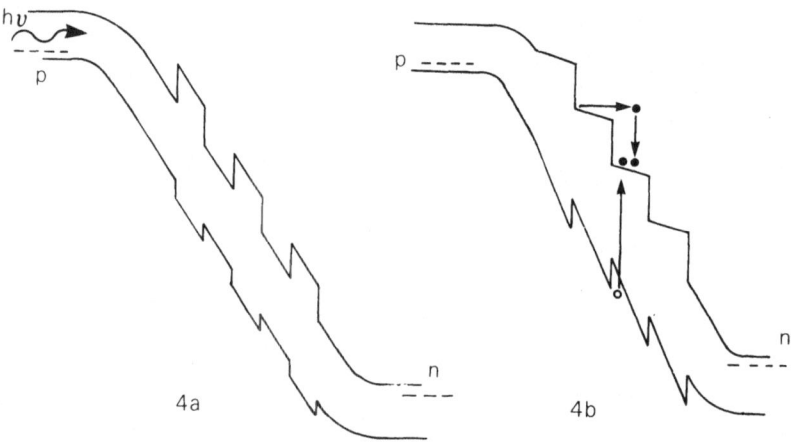

Fig.4   Quantum well (a) and staircase (b) APD band edge structure under
        reverse bias.

Monte Carlo simulation seems ideally suited to the evaluation of impact
ionization rates. The process is intrinsically random and involves initial
states which are high in the bands, requiring the incorporation of realistic
band structure. Fields are high and non equilibrium transport will
certainly be important since carrier mean free paths are comparable with the
scale of quantum well and wall widths. Indeed suitably sophisticated Monte
Carlo simulation seems to give excellent agreement with measurement both for
bulk and quantum well aided impact ionization rates[19]. This is all very
encouraging and provided the technique allows us to make predictions for
design purposes as well as to interpret measurements after the fact then it
must be seen as another significant achievement for the Monte Carlo
technique.

DISCUSSION

This brief survey constitutes a personal appraisal of the contribution of modelling to some existing heterostructure devices. There is much work that I have omitted, either in the interests of brevity or out of ignorance and I apologise for that. A number of points emerge, however, from the foregoing material.

Optical device modelling seems on firmer ground than microwave device modelling. This is partly because each contributing process is checked more easily for optical devices. For microwave devices the checkable results are largely the DC transfer characteristics and the bias dependent S-parameters and the latter can often be dominated by parasitics which are difficult to strip. The technique of on-chip electrooptic sampling[20] can avoid parasitic masking and may be refined to probe the interior of a device with an ultimate resolution equal to the wavelength of the light.

Moreover the two dimensional nature of a MODFET is unavoidable, as is the importance of self consistency in calculating electric fields from Poisson's equation. These factors have haunted MESFET modelling and the inclusion of quantum confinement effects in the channel of a MODFET make its simulation a particularly difficult task.

The transport of hot carriers is the key feature of high frequency microwave devices and whilst the classical description of these processes is best performed with Monte Carlo techniques these can be enormously expensive of computer time, especially if self consistency and time dependence in two or three dimensions are needed. Development of techniques of equivalent accuracy but higher efficiency such as those described by Herbert (these procedings) will ease the burden on computing time.

Finally, use of such classical transport descriptions over distances of the order of quantum well widths must be suspect. The observed sharp bound states in a quantum well result from the phase coherence of the wave function. Transport over a barrier of comparable width, as in a quantum well APD, must also be similarly phase coherent and a quantum transport description, which would include the effect of reflection at the 'step down' junction and of distributed scattering would provide a firmer basis for classical descriptions. The Wigner Function technique described by Frensley (these procedings), or other such techniques which go beyond the classical description, will be needed as quantum transport effects become more apparent. This method is essential for modelling scattering in such devices as resonant tunnelling structures which are inherently nonclassical. It may also provide a sounder base for modelling some aspects of more classical devices where quantum effects are lurking, for example, in the short base of an HBT or hot electron transistor or under the short gate of a MESFET or MODFET.

ACKNOWLEDGEMENTS

It is my pleasure to thank Tony Holden and David Robbins for their experienced advice on microwave and optical device modelling.

REFERENCES

1. R. Dingle 'Optical properties of semiconductor superlattices' Proc 13th Int.Conf. Phys. Semicond., F.G. Fumi, ed., 965 (1976)

2. J. Nagle and C. Weisbuch 'Design rules for very low threshold quantum well lasers', 13th European conf. Optical Comm. R.Y. Sähköinsinööriliitto, ed., 2, 25 (1987).

3. D. J. Robbins, 'Lifetime broadening in quantum well lasers' SPIE Conf., Novel optoelectronic devices, the Hague, 800, 34 (1987).

4. R. I. Taylor, R.A. Abram, M.G. Burt and C. Smith 'Auger recombination in a quantum-well heterostructure laser' IEE Proc 132J, 364 (1986).

5. G. Childs, 'Effects of high carrier concentrations on some optical properties of semiconductors', PhD thesis, Department of Applied Physics and Electronics, Durham University, UK, p130 (1987).

6. M. Osinsky and M.J. Adams, 'Gain spectra of quaternary semiconductors', Proc IEE I129. 229 (1982).

7. A. Adams, 'Band structure engineering for low-threshold, high efficiency semiconductor lasers' El.Lett., 22. 249 (1986).
   E. P. O'Reilly, K.C. Heasmann, A.R.Adams and G.P. Witchlow 'Calculation of the threshold current and temperature sensitivity of a GaInAs strained quantum well laser operating at 1.55 µm' Superlatt. and Microstr.3, 99 (1987).

8. E. Yablonovitch and E.O. Kane 'Reduction of lasing threshold current density by lowering of valence band effective mass' IEEE LT4, 504(1986).

9. J. J. Harris, C. T. Foxon, K.W.J. Barnham, D.E. Lacklison, J. Hewett and C. White, 'Two dimensional electron gas structures with mobilities in excess of $3 \times 10^6 cm^2/Vsec$', J.Appl.Phys, 61, 1219 (1987).

10. B. Vinter, 'Subbands and charge control in a two-dimensional electron gas field effect transistor', Appl.Phys. Lett., 44, 307 (1984).

11. D. Widiger, I.C. Kizilyalli, K. Hess and J.J. Coleman, 'Two-dimensional transient simulation of an idealized high electron mobility transistor' IEEE ED32, 1092 (1985).

12 J. Y. F. Tang 'Two dimensional simulation of MODFET and GaAs gate heterojunction FETS', IEEE ED32, 1817 (1985)

13. W. A. Hughes and C. M. Snowden 'Nonlinear charge control in AlGaAs/GaAs modulation doped FETS', IEEE ED34, 1617 (1987)

14. L. S. Nguyen, W. J. Schaff, P.J. Tasker, A. N. Lepore, L. F. Palmateer, M. C. Foisy and L. F. Eastman, 'Charge control, DC and RF performance of a 0.35µm pseudomorphic AlGaAs/InGaAs modulation doped FET' IEEE ED30, 139 (1988)

15. K. S. Yoon, G.B. Stringfellow and R.J. Huber 'Transient transport in bulk GaInAs/AlInAs' J. Appl. Phys. 63, 1126(1988).

16. C. M. Maziar, M.E. Klousmeier-Brown, S. Bandyopadhyay, M.S. Lundstrom and S, Datta, 'Monte-Carlo evaluation of electron transport in heterojunction bipolar transistor base structures' IEEE ED33, 881(1986).

17. R. J. McIntyre 'Multiplication noise in uniform avalance diodes', IEEE ED13, 164 (1966).

18. R. Chin, N. Holonyak, G. E. Stillman, J. Y. Tang and K. Hess, 'Impact ionization in multilayered heterojunction structures' El.Lett.16, 467 (1980).

19. K. Brennan, 'Theory of electron and hole impact ionization in quantum well and staircase superlattice avalanche photodiode structures' IEEE ED32, 2197 (1985).

20. K. J. Weingarten, R. Majidi-Ahy and D. M. Bloom, 'GaAs Integrated circuit measurements using electro-optic sampling' Proc. GaAs IC symposium, 11 (1987).

PARTICIPANTS

Abram, R.A.
    Applied Physics Group
    School of Engineering and
        Applied Science
    University of Durham, DHI 3LE, UK

Adams, A.R.
    Department of Physics
    University of Surrey, Guildford
    Surrey, GU2 5XH, UK

Andreoni, W.
    IBM Zurich Research Laboratory
    CH-8803, Ruschlikon, Switzerland

Arent, D.J.
    IMEC-VZW, Kapeldreef
    75 B-3030, Leuven, Belgium

Bachelet, G.B.
    Centro Cnr - Dipartimento di Fisica
    I-38050, Povo, Trento, Italy

Bajaj, K.K.
    Avionics Laboratory, AFWAL/AADR
    Wright-Patterson Air Force Base
    Ohio 45433, USA

Baroni, S.
    SISSA - ISAS Strada Costiera 11
    I-34014, Trieste, Italy

Brugger, H.
    IBM Zurich Research Laboratory
    CH-8803, Ruschlikon, Switzerland

Burt, M.G.
    British Telecom Research Laboratory
    Martlesham Heath
    Ipswich, IP5 7RE, UK

Chang, Y.C.
    Department of Physics
    University of Illinois
    1110 W. Green Street
    Urbana, IL 61801, USA

Christen, J.
    Technische Unersitat Belin
    Institut fur Festkorperphysik
    Sekr PN 5-2, Hardenbergstrasse 36
    D-1000 Berlin 12

Ciraci, S.
    Department of Physics
    Bilkent University
    Bilkent 06533, Ankara, Turkey

Claessen, L.M.
    Max-Planck Institut fur
    Festkorperforschung
    25 Av des Martyrs
    F-38042, Grenoble, France

Eaves, L.
    Department of Physics
    University of Nottingham
    Nottingham, NG7 2RD, UK

Eppenga, R.
    Phillips Research Laboratories
    PO Box 80000, 5600 JA Eindhoven
    The Netherlands

Fasol, G.
    Cavendish Laboratory
    University of Cambridge
    Madingley Road
    Cambridge, CB3 OHE, UK

Fasolino, A.
    SISSA-ISAS, Strada Costiera 11
    I 34014, Trieste, Italy

Faurie, J.P.
    Department of Physics
    University of Illinois at Chicago
    PO Box 4348 M/C 273
    Chicago, Illinois 60680, USA

Flores, F.
    Departamento de Fisica de la
    Materia Condensade
    Universidad Autonoma
    28049 Madrid, Spain

Frensley, W.R.
    Texas Instruments, P.O. Box 655936
    MS154 Dallas, Texas 75265 USA

Göbel, E.O.
    Fachbereich Physik
    Philipps University, Renthof 5, 3550
    Marburg, Fed. Rep. Germany

Heiblum, M.
    IBM Research Centre, PO Box 218
    Yorktown Heights, N.Y. 10598 USA

Herbert, D.C.
    Royal Signals and Radar
    Establishment, St. Andrews Rd.
    Great Malvern, Worcester
    WR14 3PS, UK

Jaros, M.
    Department of Physics
    The University
    Newcastle-upon-Tyne
    NEL 7RU, UK

Kelley, M.J.
    GEC Research, East Lane
    Wembley, HA9 7PP, UK

Lazzarini, L.
    MASPE-CNR Institute
    Via Chiavari 18/A
    43100 Parma, Italy

Lugli, P.
    Department of Mechanical Engineering
    The Second University of Rome
    Via Raimonde, 00173 Rome, Italy

Malik, R.J.
    AT&T Bell Labs. Rm IC-318
    600 Mountain Ave
    Murray Hill, N.J. 07974 USA

Martin, R.M.
    Departmemt of Physics
    University of Illinois
    1110 W. Green St.
    Urbana, IL 61801 USA

Nazare, M.H.
    Universidade de Aveiro
    Departamento de Fisica
    Aveiro, Portugal

Newman, K.E.
    Department of Physics
    University of Notre Dame
    Notre Dame, IN 46556 USA

Pepper, M.
    Cavendish Laboratory
    University of Cambridge
    Madingley Road
    Cambridge, CB3 OHE, UK

Polatoglou, H.M.
    Department of Physics
    University of Thessaloniki
    Thessaloniki 54006, Greece

Rees, C.J.
    Plessey Research Caswell
    Towcester, Northants, NN12 8EQ UK

Sermage, B.
    CNET, 196 Ave. Henri Ravera 92220
    Bagneux, France

Skolnick, M.S.
    Royal Signals and Radar
    Establishment, St. Andrews Road
    Malvern, Worcestershire
    WR14 3PS UK

Steiner, T.W.
    Department of Physics
    Simon Fraser University, Burnaby
    British Columbia, V5A 1S6, Canada

Vanzetti, L.
    MASPEC-CNR, Via Chiavari 18/A
    43100 Parma, Italy

Voos, M.
    Groupe de Physique des Solides
    de l'Ecole Normale Superieure
    24 Rue Lhomond
    75005 Paris, France

Wolford, D.
    IBM Watson Research Centre
    PO Box 218, Yorktown Heights
    New York 10598 USA

INDEX

Ab-initio calculations, 2, 34-49,
    51-60, 129-135, 359-365
Absorption measurements, 12-14,
    64-68, 73-77, 84, 86-88,
    250-252, 280-292, 322,
    347-356
Alloy fluctuations, 234-236
Alloy composition grading, 217-218,
    220, 379
Anderson localisation, 138
Anderson model of heterojunctions,
    24
Auger processes, 256-258, 281-282,
    285-289, 291, 293, 296,
    376-377, 380
Augmented spherical wave (ASW)
    method, 359
Avalanche photodiode, 379-380

Ballistic electrons, 188, 198,
    201, 206, 220
Ballistic holes, 167-175
Ballistic transport, 167-175, 188,
    198, 201, 206, 220
Band gap bowing, 119, 125
Band line-ups and offsets
    ab-initio calculations, 2
    charge neutrality level concept,
        16-17, 22
    common anion rule, 63, 77, 82-84
    dielectric midgap energy method,
        17-18
    electrical measurements, 7, 93-94
    electron affinity rule, 15
    Harrison's atomic orbital
        theory, 15-16, 82-83
    interface orientation
        dependence, 2, 51-60, 92-93
    matching of hybrids, 18
    optical measurements, 7, 9, 84,
        94-96
    polar interfaces, 3, 21-31,
        33-49
    pressure dependence, 7-20
    pseudopotential theory, 10

Band line-ups and offsets (cont)
    Raman measurements, 77, 84
    Schottky barrier measurements,
        77
    simple models, 2
    simplified electronic
        Hamiltonians, 2
    strain effects, 14
    tight binding calculations, 25
    transitivity relations, 3, 24,
        52, 89
    UPS measurements, 8
    XPS measurements, 8, 77, 82, 84,
        88-96
Band structure calculations (see
    under entries for
    individual methods)
Bipolar transistor, 225-227, 375,
    379
Bloch oscillations, 203
Boltzmann equation, 202-207
Bond-orbital method, 343
Boundary conditions at an inter-
    face, 99, 106 (see also
    Band line-ups and offsets)

Cathodoluminescence, 269-278
Charge neutrality level concept,
    16-17, 22
Coherent backscattering, 139
Common anion rule, 63, 77, 82-84
Conduction band non-parabolicity
    measurements, 335
Conductivity corrections, 138
Configuration interaction methods,
    131
Cross-interface recombination,
    253-258

Delta doping, 43, 227
Density functional theory, 34-35,
    53-55, 130, 360
Density matrix, 177
Devices, 143-146, 168, 187-188,
    193-195, 225-231, 247,
    279-301, 318, 322, 375-382

Dielectric midgap energy method, 17-18
Diffusion loop, 138
Dimensionality transitions, 137-147
Disordered anion alloy, 119
Disordered cation alloy, 119

Effective mass equation for multilayer structures, 106, 343-344
Effective mass theory, 99, 102, 343
Electron affinity rule, 9, 15
Electron beam evaporation, 217-223
Electron-electron scattering, 188-199
Electron-plasmon scattering, 188-199, 201, 207
Electro-optic devices, 318, 322
Electrostatic squeezing, 144-146
Energy dispersive spectroscopy, 85
Envelope function formalism, 64, 68, 82, 96, 99-109, 294-295, 312-322, 328-331, 342-347, 368-373
  boundary conditions, 106
  exact equations, 99-109
  non-lattice matched structures, 103-105
Esaki diode, 228
Excitons
  binding energy, 312-318, 345-355
  electric field effects, 318-323
  in quantum wells, 233-238, 250-252, 263-264, 311-324
  in superlattices, 350-356
Extended X-ray absorption fine structure (EXAFS) measurements, 121, 123
Fermi energy edge singularity, 238-241
Flux quantum, 139
Formation energies of superlattices, 33-49

Gunn devices, 227

Harrison's atomic orbital method, 9, 15-16, 82-83
Harrison's model of strain energy, 121
Heterojunction bipolar transistor (HBT), 226-227, 375, 379
High electron mobility transistor (HEMT), 187, 226 (see also MODFET)
Hole
  levels in a magnetic field, 367-373
  localisation, 234, 236, 239-240, 242

Hole (continued)
  wavefunction determination, 336
Hot electron spectroscopy, 167, 220
Hot hole spectroscopy, 167-176
Hot hole transistor, 168
Hydrostatic pressure experiments, 7-20, 253-258, 279-292

Induced density of interface states (IDIS) model, 16, 22-25, 83
Interface
  dipoles, 1-5, 16-17, 21-31, 51-60, 83
  effects in a quantum well, 259-278
  linear response theory, 53-55
  metal-semiconductor, 21-27
  recombination, 259-268
  states in Type III structures, 64
  structure, 269-278
Intervalence band absorption (IVBA), 280-281, 285-289, 291, 293, 295, 376-377
Intralayer, 21, 29
Inversion layer, 142

Keating's model, 121

Landau-Baber scattering, 141
Linear combination of atomic orbitals (LCAO) method, 82-83, 110
Linear muffin tin orbital (LMTO) method, 83, 110, 359
Linear response theory for interfaces, 53-55
Liouville
  equation, 178
  super-operator, 178
Local density approximation, 34-35, 53-55, 130, 360
Localisation (weak and strong), 139

Magnetoquantum transport measurements, 150, 155, 158-165
Magnetoresistance, 139-146, 158-165
  negative, 139-146
Magnetospectroscopy, 10, 61-80, 84, 86-88, 241-244, 367
Mass reversal at an interface, 63, 96
Matching of hybrids, 18
Metal base and related transistors, 227

386

Metal-insulator transition, 48
Metal-oxide-semiconductor field
        effect transistor (MOSFET),
        144-146
Metal-semiconductor field effect
        transistor (MESFET) 143-
        146, 187, 225
Metal-semiconductor interface, 21
  Bardeen model, 23
  charge neutrality, 22-23
  IDIS model, 23
  Schottky model, 22
  tight binding theory, 27
Microclusters, 129-135
Miscibility gap, 120
Modelling of semiconductor devices,
        193-199, 201-202, 211-213,
        280-298, 375-382
Modulation doped field effect
        transistor (MODFET), 226,
        375, 377-378
Molecular beam epitaxy (MBE), 85,
        217-222, 304
Molecular dynamics, 129-135
Monte Carlo simulation, 187-199,
        202, 379-380, 390
  carrier energy loss, 193-199
  quantum effects, 196-198

Negative differential conductivity
        and mobility, 150-158, 162,
        179-183, 204, 207

Open system, 177-179
Ordered alloys, 119-127
  stability of, 119-127
  strain in, 119-127

Particle in a box model, 99, 107,
        341
Persistent photoconductivity, 236
Perturbation method for band
        structure, 111-117
Phase relaxation time, 139
Photoluminescence, 84, 233-278,
        303-310, 317-318, 322, 359
Photoreflectance, 303-310.
Planar doped barrier (PDB), 187-
        188, 193-195, 226-227
Plasmon-phonon coupling, 192-193
Pseudopotential calculations, 16,
        33-60, 129-135, 255

Quantum interference, 137-147,
        167-168, 174, 178
Quantum kinetic transport theory
        (see Wigner function theory
        and Weak coupling quantum
        kinetic equation)

Quantum well
  absorption measurements,
        250-252, 280-292, 322,
        347-356
  avalanche photodiode, 379-380
  carrier trapping in, 247-250
  cathodoluminescence, 269-278
  electronic structure
        calculations, 312-318,
        342-346
  enhanced exciton-phonon
        coupling, 233-238
  excitons, 233-238, 250-252,
        263-264, 311-324
  Fermi energy edge singularity,
        238-241
  hole localisation, 234
  interface effects, 259-278
  interface recombination, 259-268
  laser, 229, 247, 279-301,
        375-377
  magnetospectroscopy, 241-244
  optical properties calculations,
        240-241, 292-297, 311-324,
        344-350
  photoluminescence, 233, 278,
        317-318
  quantum confined Stark effect,
        250-251, 318-323
  Schottky gated structure,
        237-238, 250
  screening of the Frohlich
        interaction, 233-238
  strain and pressure effects,
        279-301, 303-310
  structure of interfaces, 269-278
  time resolved spectroscopy,
        247-252, 259-268, 276

Reflection high energy electron
        diffraction (RHEED), 270
Resonant Raman scattering, 77, 84,
        325-339
Resonant tunnelling
  coherent theory, 150, 151
  current bistability, 150-158
  double barrier diode, 228
  electron scattering, 150,
        158-162
  intrinsic bistability, 150-158
  magnetotransport, 150, 155,
        158-165
  negative differential
        conductivity, 150-158,
        162, 179-183
  sequential theory, 150-158
  skipping states, 150, 163-165
  space charge, 150-158
  wide well case, 162-165
  Wigner function theory, 177-185

Scattering rates in superlattices, 207-211
Schottky barrier, 22, 24, 27, 228, 237-238, 250
Segregation, 134
Self consistent tight binding (SCTB) theory, 25-28
  of heterojunctions, 28, 82-84
  of metal-semiconductor interfaces, 25,
Semiconductor laser, 229, 247, 279-301, 375-377
  Auger recombination, 281-282, 285-289, 291, 293, 296, 376-377
  distributed feedback, 290-292
  carrier leakage, 282-283, 287
  graded index, 247, 376
  hydrostatic pressure effects, 279-301
  IVBA, 280-281, 285-289, 291, 293, 295, 376-377
  modelling, 376-377
  optical confinement, 247, 376
  short wavelength devices, 298
  strain engineering, 292-297, 377
  T$_o$ problem, 279-283, 377
  threshold current, 279, 283, 290, 293, 295, 298, 377
Simulated annealing, 129-135
Single particle density matrix, 177
Stability of superlattices, 33-49
Strained layers, 33, 38-47, 90-94, 103-105, 292-297, 303-310, 376, 378
Superlattice
  avalanche photodiode, 380
  band structure calculations, 33-60, 111-117, 342-344, 359-373
  carrier scattering rates, 207-211
  cross-interface recombination, 253-258
  formation energies, 33-49
  hole levels in a magnetic field, 367-373
  magnetospectroscopy, 10, 61-80, 84, 86-88, 367
  pressure effects, 7-20, 253-258
  stability, 33-49
  strained layer, 33, 38-47, 90-94, 103-105
  vertical transport, 211-214, 380

Tight binding calculations, 25-27, 34, 82-84, 96, 121-122, 343
Time irreversible behaviour, 177
Time resolved spectroscopy, 247-268
Transitivity rule for heterojunctions, 3, 24, 52, 89

Tunnelling hot electron transfer amplifier (THETA), 187-188, 195-199
Type II recombination in superlattices, 253-258

Ultra-violet photoemission spectroscopy (UPS), 8

Wave packet model of transport, 206
Weak coupling quantum kinetic equation, 202
Wigner distribution function, 178, 382
Wigner function theory of resonant tunnelling, 177-185

X-ray photoemission spectroscopy (XPS), 8, 77, 82, 84, 88-96

Zener tunnelling, 207